D0161558

Capital Cities around the World

Capital Cities around the World

An Encyclopedia of Geography, History, and Culture

ROMAN ADRIAN CYBRIWSKY

 ABC-CLIO

Santa Barbara, California • Denver, Colorado • Oxford, England

Library of Congress Cataloging-in-Publication Data

Cybriwsky, Roman Adrian.
   Capital cities around the world : an encyclopedia of geography, history, and culture / Roman Adrian Cybriwsky.
     p. cm.
   Includes bibliographical references and index.
   ISBN 978-1-61069-247-2 (hardcopy : alk. paper) — ISBN 978-1-61069-248-9 (ebook)   1. Cities and towns.   2. Capitals.   I. Title.
   G140.C93   2013
   909'.09732—dc23   2012046346

ISBN: 978-1-61069-247-2
EISBN: 978-1-61069-248-9

17 16 15 14 13   1 2 3 4 5

This book is also available on the World Wide Web as an eBook.
Visit www.abc-clio.com for details.

ABC-CLIO, LLC
130 Cremona Drive, P.O. Box 1911
Santa Barbara, California 93116-1911

This book is printed on acid-free paper ∞

Manufactured in the United States of America

Contents

List of Capital Cities

Abu Dhabi
United Arab Emirates
Abuja
Nigeria
Accra
Ghana
Addis Ababa
Ethiopia
Algiers
Algeria
Amman
Jordan
Amsterdam
The Netherlands
Andorra La Vella
Andorra
Ankara
Turkey
Antananarivo
Madagascar
Apia
Samoa
Ashgabat
Turkmenistan
Asmara
Eritrea
Astana
Kazakhstan
Asunción
Paraguay
Athens
Greece
Baghdad
Iraq
Baku
Azerbaijan
Bamako
Mali
Bandar Seri Begawan
Brunei

Bangkok
Thailand
Bangui
Central African Republic
Banjul
The Gambia
Basseterre
Saint Kitts and Nevis
Beijing
China
Beirut
Lebanon
Belfast
Northern Ireland
Belgrade
Serbia
Belmopan
Belize
Berlin
Germany
Bern
Switzerland
Bishkek
Kyrgyzstan
Bissau
Guinea-Bissau
Bloemfontein
South Africa
Bogotá
Colombia
Brasília
Brazil
Bratislava
Slovakia
Brazzaville
Republic of the Congo
Bridgetown
Barbados
Brussels
Belgium

Phnom Penh
  Cambodia
Podgorica
  Montenegro
Port-au-Prince
  Haiti
Port Louis
  Mauritius
Port Moresby
  Papua New Guinea
Port-of-Spain
  Trinidad and Tobago
Port Vila
  Vanuatu
Porto-Novo
  Benin
Prague
  Czech Republic
Praia
  Cape Verde
Pretoria
  South Africa
Pristina
  Kosovo
Pyongyang
  North Korea
Quito
  Ecuador
Rabat
  Morocco
Reykjavik
  Iceland
Riga
  Latvia
Riyadh
  Saudi Arabia
Rome
  Italy
Roseau
  Dominica
San José
  Costa Rica
San Marino
  San Marino
San Salvador
  El Salvador
Sana'a
  Yemen

Santiago
  Chile
Santo Domingo
  Dominican Republic
São Tomé
  São Tomé and Principe
Sarajevo
  Bosnia and Herzegovina
Seoul
  South Korea
Singapore
  Singapore
Skopje
  Macedonia
Sofia
  Bulgaria
South Tarawa
  Kiribati
St. George's
  Grenada
St. John's
  Antigua and Barbuda
Stockholm
  Sweden
Sucre
  Bolivia
Suva
  Fiji
Taipei
  Taiwan
Tallinn
  Estonia
Tashkent
  Uzbekistan
Tbilisi
  Georgia
Tegucigalpa
  Honduras
Tehran
  Iran
Thimphu
  Bhutan
Tirana
  Albania
Tokyo
  Japan
Tripoli
  Libya

Tunis
  Tunisia
Ulaanbaatar
  Mongolia
Vaduz
  Liechtenstein
Valletta
  Malta
Vatican City
  Vatican City (Holy See)
Victoria
  Seychelles
Vienna
  Austria
Vientiane
  Laos
Vilnius
  Lithuania
Warsaw
  Poland

Washington, DC
  United States of America
Wellington
  New Zealand
Windhoek
  Namibia
Yamoussoukro
  Côte d'Ivoire
Yangon
  Myanmar
Yaoundé
  Cameroon
Yaren
  Nauru
Yerevan
  Armenia
Zagreb
  Croatia

List of Capital Cities by Country

Afghanistan
  Kabul
Albania
  Tirana
Algeria
  Algiers
Andorra
  Andorra la Vella
Angola
  Luanda
Antigua and Barbuda
  Saint John's
Argentina
  Buenos Aires
Armenia
  Yerevan
Australia
  Canberra
Austria
  Vienna
Azerbaijan
  Baku
The Bahamas
  Nassau
Bahrain
  Manama
Bangladesh
  Dhaka
Barbados
  Bridgetown
Belarus
  Minsk
Belgium
  Brussels
Belize
  Belmopan
Benin
  Porto-Novo

Bhutan
  Thimphu
Bolivia
  La Paz (administrative);
  Sucre (judicial)
Bosnia and Herzegovina
  Sarajevo
Botswana
  Gaborone
Brazil
  Brasília
Brunei
  Bandar Seri Begawan
Bulgaria
  Sofia
Burkina Faso
  Ouagadougou
Burundi
  Bujumbura
Cambodia
  Phnom Penh
Cameroon
  Yaoundé
Canada
  Ottawa
Cape Verde
  Praia
Central African Republic
  Bangui
Chad
  N'Djamena
Chile
  Santiago
China
  Beijing
Colombia
  Bogotá

Comoros
  Moroni
Congo, Republic of the
  Brazzaville
Congo, Democratic Republic of the
  Kinshasa
Costa Rica
  San José
Côte d'Ivoire
  Yamoussoukro (official); Abidjan (de facto)
Croatia
  Zagreb
Cuba
  Havana
Cyprus
  Nicosia
Czech Republic
  Prague
Denmark
  Copenhagen
Djibouti
  Djibouti
Dominica
  Roseau
Dominican Republic
  Santo Domingo
East Timor (Timor-Leste)
  Dili
Ecuador
  Quito
Egypt
  Cairo
El Salvador
  San Salvador
Equatorial Guinea
  Malabo
Eritrea
  Asmara
Estonia
  Tallinn
Ethiopia
  Addis Ababa
Fiji
  Suva
Finland
  Helsinki
France
  Paris

Gabon
  Libreville
The Gambia
  Banjul
Georgia
  Tbilisi
Germany
  Berlin
Ghana
  Accra
Greece
  Athens
Grenada
  Saint George's
Guatemala
  Guatemala City
Guinea
  Conakry
Guinea-Bissau
  Bissau
Guyana
  Georgetown
Haiti
  Port-au-Prince
Honduras
  Tegucigalpa
Hungary
  Budapest
Iceland
  Reykjavik
India
  New Delhi
Indonesia
  Jakarta
Iran
  Tehran
Iraq
  Baghdad
Ireland
  Dublin
Israel
  Jerusalem
Italy
  Rome
Jamaica
  Kingston
Japan
  Tokyo

Jordan
  Amman
Kazakhstan
  Astana
Kenya
  Nairobi
Kiribati
  South Tarawa
Korea, North
  Pyongyang
Korea, South
  Seoul
Kosovo
  Pristina
Kuwait
  Kuwait City
Kyrgyzstan
  Bishkek
Laos
  Vientiane
Latvia
  Riga
Lebanon
  Beirut
Lesotho
  Maseru
Liberia
  Monrovia
Libya
  Tripoli
Liechtenstein
  Vaduz
Lithuania
  Vilnius
Luxembourg
  Luxembourg
Macedonia
  Skopje
Madagascar
  Antananarivo
Malawi
  Lilongwe
Malaysia
  Kuala Lumpur
Maldives
  Malé
Mali
  Bamako

Malta
  Valletta
Marshall Islands
  Majuro
Mauritania
  Nouakchott
Mauritius
  Port Louis
Mexico
  Mexico City
Micronesia, Federated States of
  Palikir
Moldova
  Chişinău
Monaco
  Monaco
Mongolia
  Ulaanbaatar
Montenegro
  Podgorica
Morocco
  Rabat
Mozambique
  Maputo
Myanmar (Burma)
  Rangoon (Yangon);
  Naypyidaw or Nay Pyi Taw
  (administrative)
Namibia
  Windhoek
Nauru
  no official capital; government offices in
  Yaren District
Nepal
  Kathmandu
Netherlands
  Amsterdam; The Hague (seat of
  government)
New Zealand
  Wellington
Nicaragua
  Managua
Niger
  Niamey
Nigeria
  Abuja
Norway
  Oslo

Oman
  Muscat
Pakistan
  Islamabad
Palau
  Melekeok
Panama
  Panama City
Papua New Guinea
  Port Moresby
Paraguay
  Asunción
Peru
  Lima
Philippines
  Manila
Poland
  Warsaw
Portugal
  Lisbon
Qatar
  Doha
Romania
  Bucharest
Russian Federation
  Moscow
Rwanda
  Kigali
Saint Kitts and Nevis
  Basseterre
Saint Lucia
  Castries
Saint Vincent and the Grenadines
  Kingstown
Samoa
  Apia
San Marino
  San Marino
São Tomé and Principe
  São Tomé
Saudi Arabia
  Riyadh
Senegal
  Dakar
Serbia
  Belgrade
Seychelles
  Victoria

Sierra Leone
  Freetown
Singapore
  Singapore
Slovakia
  Bratislava
Slovenia
  Ljubljana
Solomon Islands
  Honiara
Somalia
  Mogadishu
South Africa
  Pretoria (administrative)
  Cape Town (legislative)
  Bloemfontein (judiciary)
South Sudan
  Juba (Relocating to Ramciel)
Spain
  Madrid
Sri Lanka
  Colombo; Sri Jayewardenepura
  Kotte (legislative)
Sudan
  Khartoum
Suriname
  Paramaribo
Swaziland
  Mbabane
Sweden
  Stockholm
Switzerland
  Bern
Syria
  Damascus
Taiwan
  Taipei
Tajikistan
  Dushanbe
Tanzania
  Dar es Salaam; Dodoma
  (legislative)
Thailand
  Bangkok
Togo
  Lomé
Tonga
  Nuku'alofa

Preface

From Abu Dhabi, the soaring-skyline capital of the United Arab Emirates, to Zagreb, the millennium-old capital of the Republic of Croatia, this book takes readers in alphabetical order through the 202 capital cities of all 198 countries in the world. (Some countries have more than one capital.) No book is so comprehensive. This encyclopedia covers the smallest capital cities such as Vatican City (population 832), exotic Funafuti, a coral atoll with a population of 4,492 that is the center of government of the Pacific island nation of Tuvalu, a country that has only 10,544 inhabitants, and Vaduz, population 5,100, the capital of the tiny European principality of Liechtenstein, as well as the world's largest capital cities: Tokyo, Japan (metropolitan population 32.4 million); Seoul, Republic of Korea (20.6 million); and Mexico City, the bustling capital of the Republic of Mexico (more than 20.5 million). The work covers as well the world's smallest countries in size (Vatican City again, Monaco, Nauru in the Pacific Ocean, and Tuvalu again), and the size giants of world—the Russian Federation, Canada, and the People's Republic of China. The United States is also featured, with its capital of Washington, the District of Columbia, and the Washington Metropolitan Area that spills on either side of the Potomac River into the neighboring states of Maryland and Virginia.

The range in cities and countries is enormous, with remarkable contrasts across the map of the world in cultures, ways and standards of living, systems of government, economy, and society, and the look of the built environment. There are ancient cities that are still national capitals, or are capitals once again in their history, such as Baghdad, Iraq; Damascus, Syria; and Jerusalem, the contested capital of Israel, as well as cities that have only recently become capitals either because their countries had only recently become born (e.g., Juba in South Sudan and Dili in East Timor) or because the capital of an existing country has recently been moved there (Naypyidaw, Myanmar, the country that is also known as Burma, replacing the former capital of Yangon, the large city that is also known as Rangoon). Other contrasts are between capital cities that are extremely poor (e.g., N'Djamena in Chad, Bangui in the Central African Republic, and Brazzaville, the capital city of the Republic of the Congo); those that are generally prosperous with high standards of living (e.g., Tokyo and Singapore in Pacific Asia and Bern and Oslo in Switzerland and Norway, respectively); those that are presently torn by civil war or other strife (Damascus, Baghdad, and Kabul in Afghanistan); and those that have been developed in recent times to showcase the achievements and the potentials of their respective countries (e.g., Abuja in Nigeria, Astana in Kazakhstan, and Doha, the showy capital of oil-rich Qatar).

This book is written by one author in order to maintain a uniform standard of content and organization of the various essays about capital cities, and to keep the writing uniform. Each essay has four parts: (1) An overview that positions each city and its country in geographic space, and gives basic information such as population data, the origin of the city name, the basis of the city's economy in addition to government administration, and other defining characteristics; (2) an overview of historical development from founding to when the city became capital if not at founding, to current trends in urban growth, economic development, and other characteristics; (3) a description of the city's major landmarks, including those associated with its role as national capital, important tourist or visitor attractions, places of historical interest, and in many cases, prominent new structures and development projects that refocus the image of a city and that carry it into the new millennium; and (4) a brief profile of the social and cultural characteristics of the population, including comments about comparative wealth, ethnic and language groups, major religions, and ways of work, life, and recreation. For every city there is also a carefully selected list of up to four or five additional readings, always in English and as much as possible, accessible to general readers, although for some smaller or more remote places, only one or two reading links are provided because information is just that scarce, even on the Internet where it is said that one can find everything. For many cities, there are also representative or iconic photographs, some from the extensive global travels of your geographer-author, and others from other photographers.

In addition to the 202 capitals, this book also showcases a wide selection of cities that had been capitals in the past, as well as some examples of cities in which there are broad-based movements for separatism from the countries in which they are located and that could conceivably become national capitals (or return once again to being national capitals) in the future. Examples of past capitals include ancient archaeological sites such as Babylon (Mesopotamian civilization), Thebes (Egypt), and Tenochtitlan, the ruined Aztec capital whose remains are under the site of present-day Mexico City; as well as cities that continue to be prominent in today's world such as St. Petersburg in Russia, Kyoto in Japan, Salvador and Rio de Janeiro in Brazil, and Philadelphia in the United States. With due respect to the countries in which they are located, examples of possible would-be capitals include Lhasa in Chinese-occupied Tibet, Grozny in Chechnya, a staunchly anti-Russian region of the Russian Federation, and San Juan, the capital of Puerto Rico, a Caribbean Sea territory of the United States in which voters periodically consider (and have so far consistently rejected) the option of breaking ties with the United States and becoming an independent nation. Other text amplifies information about prominent world capitals: about the design of Washington, DC, details about ancient Athens, ancient Rome, and the Forbidden City of Beijing, and the 19th-century reconstruction and modernization of Paris and Cairo.

Other features of this book include a bibliography of books and articles in English about groupings of cities (e.g., cities of Europe, cities of Africa) as opposed readings about individual cities which are appended to the essays about individual capitals, and books and articles about what is a capital city, the kinds of capital cities that are found in the world, and the roles of capital cities in general. The introduction to this book is an essay on these subjects, as well as a listing of capital cities according to various superlatives: biggest, highest, nearest to one another, furthest away, and so on. No other book has this mix of features, so this text is unique and original. I think that readers will find it useful and enjoyable.

Now comes the question of who to thank and acknowledge for help with this book. As I did almost all this completely on my own, this paragraph can be short. However, it is important to say thank you to my editor at ABC-CLIO Kaitlin Ciarmiello who offered me this enjoyable project and then worked with me to see it to fruition, to James Dare, the media editor at ABC-CLIO with whom I worked on the photographs for this book, and to my sharp-eyed copy editor Sharmila Krishnamurthy with whom I also enjoyed working. I am thankful as well to the Coffee House branches on Khreschatik and on Sahaydachnoho Streets in Kyiv, the capital of Ukraine, the Art Hotel Baccara on a ship in the Dnipro River in Kyiv, and Rocket Cat Café in hometown Fishtown, Philadelphia for allowing me to work undisturbed for hours, and for pleasant service. My friends Lyudmyla Males, Rachel Wandless, Nastia Aleksiayuk, George Baran, Elena Shpak, and Nadia Parfan looked after me and my health as I worked, as did my young relatives Anna-Maria Yanchak-san and Olha Yanchak, and my three grown-up children Adrian, Alex, and Mary Cybriwsky, and Adrian's bride Natalia. Finally, I should go back to a time long, long ago when the world was still recorded in black and white and Technicolor was a new marvel at the movies, and thank my fourth grade teacher in a parochial school in Louisville, Kentucky, the stern Sister Mary Kenneth, who tried to make me learn all the capitals of states and countries only to be shocked that I already knew them well and could recite them quite rapidly, without error, in alphabetical order. There were only 48 states then and many fewer countries. I also have a fond memory of my fifth grade teacher, Katie Elder, who once said, "Roman, some day you will be a famous geographer." That's not exactly what I wanted to hear from someone I had a crush on, but it was good advice nonetheless.

Introduction

The word "capital" comes from the Latin *capit-*, a stem of the word *caput* meaning "head." It is related to *capitālis* meaning "of the head" and to *capitāle* referring to wealth. For us, then, it is good to know that a capital city is the head city of a given territory over which it has jurisdiction, and that it exerts influence over that territory's "capital" as in wealth. There might be other, larger cities within that same territory with more wealth and enormous roles in economy that may even extend beyond the geographical limits of the territory, but these larger cities are within the political jurisdictions of their respective capital cities and at least officially report to them. Whether a capital city is large or small, and whether another city or cities within its jurisdiction is more dominant in economy, cultural life, the media, or other urban function, almost all capitals have legal standing. That is, they were designated to be capitals by law or constitution, and normally remain capitals until the government itself decides to be headquartered elsewhere. There are exceptions to that pattern, however, as in the case of revolution or warfare, or when states cease to exist or are taken over by others.

This book is about the capital cities of nations or countries. There are other kinds of capitals too, such as those of states or provinces within many countries, and of so-called "autonomous regions" within large countries such as the People's Republic of China and the Russian Federation, as well as capitals of much smaller territories such as counties in the United States (these capitals are called "county seats") or other subdistricts, but the focus of this book is on those cities that head entire countries. That is a large topic in itself, because there are 198 countries in the world, give or take one or two depending on just how one defines what a country is and whether the news events of a day cause a country to crash or bring a new country into being. Our 198 countries have 202 capital cities because some countries have more than one administrative center, but that number too is subject to refinement because sometimes the work of national government is done in cities without legal designation as capital or is so decentralized within a country that it is hard to pinpoint just how many capital cities that country has. Most countries by far, however, have just one capital city about which there are no ambiguities. In the case of multiple capitals per country, the location of the legislature (i.e., where laws are made) provides a rule of thumb for identifying the one *main* capital city, even though the head of state (executive functions), the highest court (judicial functions), and foreign embassies (diplomatic functions) might be elsewhere. One country, Nauru, a tiny Pacific Island nation with only 9,322 inhabitants, actually has no capital. Its government meets in a district called Yaren, not really a city or town, which is, then, Nauru's de facto capital. The United Nations refers to Yaren is the "main district" of Nauru instead of capital.

Capital Superlatives

Having introduced Nauru and its phantom Yaren, we can segue to some details about superlatives relating to capital cities. After Yaren, the smallest city to be a capital city is Vatican City, capital of a coterminous city-state with a population of only 832, none of whom is a permanent resident. After Vatican City, the smallest capital city seems to be Funafuti, a coral atoll with 4,492 inhabitants and the officially designated capital of Tuvalu, a Pacific Island nation of 10,544 inhabitants, with only slightly more than the population of Nauru. At the other extreme, is Tokyo, the capital of Japan, with approximately 32.4 million inhabitants in its metropolitan area. It would not be too great an exaggeration to say, in fun, that the populations of Funafuti and Vatican City combined are about the same as that of a Tokyo subway train laden with early morning commuters. After Tokyo, the next largest capital city-metropolitan area is Seoul (population nearly 21 million), the heart of neighboring Korea (Table 1). Its subways are also enormously crowded.

In terms of population density, that is numbers of people per unit area, the most crowded cities (non-capitals) and the most crowded capital cities tend to be in Asia. Mumbai (formerly Bombay) and Kolkata (formerly Calcutta) in India and Karachi in Pakistan rank 1, 2, and 3 in population density in the world among very large cities, while Seoul (Korea) and Taipei (Republic of China or Taiwan) rank 6 and 7, respectively, and are the most densely inhabited capital cities. Seoul has 6,448 residents per square mile (16,700 per sq km) and Taipei has 5,869 per square mile (15,200 per sq km). Table 2 lists the 10 most densely settled capital cities in the world.

In terms of land area, we have Vatican City again as the world's smallest capital. It measures only 0.17 square miles (0.44 sq km). Identifying the world's largest capital cities in area is problematic because it is always difficult to compare cities across

Table 1 Largest Capital Cities by Population

Capital City (metro area)	Country	Population (× 1 million)	Rank among Capitals	World Rank
Tokyo	Japan	32.4	1	1
Seoul	Republic of Korea	20.6	2	2
Mexico City	Mexico	20.5	3	3
Jakarta	Indonesia	18.9	4	6
Delhi	India	18.6	5	8
Manila	Philippines	16.3	6	11
Moscow	Russian Federation	15.0	7	15
Cairo	Egypt	15.0	8	16
Buenos Aires	Argentina	13.2	9	17
London	UK (and England)	12.9	10	18

(*Source*: Forstall, Greene, and Pick, 2009)

Table 2   Most Densely Populated Large Capital Cities

Capital City (metro area)	Country	Population per Square Mile	Population per Square km	World Rank
Seoul	Korea	6,448	16,700	6
Taipei	ROC (Taiwan)	5,869	15,200	7
Bogotá	Colombia	5,212	13,500	9
Lima	Peru	4,537	11,750	11
Beijing	PRC	4,440	11,500	12
Delhi	India	4,266	11,050	13
Kinshasa	DR Congo	4,112	10,650	14
Manila	Philippines	4,073	10,550	15
Tehran	Iran	4,073	10,550	16
Jakarta	Indonesia	4,054	10,500	17

Note: Mumbai, Kolkata, Karachi, Lagos, and Shenzhen, all non-capitals, rank 1–5.
http://www.citymayors.com/statistics/largest-cities-density-125.html

the world because of differences in how cities are defined legally from country to country. Therefore, comparisons are usually made instead between metropolitan areas, that is, between large cities and their highly urbanized surroundings, including their suburbs, neighboring cities, and towns, and the suburbs of those cities and towns. Despite measurement problems, it seems certain that the largest capital city-metropolitan area in the world by area is that of Tokyo again. It measures 2,700 square miles (6,993 sq km), second is that of New York City (3,353 square miles; 8,683 sq km), which is not a capital city. Other than Tokyo, all the other top 13 metropolitan areas in size in the world are in the United States. This is because of American patterns of urban sprawl, because the United States is a suburban society, and because of peculiarities about how it formally defines "metropolitan area" using entire counties as building blocks. Thus, the second-largest capital city-metropolitan area in the world after Tokyo is Washington, DC, ranking 11th among the world's metropolitan areas with 1,157 square miles (2,996 sq km). Paris, capital city of France, ranks 14th (1,051 square miles; 2,723 sq km, see Table 3).

We can also rank capital cities according to the numbers of tall buildings (Table 4). This can be an index of how "urban" a capital city looks. The Web site of Emporis (http://www.emporis.com), a clearing house for date about architecture and construction, indicates that the capital city with the most high-rise buildings (defined as at least 115 ft (35 m) or 12 stories in height) is Singapore, the capital of the small Southeast Asian country with the same name. It has 4,767 high rises. After Singapore it is Caracas, the capital of Venezuela, with 3,864 high rises. These two cities rank 4 and 5 in the world in numbers of high rises, after non-capitals Hong Kong, China (7,896), New York City (6,504), and Sao Paulo, Brazil (6,467). The other extreme is not worth considering, as there are a great many of

Table 3   Largest Capitals by Land Area

Capital City (metro area)	Country	Area, Square Miles	Area, Square km	World Rank
Tokyo	Japan	2,700	6,993	2
Washington, DC	United States	1,157	2,996	11
Paris	France	1,051	2,723	14
Buenos Aires	Argentina	875	2,266	21
Moscow	Russian Federation	830	2,150	23
Mexico City	Mexico	800	2,072	27
London	United Kingdom (also England)	627	1,623	37
Kuala Lumpur	Malaysia	620	1,606	38
Manila	Philippines	540	1,399	44
Jakarta	Indonesia	525	1,360	47

http://www.citymayors.com/statistics/largest-cities-area-125.html

the world's 202 national capitals with not one high-rise building. In terms of super high-rise buildings, which Emporis calls skyscrapers and defines as at least 492 ft (150 m) tall, the capital city with the most is Tokyo with 109. It ranks sixth in world behind non-capitals Hong Kong, New York City, Dubai (United Arab Emirates), Shanghai (China), and Chicago (United States). Even more capitals in the world have no skyscrapers.

Another dimension for comparison is elevation. There are a great many capital cities at sea level, as many capitals such as Wellington, New Zealand, Buenos Aires, Argentina, Lomé, Togo, Helsinki, Finland, and Singapore, to name some widely scattered examples, are ports. Some capitals, however, have little or no elevation above sea level (i.e., no hill areas inland) and can be thought of as lowest in elevation. That is famously the case for Malé, the capital of the Indian Ocean island nation of the Maldives, which along with the country as a whole is in danger of inundation from rising sea levels because of global warming. At the other extreme, the highest altitude capital city is La Paz, one of the two capitals of South American country of Bolivia. Its average elevation is 11,942 ft (3,640 m). Number 2 is Quito, Ecuador, also in South America (9,350 ft; 2,850 m), and number 3 is Sucre, Bolivia's other capital (9.022 ft; 2,750 m). The highest capital city outside South America is Thimpu (8,688 ft; 2,648 m) in Asia, the capital city of the Himalayan Mountain kingdom of Bhutan. Table 5 lists the 10 highest capital cities by altitude.

Other distinctions for capital cities have to do with coordinates on the earth's grid of latitude and longitude. The northernmost capital city in the world is Reykjavik, Iceland at 64°08′N latitude, with the Scandinavian capitals of Helsinki, Finland (60°10′N), Oslo, Norway (59°10′N), and Stockholm, Sweden (59°19′N) being a close 2–3–4. The southern-most capital is Wellington, New Zealand

Table 4   Capital Cities by Numbers of Tall Buildings

Capital City (metro area)	Country	Number of High Rises*	World Rank
Singapore	Singapore	4,764	4
Caracas	Venezuela	3,864	5
Moscow	Russian Federation	3,754	6
Seoul	Korea	2,955	7
Tokyo	Japan	2,779	9

Capital City (metro area)	Country	Number of Skyscrapers**	World Rank
Tokyo	Japan	556	3
Bangkok	Thailand	355	6
Seoul	Korea	282	9
Kuala Lumpur	Malaysia	244	10
Singapore	Singapore	238	11

*Minimum 115 feet (35 m) or 12 stories in height

**Minimum 492 feet (150 m) in height

http://www.emporis.com

Table 5   Highest Capital Cities by Altitude above Sea Level

Rank	City	Country	Altitude (ft)	Altitude (m)
1	La Paz	Bolivia	11,942	3,640
2	Quito	Ecuador	9,350	2,850
3	Sucre	Bolivia	9,022	2,750
4	Thimphu	Bhutan	8,688	2,648
5	Bogotá	Colombia	8,612	2,625
6	Addis Ababa	Ethiopia	7,726	2,355
7	Asmara	Eritrea	7,628	2,325
8	Sana'a	Yemen	7,382	2,250
9	Mexico City	Mexico	7,350	2,240
10	Nairobi	Kenya	5,889	1,795

(41°17′S), with Canberra, Australia being second-most south (35°18′S). The capital city nearest the equator is Quito, Ecuador, a city in a country that is named for the equator. Its latitude is 0°14′S. Other equatorial capitals are in Africa: Kampala, Uganda (0°19′N); São Tomé, the capital of São Tomé and Principe at 0°20′N; and Libreville, Gabon (0°30′N). The capital city closest to the Prime Meridian

(Greenwich; the 0°N/S line) is London, England, the site of the Greenwich Observatory. The center of the city is at 0°7′W. The next closest capital to Prime Meridian is Accra, Ghana at 0°12′W. On the other side of the globe, the eastern-most capital city (that is nearest the 180° line and with an East longitude designation is Funafuti, Tuvalu, 179°13′E, while second-closest is Suva, Fiji, 178°44′E. The western-most capital city, also in the mid-Pacific, is Apia, the capital of Samoa, at 171°45′W. You will never guess the second western-most capital city in the world. It is some distance away from Samoa: Mexico City, at 99°08′W.

The two capital cities that are near one another are Vatican City and Rome, Italy, as the former is an enclave contained wholly within the municipal limits of the latter. Otherwise, the closest capital cities are Kinshasa, Democratic Republic of the Congo, and Brazzaville, Republic of the Congo, located about 1 mile apart (1.6 km) on opposites of the lower Congo River. In Europe, Vienna, Austria and Bratislava, Slovakia, 34 miles (55 km) apart, are famous as neighboring capital cities. The two capital cities that are farthest apart on the globe are Wellington, New Zealand and Madrid, Spain. They are 12,353 miles (19,880 km) apart, only 99 miles (160 km) from being antipodes at the maximum distance possible on the globe. The greatest distance between two capital cities whose countries share a border is the 3,991 miles (6,423 km) between Moscow, Russia and Pyongyang, North Korea. The longest distance from one capital city to the one closest to it is between Wellington and Canberra, 1,448 miles (2,330 km).

Types of Capital Cities

By definition, a national capital city engages in administration of government as a key economic function. However, virtually all cities, no matter how specialized they are in economy, engage in multiple economic roles, so one dimension of distinguishing between different types of capital cities is whether they are identified overwhelmingly as government centers primarily or are also known for production of other services and goods. The eminent urban scholar Peter Hall (1993, p. 69) has expressed this as a distinction between "*political capitals*" which were created specifically to be seats of government such as The Hague in the Netherlands, Washington, DC, Canberra, Ottawa in Canada, and Brasilia, the special-purpose capital of Brazil, and "*multi-function capitals*" that combine all or most high-end national functions in addition to the function of government administration. The examples that he lists are London, Paris, Madrid, Stockholm, Moscow, and Tokyo, to which I would add Seoul, Cairo, Buenos Aires, Lima, Mexico City, and Bangkok, among other cities. There are also capital cities that Hall calls "*global capitals*"—large cities that are political capitals within their own countries that dominate other aspects of the national economy and that have considerable global influence as well. London and Tokyo are the two most obvious examples, as identified by Hall, but Paris and perhaps Singapore, Rome, and maybe Moscow could be added to the list too.

Even if a city such as Washington, DC, is established specifically to be capital, it develops other branches of economy with time, especially if that city grows in size and prominence as did Washington. Not only are there thousands upon thousands of government employees in various branches of the capital economy, but there are also various ripples of other economy that develop: government-funded research and development, lobbying of government officials, headquarters of trade associations and professional associations, news media agencies, higher education, tourism and the hospitality industry, and others. While Washington was never noted as a manufacturing city, as was the case for so many of the other larger urban centers in the United States Northeast and Atlantic coast, it did grow an industrial economy as government grew: manufacturing in the printing and publishing industry such as of government reports, tax forms, and other paper production of an enormous bureaucracy. To a lesser extent, the same observations can be made for the other "political capitals" such as Ottawa and Canberra.

Overlapping these distinctions is the notion of a *primate city.* The term was coined in 1939 by the geographer Mark Jefferson to describe a city that dominates its own country as not only the officially designated political-administrative center, but also in many other critical parts of the nation's economy and life: banking and finance, industry, transportation, media, cultural institutions, education, technology, and others or some prominent combination thereof (Jefferson, 1939; Pacione, 2005, p. 83). The alternative pattern to a primate city pattern is one in which two or more cities divide up the work of the nation, and no one city dominates although only one is capital. As an example, the United States does not have a primate city pattern because, while Washington is the capital, New York is the leading financial center, Chicago and Los Angeles are also important in finance and other branches of the economy, Houston commands energy, Las Vegas has more casinos than any other city, and a host of other cities makes steel, cars, airplanes, and beer. Table 6 distinguishes examples on the left of capital cities that are clearly primate cities within their respective countries from those on the right where multiple cities direct the national economy. A number of very prominent cities are deliberately in neither column, because they are problematic. London and Moscow, for example, are both national capitals, have highly diversified economies in addition to government service, and also have much influence beyond national borders, but in both countries there are also other important cities that lessen the capitals' primacy: for example, Manchester, Birmingham, and Glasgow in the United Kingdom; and St. Petersburg, Kazan, and Irkutsk in the Russian Federation.

Another category of capital is what Hall (1993, p. 70) called the *"super-capital"* These are cities that function as centers for international organizations and may or may not be national capitals as well. Examples include Brussels, the capital of Belgium, which is also the headquarters city of much of the administration of the EU and of NATO. Brussels shares EU administration with Strasbourg, a smaller city in France near the border with Germany, and with Luxembourg, another smaller city that is capital of the Duchy of Luxembourg, but these two cities hardly seem to be "super" on a global scale, although they are both very pleasant places. Geneva in Switzerland and New York City are also "super capitals" because of their many international organizations, most notably offices of the United Nations, although neither is a national capital. National capitals Nairobi, Kenya, and Addis Ababa, Ethiopia are also "super capitals" because they are headquarters for

Table 6   Primate Cities and Non-Primate Cites as Capitals: Some Examples

Primate City Capitals	Non-Primate City Capitals
North America: None	North America: USA: Washington, New York, Chicago, Los Angeles, Houston, Boston, Atlanta, etc. Canada: Ottawa, Toronto, Montreal, Vancouver, Calgary, etc.
Central and South America: Mexico: Mexico City Peru: Lima Argentina: Buenos Aires Costa Rica: San Jose	Central and South America: Brazil: Brasilia, Sao Paulo, Rio de Janeiro, Manaus, etc. Ecuador: Quito, Guayaquil Venezuela: Caracas, Maracaibo, Ciudad Guyana, etc.
Europe: France: Paris Austria: Vienna Hungary: Budapest Denmark: Copenhagen Ireland: Dublin Latvia: Riga	Europe: Italy: Rome, Milan, Turin, Naples, etc. Germany: Berlin, Frankfurt, Munich, Hamburg, Cologne, etc. Switzerland: Bern, Zurich, Geneva, Basel, etc. Spain: Madrid Barcelona, Sevilla, etc.
Middle East/North Africa: Egypt: Cairo Iran: Tehran Azerbaijan: Baku	Middle East/North Africa Turkey: Ankara, Istanbul, Izmir, etc. Israel: Jerusalem, Tel Aviv, Haifa, etc. United Arab Emirates: Abu Dhabi, Dubai
Africa: Ethiopia: Addis Ababa Angola: Luanda Senegal: Dakar Uganda: Kampala	Africa: South Africa: Pretoria, Cape Town, Bloemfontein, Johannesburg, Durban, etc. Tanzania: Dodoma , Dar es Salaam, Zanzibar, etc. Cote D'Ivoire: Yamoussoukro, Abidjan
Asia: Thailand: Bangkok Malaysia: Kuala Lumpur Japan: Tokyo Singapore: Singapore	Asia: P.R. China: Beijing, Shanghai, Hong Kong, Chongqing, Nanjing, Guangzhou, Shenzhen, etc. India: New Delhi, Kolkata, Mumbai, Chennai, Bangalore, etc. Pakistan: Islamabad, Karachi, Lahore, Hyderabad, etc.
Oceania: Papua and New Guinea: Port Moresby Samoa: Apia	Oceania: Australia: Canberra, Melbourne, Sydney, Perth, Adelaide, Brisbane, etc. New Zealand: Wellington, Auckland, Christchurch, Dundee, etc.

so many African international organizations. They are bases as well for foreign or-ganizations that provide aid in Africa. Rome, Italy can be added to this list because it, along with its Vatican City enclave, headquarters the global Roman Catholic Church. Likewise, Mecca in Saudi Arabia is a "super capital" of sorts (although not a national capital) because this holy city is the destination for the *hajj*, a religious pilgrimage that millions of Moslems from around the world make each year and that every Moslem, no matter how far from Mecca they happen to be, tries to make at least once in a lifetime.

We can also identify *imperial capital cities* and *colonial capital cities* as two other categories of world capitals. The former are cities that as capitals of their own na-tions also ruled a network of colonies abroad. This was particularly so in the age of colonialism in the 19th and early 20th centuries. The fact of being imperial capi-tals means that great wealth came to these cities from other places, with a result that such cities often have grand palaces, monuments, and boulevards, beautiful cathedrals, mosques, or temples for religious worship, and various other land-scape features of privileged cities depending on where in the world and the time in history. The most famous examples are several of the capitals of European coun-tries, most especially London, the capital of England and the United Kingdom. It once ruled an empire that spanned all the continents. India and other parts of South Asia were once British, as were many parts of Africa, Canada, Australia, and other places. Other imperial capitals in Europe were Paris, Amsterdam, Madrid, and Lisbon, as well as Brussels, the Belgian capital that once ruled a huge territory in the heart of Africa that was called the Belgian Congo. Vienna and Moscow were imperial capitals too, although in this case the colonies were not strung out over-seas but were subjugated neighboring countries: the Austro-Hungarian Empire in the case of Vienna and the Russian Empire and its successor the Soviet Union in the case of Moscow. Athens, Greece, and Rome, Italy, are examples of national cap-itals that, centuries ago, were once imperial capitals.

Outside the European world, Beijing and Tokyo are examples of current capital cities that also ruled (or still rule, in the case of Beijing) vast conquered territories. Japan lost its empire with defeat in World War II, while China has succeeded in annexing the territory of Manchus, Mongols, Uyghurs, Tibetans, and other non-Han Chinese peoples into borders that are now ruled from Beijing. By the same logic, even though Americans rarely apply such vocabulary to their own country, Washington can be considered to be an imperial or former imperial capital, be-cause the United States expanded across the continent to absorb the lands of oth-ers at the frontier and parts of Mexico. The U.S. "empire" also reached across the waters to the Hawaiian Islands and other places in the Pacific, to Puerto Rico, for a time to the Philippines, and in a way to Cuba and other places in the Caribbean and Central America that may not have been American colonies per se but whose governments were once under disproportionate American influence.

Colonial capital cities, on the other hand, are those capital cities that were once the seats of colonial government abroad. These cities stand out because some parts of their architecture and urban form resembles that of European imperial capitals, and because the cities were developed in whole or in part for the convenience, comfort, and profit of colonial elites. Sometimes, these colonial capitals were built atop an existing indigenous capital or other settlement, as in the famous case of Mexico City being built by Spanish conquistadores atop the ruins of Aztec Tenochtitlan, while at other times they were grafted onto existing settlements. Examples of the latter include Batavia (now Jakarta) which the Dutch developed beside the locals' Sunda Kelapa as their capital for the Dutch East Indies, now Indonesia, and New Delhi, India which was developed in 1911 beside "Old Delhi," already an historic city. Other colonial capitals were entirely new: Spanish Lima (Peru), English Nairboi and English Singapore, and Porto-Novo, the capital of the small African nation of Benin. That city originated with a slave port that was established by the Portuguese and later became the capital of the French colony of Dahomey. A colonial capital of a different sort is Kyiv, the historic capital of Ukraine. It eventually became part of the Russian Empire and was transformed first into a Russian city, and then into a Soviet city that reported to Moscow and took architectural and design cues from Soviet builders and planners. During Russian and Soviet rule, the city was known as Kiev, following the Russian pronunciation. Minsk, the capital of Belarus also took on colonial city characteristics after it became part of the Soviet Union, as did the capitals of other non-Russian republics of the former USSR.

Former Capitals

This opens the way for mention of various categories of *former capital cities*. In the Russian Empire, such a city was St. Petersburg which, except for a short interlude, ruled the Russian Empire from 1713 to 1918 between periods of rule by Moscow. Other former imperial capitals include Istanbul, then named Constantinople, which ruled the Ottoman Empire until the founding of modern Turkey and designation of Ankara as national capital in 1923, Lahore (now Pakistan) which once ruled the Mughal Empire, Königsberg which was capital of the Prussian Empire from the late Middle Ages until 1701 when the capital was moved to Berlin, Cusco in Peru which was the former capital of the Inca Empire, Chang'an in China, and Angkor, the once glorious capital of the Khmer Empire. St. Petersburg, Chang'an (now named Xi'an), Lahore, and Istanbul are all still prominent urban centers, while Cusco and Königsberg have more of a local influence only, and Angkor is an archaeological site and tourism attraction. Other globally prominent cities that once flourished as capital cities but no long are Kyoto in Japan, Nanjing in China, Kolkata (Calcutta) in India, Karachi in Pakistan, Yangon (Rangoon) in Myanmar (Burma), Rio de Janeiro in Brazil, and Bonn in Germany.

Further Reading

Forstall, Richard L., Richard P. Greene, and James B. Pick. "Which Are the Largest? Why Lists of Major Urban Areas Vary So Greatly," *Tijdschrift voor Economische en Sociale Geografie* 100, no. 3 (2009): 277–97.

ABIDJAN. *See* Yamoussoukro.

ABU DHABI

Abu Dhabi is the capital and second-largest city after Dubai in the United Arab Emirates (UAE), a federation of seven emirates or principalities in the Middle East on coast of the Persian Gulf on the southeastern Arabian Peninsula. It is also the capital of the Emirate of Abu Dhabi. The city is located on a small triangular island in the Persian Gulf that is just off shore from the mainland and is connected to the mainland by a short bridge. The population of Abu Dhabi is about 900,000. In addition to its role as government center for the UAE, it is also one of the world's leading producers of oil and natural gas and a major exporter. Income levels and living standards for citizens are correspondingly high. In recent years, oil revenues have been applied to diversification of the economy, such that the city is now also a leading financial and banking center, and is a center of tourism, retailing, real estate development, and industrial production.

Historical Overview. Oil was discovered in Abu Dhabi in 1958 and the commercial production began in 1962. Before then, Abu Dhabi was but a small town of local importance only. Its origins are traced to 1793 when members of the Bani Yas Bedouin tribe migrated there because of the presence of fresh water. The Al Nahyān family was part of that first migration and is the ruling family of Abu Dhabi till today. Pearling, which was an early basis of the economy, declined with the rise of the cultured pearl industry in Japan in the 1930s. The emirates were a protectorate of Great Britain from the 19th century until 1971, and have since been united into a federation of absolute monarchies government by a president chosen by the seven emirs. At that time, Abu Dhabi was chosen as the capital for the federation. Economic growth had been slow during the very conservative long rule of Sheikh Shakhbūt ibn Sultān Āl Nahyān (1928–1966), but then picked up when his younger brother Zāhid ibn Sultān took over and began investing in infrastructure to support the development of a petroleum-based economy. The city has since grown very rapidly and has one of the world's leading construction projects, with considerable development of impressive skyscrapers, commercial centers, new airport, roads, international hotels, sports facilities, and many other projects.

Major Landmarks. The Sheikh Zayed Mosque, now nearing completion, is one of the most opulent and beautiful newly built religious structures in the world. It is the world's sixth-largest mosque and can accommodate as many as 40,000 worshippers, including up to 7,000 in the main prayer hall. The waterfront of Abu Dhabi is a spectacular stretch of gleaming high-rises, shopping malls, and recreation places. There are sandy beaches as well. Khalifa Park has an aquarium, museum, and other attractions. Yas Island, located off the main island on which Abu Dhabi is situated, has a famous Formula 1 automobile racetrack and is the site of the Abu Dhabi Grand Prix.

Culture and Society. The official state religion of the UAE is Islam and Arabic is the official language. However, the population is highly mixed ethnically, so other languages are spoken and foreign residents have the freedom to practice their own faiths. Native-born Emiratis are a minority in their own country, and native-born Abu Dhabians are a minority in their emirate and its capital city. The majority of inhabitants of the UAE, Abu Dhabi Emirate, and the city of Abu Dhabi are foreign expatriate workers. Some hold such specialized, well-paying jobs as engineers and architects, but the majority of inhabitants are laborers or service workers and earn far less. More than one-half of the total population is from India and Pakistan. Other sources of workers are Sri Lanka, Bangladesh, Eritrea, Ethiopia, Somalia, Iran, the Philippines, Sri Lanka, and the United Kingdom. The enormous oil wealth of Abu Dhabi assures its own citizens a very comfortable living with many benefits provided by the government. Foreigners live well too, but are clearly in a lower class of residence. Abuses of human rights have been reported, including sexual abuse of female domestic workers from abroad, human trafficking, and perpetuation of indebtedness on the part of foreign workers, especially those from South Asia, turning them into de facto indentured servants.

Further Reading

Al Fahim, Mohammed. *From Rags to Riches: A Story of Abu Dhabi.* London: London Center of Arab Studies, 1995.

Davidson, Christopher M. *Abu Dhabi: Oil and Beyond.* New York: Columbia University Press, 2009.

Tatchell, Jo. *A Diamond in the Desert: Behind the Scenes in Abu Dhabi, the World's Richest City.* New York: Black Cat Publishing, 2009.

ABUJA

Abuja is the new capital of the Federal Republic of Nigeria, a West African country that is the continent's most populous nation. The city became the capital on December 12, 1991, replacing Lagos, and was specially built for the purpose. It is located in the very center of the country near Aso Rock, a 1,300-ft-high (400 m) feature of the natural landscape that stands as an icon of the city. It is the focus of the Federal Capital Territory (FCT), which was formed in 1976 from parts of three states in the center of Nigeria: Nasawara, Niger, and Kogi. The population of Abuja is increasing very rapidly and was 776,298 as per the 2006 census.

Historical Overview. Lagos, a port city near the southwest corner of Nigeria on the Bight of Benin of the Atlantic Ocean had been the capital of the country since 1914 when the country was still a colony of Great Britain. It was the first capital after independence was achieved in 1960. Nigeria, however, is extremely diverse in terms of ethnic groups, languages, and other aspects of culture, and is divided geographically between a mostly Christian south and an overwhelmingly Muslim north. Consequently, the leaders of Nigeria agreed that a central capital would be better for unification of the country and as a symbol of ethnic-religious neutrality in national government than the giant city in a corner of the nation. Besides, Lagos was far too big and was growing much faster than services and infrastructure could be provided. The decision to move the capital and build Abuja took shape during the 1960s and early 1970s. Ground was broken in the late 1970s, and the capital was inaugurated in 1991. Planning and construction took much longer and was considerably more expensive than was expected because of political conflict and Nigeria's endemic corruption. Three American firms (Planning Research Corporation; Wallace, McHarg, Roberts, and Todd; and Archisystems) took charge of the master plan for the new city, while the detailed design for the city center was principally the work of Japanese architect Kenzo Tange.

Major Landmarks. In addition to Aso Rock, there is also Zuma Rock, another high monolith in the north of the city. It stands 2,379 feet (725 m) above the surroundings and is depicted on the Nigerian 100 naira bill. Major architectural landmarks include the spectacular Abuja National Mosque with its golden dome and four high minarets, the National Church of Nigeria, the iconic Abuja City Gate, and Ship House, the ship-shaped headquarters building for the national Ministry of Defense. The headquarters of the Central Bank of Nigeria is also an impressive structure. The largest park in the city is Millennium Park. Millennium Tower is under construction in the center of the city's main business district. Designed by Italian architect Manfredi Nicoletti, the structure pro mises to be spectacularly dist inctive in architectural form and will be Nigeria's tallest building when completed.

Culture and Society. Nigeria is a country with great wealth from its oil resources and other parts of the economy, but the wealth is distributed very unevenly and there is great poverty. Abuja is a symbol of opportunity for all and national unity. The city has grown very rapidly due to migration from all of the country's 36 states, and reflects

The Abuja National Mosque. (Klaas Lingbeek-van Kranen/ iStockPhoto.com)

the ethnic, linguistic, and religious diversities of the nation. The city is at times a model of coexistence and cooperation across social boundaries, and at other times the scene of bloody conflict. For example, on August 26, 2011, the Nigerian jihadist group Boko Haram detonated a car bomb in front of Abuja's United Nations building that killed at least 21 individuals and injured about 60 others. As the city builds monuments and beautiful government and business buildings that are appropriate for the capital of a populous and prominent nation, new arrivals to Abuja build squatter shacks at the city's outskirts, which multiply by the thousands into enormous new slum districts. English is the official language of Nigeria and the language that binds the residents of Abuja from different regions of the country.

Further Reading

Dawam, Patrick D. "Urban Development and Population Relocation in Abuja, Nigeria," *Ekistics* 61, no. 366/367 (1994): 216–20.

Jibril, Ibrahim Usman. "Resettlement Issues, Squatter Settlements and the Problems of Land Administration in Abuja, Nigeria's Federal Capital," a paper presented at a conference in Accra, Ghana, on March 8–11, 2006 and retrieved at www.fig.net/pub/accra/papers/ts18/ts18_01_jibril.pdf.

Wright, Herbert. *Instant Cities*, 28–31. London: Black Dog Publishers, 2008.

ACCRA

Accra is the capital and largest city of Ghana, a country in West Africa on the Gulf of Guinea. It is located in the south of the country along the coast. In addition to its administrative functions, the city is the major business and financial center of Ghana and houses the headquarters of the national Cocoa Marketing Board that oversees sales of Ghana's principal export. The heart of the city was once an important port as well, but now most of the nation's seagoing imports and exports are processed through Tema, a manmade port some 16 miles (25 km) to the east. Accra's population is about 1.7 million, while that of the metropolitan area exceeds 4 million.

Historical Overview. The area of Accra was settled in the 15th century by the Ga people, one of the main ethnic groups of coastal Ghana, and the city itself started as a fishing port. Beginning with the Portuguese in 1482, Europeans arrived on the coast of Ghana, which they named the Gold Coast, and built forts to enable exploitation of natural resources and trade in slaves. In all, 27 forts were built by Europeans along the Ghana coast, including three in the territory of present-day Accra: Ussher Fort by the Dutch in 1605, Christiansborg Castle by the Danes in 1657, and James Fort by the British in 1673. The slave trade was the main occupation of Europeans along the Gold Coast until it was abolished in 1807. In 1873, the British captured the Asante capital of Kumasi, a city in the interior of Ghana and made the Gold Coast a crown colony. Accra became the capital in 1877, replacing Cape Coast, because its climate was drier and therefore safe from diseases borne by the tsetse fly. In the late 19th century, the British developed Victoriaborg as an exclusive suburb for themselves and other Europeans to the immediate east of Accra's historic core. In 1923, a critical rail link was completed to Kumasi, enabling Ghana to export cocoa. Accra remained the capital of British Gold Coast until Ghana achieved independence in 1957, when the city became a national capital. The first leader of independent Ghana, Kwame Nkrumah (prime minister, 1957–1960; president, 1960–1966), ended the British system of racial segregation in the city, and erected monuments and public spaces in Accra that promoted national unity, pride in Africa, and the philosophy of pan-Africanism.

Major Landmarks. Independence Square, also known as Black Star Square, is located at the waterfront of Accra and was built during the administration of Kwame Nkrumah to celebrate Ghana's independence and honor three young Ghanaians who were shot and killed by British authorities during the colonial period. Other monuments are Independence Arch, the Kwame Nkrumah Mausoleum, and the three historic forts, Ussher Fort, James Fort, and Christiansborg Castle. Christiansborg Castle is also known as Osu Castle; it has been rebuilt and modernized many times since its original construction, and has served for many years as the seat of government of Ghana, a role that it continues to have till today. Golden Jubilee House is the new residence and office of the president of Ghana. Important cultural sites are the National Museum, the China-built National Theater of Ghana, and the Du Bois Centre (a research library), and the gravesites of the prominent African American scholar W. E. B. Du Bois and his wife, the playwright Shirley Graham Du Bois.

Culture and Society. Ghana is home to at least 100 indigenous ethnic groups, and Accra, as the national capital, houses a representation of all of them. The main groups in the city, however, are the Ga, who are indigenous to the site of Accra, and the Akan, the most populous of Ghana's ethnic groups. English is the official language of the country, but most Ghanaians also speak at least one ethnic language. About three-quarters of Ghana's residents are Christians and 16 percent Muslims. Accra is one of the fastest-growing cities in Africa, with considerable migration from the countryside and high birth rates. Much of the city is modern, prosperous, and impressive looking, but there are also large expanses of poor squatter settlements with substandard housing and inadequate services and infrastructure.

Further Reading

Grant, R. and P. Yankson. "Accra," *Cities: The International Journal of Urban Policy and Planning* 20, no. 1 (2003): 65–74.

Grant, Richard *Globalizing City: The Urban and Economic Transformation of Accra, Ghana.* Syracuse: Syracuse University Press, 2009.

Yeboah, I. "Structural Adjustment and Emerging Urban Form in Accra, Ghana," *Africa Today* 7 (2000) 61–89.

ADDIS ABABA

Addis Ababa is the capital and largest city of Ethiopia, formally named the Federal Democratic Republic of Ethiopia, a landlocked country of 82 million people on the Horn of Africa in East Africa. It is Africa's second most populous nation and the world's most populous landlocked country. Addis Ababa is located in the center of the country at the foot of Mount Entoto, and has an elevation that ranges from 7,631 feet (2,326 m) at its lowest level to some 9,800 feet (3,000 m) in the Entoto Mountains to the north. The population of the city is about 3.4 million, while that of the metropolitan area is about 4.6 million. The name Addis Ababa means little flower, and was given to the city by Empress Taytu Betul when it was founded in 1886.

Historical Overview. Prior to the founding of Addis Ababa, the capital of Ethiopia had migrated within the kingdom until Emperor Menelik II decided to build a permanent capital on the Entoto Mountains. However, the site lacked wood for fuel and was cold, so according to the often-repeated story, Empress Taytu Betul convinced her husband to reconsider and build the capital instead at a lower elevation near mineral hot springs where she had built her own palace and enjoyed bathing in warm therapeutic waters. Other members of the court built residences nearby, and the new capital began to grow. To counter the shortage of firewood, Menelik wisely imported eucalyptus trees from Australia, which took root well in the Entoto Highlands and turned the young Addis Ababa into a green and leafy city. The new capital was laid out with wide streets, and many beautiful churches and other buildings were erected. Italian troops took over Ethiopia in 1936 in the Second Italo-Abyssinian War, and ruled the country as part of Italian East Africa until 1941 when they were defeated by the British forces in World War II. During the Italian period, Addis Ababa was the colonial capital.

Emperor Haile Selassie I, who ruled Ethiopia from 1930 until his death in 1974, reestablished the city as a national capital after the Italians were defeated, and made Addis Ababa a center of African unity and anticolonial activity among emerging African nations. The city was the headquarters of the Organization of African Unity (OAU) from its inception in 1963, and in 2002 became headquarters of the successor of the OAU, the African Union, AU. Addis Ababa is also the base of the United Nations Economic Commission for Africa. Both organizations are located in Africa Hall. Despite these distinctions for the city, Addis Ababa is quite poor and, like Ethiopia as a whole, has suffered from drought conditions, famine, revolution, repressive government, and a lost war that resulted in the breaking off from the country of Eritrea.

Major Landmarks. Addis Ababa has many important landmarks, including Africa Hall, the Parliament Building, the National Palace (now residence of the president of Ethiopia), King Menelik's Palace (now the seat of the government of Ethiopia), and Addis Ababa University. Principal places of worship are the Holy Trinity Orthodox Cathedral, Medhane Alem (meaning "Savior of the World"), St. George's Cathedral (the second-largest church on the African continent),, and the Anwar Mosque. Important museums include the Ethiopian National Museum in which there is a famous replica of Lucy, an early hominid, the Ethiopian Ethnological Museum, the Addis Ababa Museum, and the Red Terror Museum, a museum that commemorates the suffering of Ethiopia's people during the 1977–1978 violent political campaign by Communist ruler Mengistu Haile Mariam. The Arat Kilo Monument stands in the center of a busy traffic circle in the city and commemorates Ethiopia's struggle against Italian occupation. Outside the city is the summit of the Entoto Mountains (10,827 feet; 3,300 m) and two old stone churches, St. Mary and St. Raguel, that were built by Menelik II, as well as a smaller royal palace.

Culture and Society. The Ethiopian society consists of many ethnic and linguistic groups and Addis Ababa, as the capital city, reflects the mix. In the city, the main ethnic groups are the Amhara people (47% of the population), followed by Oromo (19.5%), Gurage (16%), Tigray (6%), Silt'e (3%), and Gamo (nearly 2%). Amharic is the main language, spoken by 71 percent of the population, followed by other ethnic languages. About three-quarters of the population belongs to the Ethiopian Orthodox Church, the principal religious denomination of the country, while some 16 percent comprises Muslims and 8 percent comprises Protestants. Many Ethiopians have distinguished themselves globally as champion long distance runners. The mountains outside the city are said to be excellent training grounds.

Further Reading

Angélil, Marc. *Cities of Change: Addis Ababa.* Basel: Birkhäuser Architecture, 2009.

Benti, Getahun. *Addis Ababa: Migration and the Making of a Multi-Ethnic Metropolis, 1941–1974.* Trenton, NJ: Red Sea Press, 2007.

Reminick, Ronald A. *Addis Ababa: The Evolution of an Urban African Cultural Landscape.* Lewiston, NY: Edwin Mellen Press, 2010.

ALGIERS

Algiers is the capital and largest city of the People's Democratic Republic of Algeria, usually referred to as simply Algeria, a country in the Maghreb region of north-western Africa that stretches from the Mediterranean Sea in the north into the heart of the Sahara Desert in the south. The city is located on the Mediterranean coast on the west side of the Bay of Algiers. The word "Algiers" comes from the Arabic for "the islands," referring to four landmark islands in the bay just offshore. The city is sometimes referred to as "Algiers the White" (more commonly *Alger la Blanche* in French) because its white buildings offer a contrast against the blue sea and blue sky. The population of Algiers is estimated to be about 2.9 million, while that of the metropolitan area is 5.5 million.

Historical Overview. The site of Algiers was once a Phoenician port city that was later incorporated into the Roman Empire and was named Icosium. The present-day city was founded in 944 by Bologhine ibn Ziri. In 1159, the city came under the dominion of the Almohad Caliphate, and in the 13th century the city was ruled by sultans from Tlemcen, a coastal city in northwest Algeria. In 1510, after defeating the Moors on the Iberian Peninsula, Spain began to exert authority over the north coast of Africa, and constructed a fortress on the rocky island of Peñon in Algiers's harbor. By 1524, the Spaniards were defeated and the Algiers territory was annexed to the Ottoman Empire, although the Ottomans exercised little control over the city. The city became a base for piracy against Mediterranean shipping and ports. Other centers of the Barbary pirates were Tunis and Tripoli, the capitals of today's neighboring countries of Tunisia and Libya, respectively. European powers allied against the pirates, and during 1801–1805 and in 1815, the United States fought the First and Second Barbary Wars, respectively, to make the Mediterranean safe for ships. Algiers was bombarded by a British squadron in 1816–1817, destroying the corsairs in its harbor. In 1830, a French army defeated

Algiers, opening the way for French takeover of Algeria as a colonial possession with Algiers as the capital city. Many French and other Europeans settled in Algeria afterward. In 1954, the Algerian National Liberation Front (FLN) launched a bloody war for independence. It was finally won in 1962, resulting in the sudden repatriation of 1 million or so of so-called *Pieds-Noirs,* the French residents of North Africa, to France. Unrest continued to mark Algeria after independence. In 1988, Algiers was the focus of a prodemocracy movement in Algeria called the "Spring of Algiers" that led to a new constitution in 1989. In the 1990s, Algiers saw considerable conflict with radical Islamists and the start of a 10-year civil war. During 2010–2012, Algiers and other cities saw mass popular protests about poor economic conditions and political repression in the country.

Major Landmarks. The Casbah or Citadel sits on a 400 ft (122 m) hill that overlooks the city. Nearby is the old quarter of Algiers. The citadel was founded on the ruins of the Roman town Icosium. Makamelchahid, or the Martyrs' Memorial, is an iconic concrete structure that was built in 1982 on the 20th anniversary of Algerian independence in honor of those who gave their lives in the fight against French colonialism. The Great Mosque dates back to at least 1097 and is the oldest mosque in Algiers; its minaret was built in 1324. The Ketchaoua Mosque has an interesting history. It was made into a mosque with Algerian independence in 1962; before then it was the Roman Catholic Cathedral of St. Philippe. It had been a mosque even before, however, erected in 1612, and was converted into a church in 1845 soon after France gained control of Algeria. The church Notre Dame d'Afrique (Our Lady of Africa) was built by the French in 1872 on a cliff overlooking the city and formerly could be reached only by a cable car. There is an inscription on the apse that reads in French: "Our Lady of Africa, pray for us and for the Muslims." The contemporary side of Algiers is represented by the Centre Commercial Al Qods, a popular modern shopping mall.

Culture and Society. The population of Algiers is mostly ethnically Berber and overwhelmingly Sunni Muslim in religious faith. The official language is Arabic. During the last few years there has been a conflict between those who want Algeria to be a fundamentalist Islamic society under strict *sharia* law and those who prefer a more liberal society. The government is authoritarian, and there have been mass popular protests in recent years in support of political freedoms and democracy.

Further Reading

Celik, Zeynep. *Urban Forms and Colonial Confrontations: Algiers under French Rule.* Berkeley: University of California Press, 1997.

Celik, Zeynep, Julia Clancy-Smith and Frances Terpak, eds. *The Walls of Algiers: Narratives of the City through Text and Image.* Seattle: University of Washington Press, 2009.

AMMAN

Amman is the capital and largest city in the Hashemite Kingdom of Jordan, more commonly referred to as simply Jordan, a small landlocked country in the Middle East, to the immediate east of Israel. The city is located in a hilly region of northwest

Jordan and was originally built on seven hills. Growth of the urban area, however, has expanded the city well beyond the original territory. The straight-line distance between Amman and Israel's capital Jerusalem is about 44 miles (70 km). The population of Amman is about 2.1 million and 2.9 million for the wider metropolitan area.

Historical Overview. Archaeological evidence indicates that the site of Amman has been settled for more than 10,000 years. The city was called Rabbath Ammon or Rabat Amon by the Ammonites in the 13th century BC, and is referred to as Rabbat 'Ammon in the Hebrew Bible. Over the centuries, the city was conquered by Assyrians, Persians, and Macedonians. In the third century BC, the Macedonian ruler of Egypt Ptolemy II Philadelphus renamed the city Philadelphia. During the first century AD, the city was ruled by Nabataeans, and then in 106 AD it became part of the Roman Empire. With the start of the Eastern Roman Empire based in Constantinople in 395, the city was designated as the site of a Christian bishopric and was later renamed Amman. It declined by the 10th century and was of local importance only until Circassian refugees from the Russian Empire's conquest of their lands in the Caucuses Mountains settled there in 1878. Later, the city became a major stop on a rail line that was built between Damascus (Syria) and Medina (Saudi Arabia). In 1921, when the Emirate of Transjordan was established, Amman became its capital city.

Since the middle of the 20th century, Amman's population swelled with refugees from nearby territories. First, there was the arrival of displaced Palestinians with the establishment of the state of Israel in 1948. Then, more refugees flocked to the city after Jordan lost Jerusalem and territory on the West Bank of the Jordan River in the Six Day War of 1967. In 1991, Palestinian settlers of Kuwait fled to Jordan in the context of the Gulf War. In 2003, tens of thousands of Iraqi citizens resettled in Amman after invasion of their country by the United States and the start of a prolonged war. As a result of such population growth, Amman is rapidly constructing new districts and new housing developments. On November 9, 2005, the center of the city was the scene of three coordinated hotel bombings by Al Qaeda terrorists.

Major Landmarks. The Citadel of Amman is on the city's highest hill at the site of Rabbath Ammon. The Umayyad Palace dating from about 720 AD is located in the northern part of the Citadel and is the principal landmark. The remains of the Roman Temple of Hercules and its towering columns are also on the citadel hill. Other landmarks in Amman are the Roman Theater, the Roman Forum, and the National Archaeological Museum. The King Abdullah I Mosque built between 1982 and 1989 is the largest mosque in Jordan. Following the death of King Abdullah I in 1999, his son and successor, Abdullah II, opened the Royal Automobile Museum to showcase his father's collection of fine automobiles. Landmarks of a different sort in Amman are its various gleaming shipping malls such as City Mall, Mecca Mall, and Abdoun Mall, and its upscale pedestrian shopping street Wakalat Street. There are quite a few new high-rise buildings in Amman, and a large construction project called Abdali New Downtown is underway.

Culture and Society. Amman is a generally prosperous city given its history of refugee settlement and accommodation, and the relatively small size of Jordan. The city has evolved into a major business center for its region and a popular destination for tourists, especially from Arabic-speaking countries in the Middle East. The city has a reputation for being open and tolerant. Jordan itself is a constitutional monarchy and is relatively stable politically. Most Jordanians are Muslims and the official language is Arabic.

Further Reading

Alon, Yoav. *The Making of Jordan: Tribes, Colonialism and the Modern State*. London: I. B. Tauris, 2009.
Kadhim, M. B. and Y. Rajjal. "Amman," *Cities: The International Journal of Urban Policy and Planning* 5, no. 4 (1988): 318–25.
Robins, Philip. *A History of Jordan*. Cambridge: Cambridge University Press, 2004.

AMSTERDAM

Amsterdam is the capital and largest city of the Netherlands, a small country on the North Sea in Western Europe. It is located in the western part of the country in the province of North Holland. The Amstel River runs through the heart of the city. The city's population is about 783,000, while the metropolitan area has about 2.2 million inhabitants. The Amsterdam metropolis comprises the northern part of the Randstad, a conurbation of close-together Dutch urban areas that totals some 7.1 million inhabitants. Other major cities of the Randstad include Rotterdam, The Hague, and Utrecht.

Amsterdam is the official capital of the Netherlands as stipulated in the August 24, 1815, national constitution and its successors. However, the seat of government, that is, the meeting places of the national legislature (the States-General), the executive office, and the Supreme Court, are all in a city named The Hague, the third-largest city in the Netherlands with a population of about 500,000. The Hague has been Holland's seat of government since 1588.

Historical Overview. Amsterdam traces its beginnings to a dam and bridge that was built across the Amstel River in 1275. It is from that beginning that the name of the city is thought to derive. In either 1300 or 1306, Amsterdam was granted the rights of a city. Later in the 14th century, it traded with the Baltic and North Sea ports of the Hanseatic League and evolved into a prosperous city. As a result of the Eighty Years' War of 1568–1648 in which the Protestant Dutch rebelled about matters of religion and taxation against the Roman Catholic King Philip II of Spain, the regent of the Hapsburg Netherlands, the Netherlands gained its independence. The city welcomed religious refugees from the Spanish Inquisition and Huguenots from France, and became known as a place of social and religious tolerance. In the latter part of the 17th century, Dutch ships began to ply the world for trade, and Amsterdam grew into a rich city and a leading financial center. The city's stock exchange, established in 1602, is the oldest stock exchange in the world. Colonies in South and Southeast Asia and in the West Indies produced enormous wealth for Dutch merchants. This period is called the Dutch Golden Age. In Amsterdam it is reflected in architecture and urban design, most prominently the city's famous network of canals.

The city lost some ground to repeated episodes of disease in the 17th century, and then warfare with England and Napoleonic France in the 18th and 19th centuries, but after the establishment of United Kingdom of the Netherlands in 1815, Amsterdam entered a second Golden Age. In addition to prospering form the Dutch colonial holdings in Central and South America, Africa, and South and Southeast Asia (most notably the Dutch East Indies, the large archipelago that would later become independent Indonesia) that it administered, Amsterdam became an industrial city, and a leading center of learning and culture. Again, new architecture and urban design expressed the city's wealth and prestige. Nazi Germany occupied the Netherlands in World War II and murdered tens of thousands of Dutch Jews. The most famous victim was the young Jewish girl, Anne Frank, whose diary about a life in hiding until she herself fell victim to the Holocaust is one of the world's most widely read books. The Netherlands lost most of its overseas colonies to the independence movements that swept across the colonial world after World War II, and engaged in a costly war in the late 1940s in a futile attempt to keep possession of Indonesia. Amsterdam is now an increasingly diverse city socially, in part because of immigration from former colonies, as well as a leading world financial, industrial, cultural, and educational center. Foreign tourism to Amsterdam is an important part of the urban economy as well.

Major Landmarks. The most distinguishing feature of Amsterdam is the *grachtengordel,* the central city's historic network of canals. They were dug during the Dutch Golden Age in the 17th century and form concentric belts around the inner city. The three main canals are called Herengracht, Prinsengracht, and Keizersgracht. The Singel is an old canal that once encircled the city as a defensive moat. All told, there are more than 60 miles (100 km) of canals, forming about 90 islands, and some 1,500 bridges. Alongside the canals are some 1,550 historic buildings. The entire complex is registered as a UNESCO World Heritage Site.

There are many excellent museums in Amsterdam, including the Anne Frank Museum, The Rijksmuseum, and the Van Gogh Museum. The Rijksmuseum is famous for works by Rembrandt and other Dutch Masters. Other landmarks in Amsterdam are Zuidas (translated as "south axis"), an area of tall, modern, and post-modern buildings that comprises the city's principal center of finance and business, and De Wallen, an area of narrow streets and back lanes set aside for legalized prostitution. It is also called the Amsterdam Red Light District, and is a popular tourist attraction if only for just a look. Dam Square, or simply "the Dam" is where Amsterdam was founded. It is a prominent public plaza with landmarks such as the Royal Palace and the National Monument, a memorial to those killed in World War II. On May 7, 1945, the Dam was the site of a particularly senseless machine gunning of Dutch civilians by Nazi troops. The Damrak is a main street in the center of Amsterdam that follows the course of a canal that had been filled in, connecting the city's central railway station and Dam Square. The old stock exchange building, Beurs van Berlage, is on this avenue. Rembrandt Square is a popular place for nightlife and cafés. In the center of the square is a statue of the famous 17th-century Dutch painter Rembrandt van Rijn.

Culture and Society. Amsterdam has long been known as a haven for persecuted religious groups. Its reputation for tolerance and openness has applied as well to views about drugs usage, especially marijuana, and to prostitution. Also, the city is among the most gay-friendly cities in the world. Because of stepped-up immigration since the latter part of the 20th century, Amsterdam now has a very diverse ethnic mix. Dutch natives comprise only about one-half of the city's population, with immigrants making up the rest. European immigrants make up about 15 percent of the total and non-European immigrants make up about 35 percent. The largest non-Dutch populations are Turks, Moroccans, and Surinamese. The rapid growth in non-western (and Islamic) population in the Netherlands (not just in Amsterdam) has strained the Dutch reputation for tolerance, as anti-immigrant discrimination is on the rise and politicians have spoken openly about instituting tighter controls on who may live and work in the country.

Further Reading

Kahn, Dennis and Gerrit van der Plas. "Amsterdam," *Cities: The International Journal of Urban Policy and Planning* 16, no. 5 (1999): 371–81.

Mak, Geert. *Amsterdam.* Cambridge: Harvard University Press, 2002.

Musterd, Sako and Willem Salet, eds. *Amsterdam Human Capital.* Amsterdam: Amsterdam University Press, 2003.

Nell, Liza and Jan Rath, eds. *Ethnic Amsterdam: Immigrants and Urban Change in the Twentieth Century.* Amsterdam: Amsterdam University Press, 2010.

ANDORRA LA VELLA

Andorra la Vella, meaning Andorra the Old, is the capital and largest city of the Principality of Andorra, a tiny landlocked country in western Europe in the Pyrenees Mountains at the border between Spain and France. It is located in the southwest of Andorra where two mountain streams, the Valira del Nord and the Valira del Orient come together to form the Gran Valira. At an elevation of 3,356 feet (1.023 m), the city is the highest capital city in Europe. Tourism, especially winter season skiing, is a major part of the economy, accounting for about 80 percent of GDP. The population of Andorra la Vella is 24,574, not even one-third of Andorra as a whole.

Historical Overview. Andorra's history begins in the early ninth century when Charlemagne, the King of the Franks and Emperor of Rome, granted residents of the Andorra Valley a charter in return for their fight against the Moors in Spain. In 1278, Andorra became independent from the Crown of Aragon, and has been governed since as a coprincipality by the Bishop of Urgell, the head of a Roman Catholic Diocese in Catalonia, Spain, and the head of state of France (now the president of the French Republic; formerly the king of France). Andorra la Vella has been capital since 1278. In the 1930s, a Russian Lithuanian adventurer named Boris Skossyreff proclaimed himself to be "Boris I, Prince of the Valleys of Andorra, Count of Orange and Baron of Skossyreff … sovereign of Andorra and defender of the faith," and attempted a coup before being arrested by Spanish authorities. In 1993, Andorra adopted a constitution that made it a parliamentary democracy.

In the 1990s as well, Andorra developed an economy as a center for banking and off-shore tax haven. Andorra la Vella bid to host the 2010 Winter Olympic Games but did not succeed.

Major Landmarks. The main landmarks of Andorra la Vella are the 12th-century Church of Saint Esteve, the Church of Saint Andreu, also from the 12th century, and Casa de la Vall, a 16th-century building that is the seat of the Andorran parliament. The Caldea Spa and swimming pool complex in the nearby town of Escaldes-Engordany is contained in a modern glass-spired structure and is heated by natural hot spring water. Also outside Andorra la Vella, in the village of La Margineda, is a beautiful old stone bridge from the 12th century that is almost perfectly preserved.

Culture and Society. Native Andorrans are ethnically Catalans and make up about 37 percent of the population. Other numerous ethnic populations are Spaniards, Portuguese, and French. Catalan is the official language of the country, but Spanish, Portuguese, and French are spoken as well. Approximately 90 percent of the population is Roman Catholic. The country is generally prosperous because of the tourism economy and banking and boasts of an unemployment rate that is almost 0 percent.

Further Reading

Eccardt, Thomas M. *Secrets of the Seven Smallest States of Europe: Andorra, Liechtenstein, Luxembourg, Malta, Monaco, San Marino and Vatican City.* New York: Hippocrene Books, 2004.

Ticktin, Michael. "Andorra Shows the Way," *Economist* 375, no. 8431 (June 18, 2005): 18.

ANKARA

Ankara is the capital and second-largest city of Turkey, a country that spans Europe and Asia at the eastern margins of the Mediterranean Sea. The city is located near the center of the country in a region known as Anatolia, and is an important hub of transportation in Turkey, as well as a commercial and industrial center. The region around the city is rich in agriculture, which underpins part of the Ankara economy. Grapes and other fruits, and wines, have been important products. In the past the region was known for raising long-haired Angora goats prized for mohair wool, and the city itself was named Angora. The population of Ankara is about 1.6 million, while that of the wider metropolitan area is nearly 4.5 million.

Historical Overview. The site of Ankara has been settled for millennia, but it began to grow into urban form in about 1000 BC as a Phrygian trade center. Persians eventually succeeded the Phrygians as rulers of the city, who then lost the city to Alexander the Great who conquered Ankara in 333 BC. In 278 BC, Ankara was occupied by a Celtic group, the Galatians. In 25 BC, the city was conquered by the Roman Emperor Augustus and annexed to the Roman Empire. It was designated as the capital of the Roman province of Galatia, and became a key center of trade and administration in the region. Ankara was ruled from Constantinople after that city succeeded Rome and became capital of the Eastern Roman Empire. It remained under Byzantine rule until the late 11th century when the Turkish Seljuk sultan Alp Arslan captured the city and much of Anatolia. In 1243, the Mongols defeated the Seljuks. In 1356, the city was captured by Orhan I, the second ruler of the Ottoman Empire. It remained under Ottoman control more or less continuously until the end of World War I. By 1920, the Turkish nationalist movement led by Mustafa Kemal Atatürk was centered in Ankara. The Republic of Turkey was proclaimed on October 29, 1923, with Atatürk as first president, and Ankara replaced Istanbul (formerly Constantinople) as national capital. The city has grown at a rapid pace since becoming capital.

Major Landmarks. The historic center of Ankara is a rocky hill with the remains of an ancient citadel that was built by the Galatians and then enlarged by the Romans, the Roman theater, the Temple of Augustus, and an old Roman bath. The Column of Julian was erected in 362 to honor a Roman emperor on his visit to the city. Major museums include the Ankara Ethnography Museum, the Museum of Anatolian Civilizations, and the War of Independence Museum. Anitkabir is Atatürk's mausoleum with an adjacent museum about the national leader's life. Victory Monument, erected in 1927, depicts Atatürk in uniform. Other landmarks include the residence of the president of Turkey, Kocatepe Mosque, the city's largest mosque, and Atakule Tower, a 410-ft-high (125 m) communications and observation tower, is the city's tallest structure, built over 1987–1989. The main commercial district of Ankara is around Kizilay Square.

Culture and Society. Ankara has grown very rapidly with Turkey's modernization and economic development, and its population includes many migrants from the countryside who come to the city for a better life. Because of a shortage of affordable housing, the outskirts of the city have districts of squatter housing called *gecekondu*. The city's government has many planning and urban development projects underway in order to improve housing and urban infrastructure. Ankara is known for being very clean and safe from crime. More than 90 percent of Ankara's residents are Muslims. The official language is Turkish.

Further Reading

Altaban, Ö. and M. Güvenç. "Urban Planning in Ankara," *Cities: The International Journal of Urban Policy and Planning* 7, no. 2 (1990): 149–58.

Kezer, Zeynep. "Ankara," in Emily Gunzburger Makaš and Tanja Damljanović Conley, eds., *Capital Cities in the Aftermath of Empires: Planning in Central and Southeastern Europe*, 124–40. London: Routledge, 2010.

Sarkis, Hashim, ed. *A Turkish Triangle: Ankara, Istanbul and Izmir at the Gates of Europe*. Cambridge, MA: Aga Khan Program at the Harvard University Graduate School of Design, 2009.

ANTANANARIVO

Antananarivo is the capital and largest city of the Republic of Madagascar, formerly the Malagasy Republic, an island nation in the Indian Ocean near the southeastern coast of the African continent. The country consists of one main island, Madagascar, the fourth largest island in the world, and many small islands offshore.

Antananarivo is located near the center of the main island on a high summit (4,183 feet above sea level; 1,275 m) from which it commands the surroundings. The city is connected by rail to the nation's main port, Toamasina, some 134 miles (215 km) to the east. The population of Antananarivo is about 903,000, while that of the metropolitan area is about 1.4 million. The city was once named Tananarive. It is often referred simply as Tana.

Historical Overview. Antananarivo was founded in 1625 by King Andianjaka who gave it its name, which means "City in the Thousand" in reference to the number of soldiers who were assigned to protect it. With time, the area that the city ruled was expanded and Antananarivo grew to be a populous settlement. In 1793, the city became the capital of the Merina kings and governed almost the whole island of Madagascar. The city's palace complex, called the Rova of Antananrivo, stood high on the largest peak in the city. Because of steep slopes, many parts of the city were accessible only by stairways. France invaded Madagascar in 1893 in the Franco-Hova War and annexed it as a colony in 1896, sending the Merina royal family into exile. Under French rule, Antananarivo saw construction of wide boulevards and stone buildings, and a considerable French expatriate population of planters, civil servants, and traders. The Rova of Antananarivo was made into a museum. Independence from France came in 1960. Since then, there have been four independent republics on Madagascar, each with its own constitution. The University of Madagascar was founded in Antananarivo in 1961.

Major Landmarks. The main landmark of Antananarivo is still the Rova complex at the top of the city's highest hill. It stood vacant for a number of years after Madagascar achieved independence, and then just prior to designation by UNESCO as a World Heritage Site it suffered a devastating fire on the night of November 6, 1995. The cause was ruled to be accidental, although there are many people who doubt this opinion and suspect arson linked to political infighting as the cause. Since the fire, there has been reconstruction of various parts of the complex, most notably the spectacular Manjakamiadana, the Queen's Palace. Other landmarks of note in Antananarivo are the French residency, the Anglican and Roman Catholic Cathedrals, the Museum of Ethnology and Paleontology, and the National Library.

Culture and Society. The main ethnic group of Madagascar is the Malagasy people, among which there are eight ethnic subgroups. The official language of the country is Malagasy as well, a language of Malayo-Polynesian origin. French is also spoken and has official status. Many residents of the country practice traditional religious beliefs that center on ancestor worship and building of tombs. About one-half of the population is Christian, approximately evenly divided between Roman Catholics and Protestants.

Further Reading

Brown, Mervyn. *A History of Madagascar.* Princeton, NJ: Markus Wiener Publishers, 2002.
Wright, Gwendolyn. *The Politics of Design in French Colonial Urbanism.* Chicago: University of Chicago Press, 1991.

APIA

Apia is the capital and largest city in Samoa, a small sovereign state comprised of two larger islands and eight islets in the western part of the Samoan Islands group in the South Pacific Ocean. The country's official name is the Independent State of Samoa, distinguishing it from American Samoa which is another part of the island chain and is a territory of the United States. Until 1997, it was referred to as Western Samoa. Apia is on the north coast of the island of Upolu, the smaller of the two main islands of Samoa. It is the country's only city and has a population of approximately 38,000, a little more than one-fifth of the national total.

Historical Overview. Apia grew from a small village of the same name. Its good natural harbor facilitated the landing of foreign missionaries, traders, and whaling ships. The first missionaries were from England and arrived in the 1830s. Germans came to trade in cocoa beans and copra, and established a plantation economy. In 1898 and 1899, Apia was the scene of conflict between the United States, Great Britain, and Germany about who should control Samoa, and was shelled by American and British warships in opposition to German claims. The 1899 Tripartite Convention divided Samoa between the three powers, but Britain withdrew as a result of a separate exchange of colonies with Germany, and Samoa came to be divided between American Samoa and what came to be known as German Samoa, the same territory as the independent country today. Germany lost the colony after World War I. The territory was then administered by New Zealand through the League of Nations as Western Samoa. Samoan determination for independence grew after two terrible incidents caused by New Zealanders that took Samoan lives: in late 1918 a breach of quarantine by a ship arriving from New Zealand created an influenza epidemic that killed about one-fifth of the Samoan population in the following months; and the Black Sunday incident at the end of 1929 in which New Zealand police fired a machine gun needlessly into a crowd of peaceful demonstrators, killing 11, including a popular local chief. Apia was designated as the capital of Western Samoa in 1959. Samoan independence was achieved from New Zealand in 1962 and the two countries signed a friendship agreement. In 2002, New Zealand issued a formal apology for the two incidents.

Major Landmarks. The center of Apia is marked by a landmark clock tower. The Samoan Parliament house is on the Mulinu'u peninsula in the western end of the city. Other landmarks are the Apia Courthouse, the Museum of Samoa and its top floor, and the Immaculate Conception of Mary Roman Catholic Cathedral, now being rebuilt following a 2009 earthquake. Near Apia in the village of Vailima is the former home of noted English author Robert Louis Stevenson that is now a museum in his honor, while his grave is near the summit of Mount Vaea.

Culture and Society. Samoa has distinctive cultural traditions known collectively as *fa'a Samoa,* the traditional Samoan way. There are many traditional ceremonies in Samoa, including rituals associated with the bestowal of titles of chief, old beliefs about creation and traditional gods and goddesses, and codes of behavior that call for respect in relations between people. The legend of Sima and the Eel tells the story of the first coconut tree. Other aspects of culture include traditional song, dance and drum beats, and intricate body tattooing for both men and women. The male dance of *fa'ataupati* is based on slapping various parts of one's own body and is said to derive from the slapping of insects.

More than 95 percent of the people of Samoa are ethnic Samoans, the second most populous Polynesian group after the Maori in New Zealand. The native language is Samoan. English is also an official language. As a result of missionary activity, most Samoan people are Christians. There are members of various Protestant and evangelical congregations, and Roman Catholics. There are also many Mormons. Rugby is very popular in Samoa, and national teams are able to compete against teams from much larger nations. Because of its position near the International Date Line, Samoans are among the first in the world to start each new day and to greet every new year.

Further Reading

Jones, Paul and John P. Lea. "What Had Happened to Urban Reform in the Island Pacific? Some Lessons for Kiribati and Samoa," *Pacific Affairs* 80, no. 3 (Fall 2007): 473–91.

Macpherson, Cluny and La'avasa Macpherson. *The Warm Winds of Change: Globalisation in Contemporary Samoa.* Auckland: Auckland University Press, 2010.

Meleisea, Malama. *The Making of Modern Samoa: Traditional Authority and Colonial Administration in the Modern History of Western Samoa.* Suva: Institute of Pacific Studies, University of the South Pacific, 1987.

ASHGABAT

Ashgabat is the capital and largest city of Turkmenistan, a country in Central Asia that borders the Caspian Sea to the west, Iran and Afghanistan to the south, and Kazakhstan and Uzbekistan to the north. The city is located in the south-central part of the country at the base of the Kopet Dag Mountain range, which demarcates the border with Iran. To the north of Ashgabat is the Kara Kum Desert. The population of Ashgabat is about 1 million.

Historical Overview. Ashgabat was but a small village until 1881 when the Russian Empire took control of the region from Persia and selected it to be an administrative center. It remained under Russian influence in 1991 when the successor of the Russian Empire, the Union of Soviet Socialist Republics, collapsed and Turkmenistan became an independent nation. Although the city is comparatively new, it is located near the archaeological remains of an ancient city that was called Nisa (also called Parthanusia), the capital of the third century BC Parthian Empire. The ruins of a Silk Road city named Konjikala that had been destroyed in the 13th century by a Mongol invasion were also nearby. The Russians built a fortress near Ashgabat to defend their interests from British imperial ambitions in Central Asia and built the city in European fashion. After the Russian revolution of 1917, Ashgabat was at first governed for Russian-Bolshevik-controlled Tashkent in today's Uzbekistan. In 1919, it was named Poltoratsk in honor of a local revolutionary figure. The city was named Ashgabat again in 1927 when the Turkmen Soviet Socialist Republic was created within the Soviet Union, and it was designated to be capital. On October 6, 1948, an enormous earthquake completely destroyed the city and killed more than 100,000 people (about two-thirds of the population). For the next five years, the site was closed to visitors. Afterward, Ashgabat was rebuilt on new lines as a modern, Soviet-style city. After Turkmenistan became independent in 1991, it continued to be ruled by Saparmurat Niyazov, who had been the leader of the Turkmen SSR, until his death near the end of 2006. Niyazov's rule was dictatorial, as seen in the "President for Life" title he assumed in 1999. He built a cult of personality around himself and many monuments to match. About Ashgabat he once said "I want Ashgabat to become the city of my dream" (Righetti, 2008).

Major Landmarks. The centerpiece of Ashgabat is the lavish Presidential Palace of white marble and a dome of gold mirrored glass. Until recently, there was also the 246-ft-high (75 m) Arch of Neutrality, constructed in 1998 to herald Turkmenistan's official position of neutrality. Because it rests on three supports, the monument was locally called "the tripod." Atop the monument was a 39-ft-high (12 m) gold-plated statue of Niyazov that rotated to always face the sun. Niyazov's successor as president, Gurbanguly Berdimuhamedow, has had the Arch of Neutrality dismantled and replaced it in 2010 with an even taller (312 feet; 95 m) "Monument to Neutrality" in another part of Ashgabat. The golden statue of Niyazov was moved to the new monument but not atop it. Other landmarks are the Earthquake Memorial, the War memorial, the Azadi Mosque, the Khezrety Mosque, the National Museum of Turkmenistan, the Museum of Fine Arts, and the Carpet Museum. The Ashgabat Flagpole is the tallest in the world at 436 feet (133 m). Outside the city are the ruins of Nisa, a UNESCO world heritage site.

Culture and Society. The population of Turkmenistan is about 85 percent Turkmen, 5 percent Uzbek, 4 percent Russian, and the rest other ethnic groups. Turkmen is the official language of the country, although Russian is still used

The three-legged Neutrality Arch in Ashgabat, also referred to as the "Tripod." (Travel-Images.com)

for interethnic communication. Nearly 90 percent of the population comprises Muslims. Although there have been moves to liberalize the country since the death of President Niyazov, such as overturning his decree banning opera and circus, the government of Turkmenistan continues to be highly autocratic. According to Reporters without Borders, Turkmenistan was reported to be the world's third-worst country in terms of freedoms of the press in 2011. The mainstay of Turkmenistan's economy is enormous reserves of natural gas, as well as oil reserves.

Further Reading

Kropf, John W. *Unknown Sands: Journeys around the World's Most Isolated Country.* Houston: Dusty Spark Publishing, 2006.

Peyrouse, Sébastien. *Turkmenistan: Strategies of Power, Dilemmas of Development.* Armonk, NY: M. E. Sharpe, Inc., 2011.

Righetti, Nicholas. *Love Me Turkmenistan.* London: Trolley Books, 2008.

ASMARA

Asmara is the capital and largest city of Eritrea, officially known as the State of Eritrea, a country in the Horn of Africa, in East Africa on the Red Sea. The city is also known as Asmera. It is located in the interior of the country near the nation's center at an elevation of 7,628 feet (2,325 m) in the Eritrean Highlands. It is linked by rail to its port city on the Red Sea, Massawa, some 40 miles (65 km) to the east. The part of the country where Asmara is located is called the Maekel Region. The population of Asmara is about 579,000.

Historical Overview. The origins of Asmara can be linked to settlements from several hundred years ago by indigenous Tigrinya and Tigre peoples. In 1900, the city became the capital of the Italian colony of Eritrea. It grew rapidly in the 1930s when Italians used the city as the base for invasion of Ethiopia. Many Italians settled in the city at that time, and constructed a palm-lined city center and residential neighborhoods that are still distinguished for Italian modern architecture. The Italians also made Asmara into an industrial center based on food processing, textiles, and leather. From 1941 to 1952, Eritrea was administered by Great Britain, and then from 1952 to 1991 it was a part of Ethiopia. From 1943 to 1977, Asmara was home to Kagnew Station, a strategic telecommunications base for the U.S. Army. The Eritrean War of Independence against Ethiopia began in 1961 and lasted for 30 years. From May 24, 1991, Asmara became a national capital.

Major Landmarks. Asmara is known for its Italian modernist architecture, especially along Independence Avenue, a street with many cafes, shops, and old cinemas. In some senses, the city is said to be frozen in time in terms of architecture. The 1937 Cinema Impero is a beautiful art deco landmark, as is the Fiat Tagliero Building. Other landmarks are the 1925 St. Joseph's Roman Catholic Cathedral, the 1937 Grand Mosque, and Enda Mariam (St. Mary's) Eritrean Orthodox Cathedral. The Eritrean National Museum showcases 6,000 years of the region's history. Outside the city is Martyrs National Park, an extensive forest and wildlife reserve with beautiful scenery.

Culture and Society. The main ethnic group of Asmara comprises the Tigrinya people, 77 percent of the population. Tigre people make up 15 percent of the total. There is also a significant population of Italian Eritreans, descendants of the Italians

An example of Italian modernist architecture in Asmara. (Jackmalipan/Dreamstime.com)

who colonized Eritrea in the early 20th century. The main language of the city is Tigrinya, although English, Arabic, and Italian are spoken as well. The religious mix includes Eritrean Orthodox (60%), Muslims (25%), and Roman Catholics (15%).

Further Reading.

Dennison, Edward, Guang Yu Ren, and Naigzy Gebremedhin. *Asmara: Africa's Secret Modernist City.* London: Merrell Publishers, 2007.
Hill, Justin. *Ciao Asmara: A Classic Account of Contemporary Africa.* London: Abacus, 2002.
Visscher, Jochen and Stefan Boness. *Asmara: The Frozen City.* Berlin: Jovis, Mul Edition, 2007.

ASTANA

Astana (population 709,000) is the capital and second-largest city of Kazakhstan, a large, oil- and gas-rich country in Central Asia that became independent in 1991 with the fall of the Soviet Union. Astana became capital in 1997 when government functions were moved from Almaty, the country's largest city. Reasons for the move include that Almaty lacked room for expansion because of mountainous surroundings, was vulnerable to major earthquakes, and was near international borders, a security issue. Astana, by contrast, is near the center of Kazakhstan in a flat steppe area with unlimited room for expansion. Critics say that that Astana is in a very lightly settled area of Kazakhstan, far from the main centers of national population, and that in winter the city is unbearably cold. It ranks as the world's second-coldest capital city, after Ulaanbaatar in Mongolia. The word "Astana" means capital in the Kazakh language. Prior to becoming capital, the city was named Akmola or Aqmola, and before that Tselinograd and Akmolinsk.

Historical Overview. The city began as a fortress city established on the Ishim River in 1824 and grew rapidly in the 20th century as a rail junction and as the hub of a large agricultural resettlement program called the Virgin Lands Campaign that was started by Soviet Union authorities in 1954. The city is also associated with the Gulag prison camps that the Soviets had established in Kazakhstan to house political dissidents. A camp for the wives of prisoners held in other camps in Kazakhstan was just outside Astana.

Major Landmarks. Kazakhstan president Nursultan Nazarbayev has been in office since before independence and has insisted that Astana should be a great and beautiful city that would rival any capital city in the world. Therefore, he has allocated considerable sums from the oil and gas wealth of the nation for construction of extraordinary monuments and buildings, and has hired for the task some of the world's best known architects, notably Sir Norman Foster from the United Kingdom and Kisho Kurokawa from Japan. The city has been one of the world's largest building projects. Among the major landmarks of Astana are the distinctive Bayterek tower, the pyramid-shaped Palace of Peace and Reconciliation, the Ak Orda ("white horde") Presidential Palace, the Khan Shatyry Entertainment Center, the Kazakhstan Central Concert Hall, and the Nur-Astana Mosque. On view inside the globe of the Bayterek Tower is the handprint of President Nazarbayev embedded in a triangle of solid gold. The Presidential Palace is flanked by enormous twin golden pillars that are thought to be Masonic symbols. The Ishim River waterfront has been made into a beautiful park.

The Ak Orda Presidential Palace. (Kristina Postnikova/Shutterstock.com)

Culture and Society. Approximately two-thirds of the population of Astana is ethnic Kazakh, with most of the rest being Russian. Many Russians had settled in Kazakhstan during the Soviet period, and in Astana in particular during the opening of surrounding farmlands. The city had a majority Russian population until after independence. Many rural Kazakhs have since settled in Astana to take construction jobs and to build careers in their national capital. It is widely assumed that Astana will be renamed in honor of President Nazarbayev after he dies. The city has gained international recognition for its winning professional cycling team, as well as for competitive teams in professional ice hockey and soccer.

Further Reading

Aitken, Jonathan. *Nazarbayev and the Making of Kazakhstan: From Communism to Capitalism.* London: Continuum Publishing Group, 2010.

Alexander, Catherine, Victor Buchli, and Caroline Humphrey, eds. *Urban Life in Post-Soviet Asia.* London: University College London Press, 2007.

Lancaster, John. "Tomorrowland," *National Geographic* 221, no. 2 (2012): 80–101.

Olcott, Martha Brill. *Kazakhstan: Unfulfilled Promise.* Washington, DC: Carnegie Endowment for International Peace, 2010.

ASUNCIÓN

Asunción is the capital and largest city of the Republic of Paraguay, a landlocked country in South America bordered by Argentina, Bolivia, and Brazil. It is located on the Paraguay River at Paraguay's border with Argentina, and is a river port city. The formal name for the city is Ciudad de Asunción, and its metropolitan area, which includes several nearby cities in the Central Department (state) of Paraguay, is called Gran Asunción. The population of the city is about 542,000, while that of Gran Asunción is more than 2 million, nearly a third of the national population.

Historical Overview. Asunción was founded in 1537 by Juan de Salazar y Espinosa and is one of the oldest cities on the South American continent. Because of its position on a navigable river in the interior, it served as a base for early Spanish explorations of the continent and the establishment of numerous colonial cities, including the second founding of Buenos Aires. When the first Buenos Aires settlement was destroyed in 1541, its Spanish population fled to Asunción. Asunción was also a base for Spanish missionary activity in South America. The First Synod of Asunción in 1603 resulted in translation of the Catholic catechism into the local Guaraní language and set guidelines for missionary work with the indigenous population. The first rebellion against Spanish rule occurred in 1831 and Paraguay gained its independence in 1811. Asunción flourished after independence, and developed as a center of industry, education, and trade. It had the first railroad service in South America. However, the 1865–1870 War of Triple Alliance between Paraguay and the combined forces of Argentina, Brazil, and Uruguay resulted in disaster for Paraguay, with nearly two-thirds of the country's population dead. Afterward the country was occupied by Brazil until 1876 and its economy stagnated. The country has had a long history of political instability and oppressive government. From 1954 to 1989, Paraguay was ruled by dictator Alfredo Stroessner. The political situation has improved more recently, and Paraguay's economy has developed considerably, with recent growth being especially strong. Asunción is increasingly a modern city with prosperous downtown, new shopping malls, and many venues for entertainment and enjoyment.

Major Landmarks. The new National Congress building was erected in 2002 as a gift from the government of Taiwan, and is an important landmark. Other landmarks are the Lopez Presidential Palace, the National Cathedral, the Municipal Museum, the Museo National de Bellas Artes, and the Panteón de los Heroes where many of Paraguay's heroes are entombed. The main commercial street is Calle Palma.

Culture and Society. The official languages of Paraguay are Spanish and Guaraní. About 90 percent of the population is Roman Catholic. The Paraguayan population includes immigrant groups and their descendants from a wide range of European and Asian countries, and is therefore unusually diverse. Much of the indigenous population of the country is poor, with unequal access to land being a major obstacle to economic progress. Many poor Paraguayans have migrated to Asunción in search of a better life. While the city has prosperous districts and a growing middle class, it also has poor slums and squatter districts. Many people work in a large informal sector of the economy, including as urban street vendors.

Further Reading

Gimlette, John. *At the Tomb of the Inflatable Pig: Travels through Paraguay.* New York: Vintage, 2005.

Horst, René Harder. *The Stroessner Regime and Indigenous Resistance in Paraguay.* Gainesville: University of Florida Press, 2007.

O'Shaughnessy, Hugh. *The Priest of Paraguay: Fernando Lugo and the Making of a Nation.* London: Zed Books, 2009.

ATHENS

Athens is the capital and largest city of Greece, a country in southern Europe on the Aegean and Mediterranean Seas. The city is one of the world's oldest urban centers, having been continuously inhabited for some 7,000 years. The city is located on the Attica Peninsula on the Aegean coast of Greece, behind which are four sizable mountains, Mount Aegaleo, Mount Parnitha, Mount Penteli, and Mount Hymettus. A basin location concentrates air pollutants above the city, particularly during times of atmospheric temperature inversion when warm air overlays cooler air at the surface. The population of the municipality of Athens is about 656,000, while that of the metropolitan area as a whole is more than 3.7 million, approximately one-quarter that of Greece.

Historical Overview. The recorded history of Athens goes back some 3,400 years. At the time the city was an important center of Mycenaean civilization defended by a fortress at the hill-top Acropolis. In 490 BC and 480 BC, a coalition of Greek city-states led by Athens and Sparta known as the Delian League defeated the Persian armies in the Battles of Marathon and Salamis, respectively. As a result, Athens became the leading city of the Greek world and the center of a great flowering of cultural and political life variously known as the Golden Age of Athens or the Age of Pericles after the celebrated orator and statesman who was the city's leader. This period is commonly dated as 480 BC–404 BC. Among the enduring intellectual lights of Athens in the fifth century BC were playwrights Aeschylus, Sophocles, and Euripides, the physician Hippocrates, historians Herodotus and Thucydides, and the philosopher Socrates. Under the expertise of sculptor Phidias and architects Ictinus and Callicrates, the Acropolis and many temples were built in Athens during this period, including the landmark Parthenon. The Peloponnesian War of 431–404 BC between Athens and Sparta resulted in Athenian defeat and decline in prestige. Even still, in approximately 387 BC, the philosopher Plato founded his Academy in Athens, where Aristotle was his student, the first institution of higher learning in the Western world.

Athens prospered once again as part of the Byzantine Empire during the Crusades of the late 11th–late 13th centuries, but was then conquered by the Ottoman Empire in 1458 and was ruled by Ottoman Turks until after the Greek War of Independence of 1821–1832. In 1834, the city was chosen to be capital of the independent Greek state, and the first King of Greece, Otto of Bavaria, commissioned architects to modernize the city. In 1896, Athens hosted the first modern Olympic Games. The city's population increased rapidly in the 20th century as a result of industrialization and because of the 1919–1922 Greco-Turkish War that resulted in expulsion of Greek residents from modern-day Turkey. In 1981, Greece became a part of the European Union. In 2004, Athens hosted the Summer Olympics once again. Since about 2009, Greece has been in an economic crisis that has rocked the entire European Union.

Major Landmarks. Athens has two UNESCO World Heritage Sites, the Acropolis on a hill in the center of the city and the Dafni (or Daphni) Monastery that was founded at the turn of the sixth century in what is now the northwest suburb of Chaidari. The Acropolis has the archaeological remains of the Parthenon (the old temple of Athena), the temple of Athena Nike, the Erechtheum with its spectacular porch of the Caryatids, the Theater of Dionysus Eleuthereus, and the Odeon of Pericles, among other landmarks. The site of the ancient Agora of Athens is to the northwest of the Acropolis. It was the focus of the city's civic life, and includes

The Acropolis of Athens

The word "acropolis" is Greek for city on an extremity and can refer to any of a number of hilltop cities in ancient Greek civilization. The Acropolis of Athens, the historic Greek capital, is far and away the most famous, such that the word Acropolis, spelled with a capital 'A', refers to this specific place only. The ruins of many historic structures are atop the Athens Acropolis, the most important of which are those of the Parthenon, the iconic temple to the patron of Athens, the goddess Athena. Later in history, the building served as a Christian Church and then as a mosque. Most of the sculptures of the Acropolis are in the Acropolis Museum in Athens, but a great many had been removed around 1800 to London under a controversial permit that was issued by Ottoman rulers in Greece, and are on display in the British Museum as the Elgin Marbles or, alternatively, the Parthenon Marbles. The Acropolis is the leading tourist site in Athens.

among other ruins the remains of the Roman Forum. Other historical sites include the ruins of the Temple of Olympian Zeus and Hadrian's Gate. The 1896 Olympic Stadium is located nearby. Major museums include the National Historical Museum in the old Parliament building, the National Archaeological Museum, the Cycladic Museum, the Benaki Museum, the New Acropolis Museum, and Agora Museum, the National Art gallery, and the City of Athens Technopolis. The Kolonaki neighborhood is known for art galleries, museums, and boutiques during the day and restaurant and bar life at night, while the Exarcheia district to the north is a busy student-focused neighborhood of cafés, bars, and book stores.

Culture and Society. Although the vast majority of residents in Athens are Greek, the city also has many immigrants from nearby countries in southern and eastern Europe such as Albania, Macedonia, Bulgaria, Romania, Ukraine, and Moldova, who were attracted by the city's stronger economy. At the same time, many Greeks have emigrated to the United States, Canada, and western Europe, where they have found better opportunities for work. The main religious faith is the Greek Orthodox Church, and the official language of the country is Greek. Greek culture goes back many centuries, and the achievements of the Athenian Golden Age are source of great pride nationwide. Foreign tourism based on Greek history and culture is a mainstay of the Athens economy.

Further Reading

Freely, John. *Strolling through Athens: Fourteen Unforgettable Walks through Europe's Oldest City.* New York: Palgrave Macmillan, 2004.

Powell, Anton. *Athens and Sparta: Constructing Greek Political and Social History from 478 B.C.* London: Routledge, 2001.

Sarrinikolaou, George. *Facing Athens: Encounters with the Modern City.* New York: North Point Press, 2004.

Smith, Michael Llewellyn. *Athens: A Cultural and Literary History.* Northampton, MA: Interlink Books, 2004.

B

BAGHDAD

Baghdad is the capital and largest city of the Republic of Iraq, a country in the heart of western Asia that has recently been at the center of considerable international military conflict, including very prominently the involvement of the armed forces of the United States. The city is on a flat plain near the center of the country and is divided into halves by the River Tigris that flows south through Baghdad to empty into the northern reaches of the Persian Gulf. The city's estimated population is 7.2 million, making it the second-largest city in the Arab world after Cairo, Egypt, and the second-largest city in West Asia, after Tehran, Iran. In the 12th century, Baghdad had a population of about 1.2 million and was probably the world's largest city.

Historical Overview. Baghdad was founded in 762 under orders of Al-Mansur, the second caliph of the Abbasid dynasty, to be the capital of the empire that he ruled. It was built from the ground up according to a master plan. When construction commenced, the caliph declared that "this is indeed the city that I am to found, where I am to live, and where my descendants will reign afterward" (Wiet, 1971). A favorable location for trade and territorial administration, as well as ample water despite the desert environment, enabled the city to grow and prosper into a leading center of Islamic culture and learning. In 1258, however, Baghdad was destroyed by Mongol invaders and then languished for centuries afterward. The city was under Ottoman rule for most of the period between 1534 and 1917. Outbreaks of cholera and plague periodically decimated the population. In 1920, Baghdad became the capital of the British Mandate of Palestine and then in 1932 of the Kingdom of Iraq. On June 14, 1958, there was a successful coup against the monarchy, and King Faisal II, members of the royal family, and other officials were killed and their bodies were dragged for effect through the streets of the city. Saddam Hussein came to power in 1979 and ruled the country with an authoritarian hand until he was deposed in 2003 in a United States-led invasion. Profits from high oil prices on the global market buoyed Iraq's economy, but costs of the Iran–Iraq War of 1980–1988 and the Gulf Wars of 1991 and 2003 took the country away from general prosperity, took many lives, and caused great damage to the city.

Major Landmarks. The Swords of Qādisīyah, also known as the Hands of Victory, is a triumphant arch-like monument that was commissioned across a busy road to commemorate Iraq's supposed victory in the Iran–Iraq War. The Monument to the Unknown Soldier is another dramatic architectural presentation that remembers that war: its form is meant to suggest a traditional shield falling from the grasp of a dying

The Swords of Qādisīyah monument in central Baghdad, also referred to as the Hands of Victory monument. (UIG via Getty Images)

soldier. The Al-Shaheed Monument is dedicated to Iraqi soldiers who perished in the Iran–Iraq War. Other landmarks are the Al-Faw Palace, also known as the Water Palace because of its location beside the River Tigris and once a favorite among the many opulent residences of Saddam Hussein, the Republican Palace that was built by King Faisal II and that Hussein used for hosting foreign dignitaries, and the historic Abbasid Palace that was built in the 12th or 13th century. The National Museum of Iraq was once a rich depository of antiquities from early Mesopotamian civilization, but it was looted during hostilities and is now rarely open. The Al Kadhimain Shrine is an important shrine for Shiite Muslims. The Umm al-Qura Mosque is a curious piece of religious architecture with minarets built to resemble the barrels of Kalashnikov rifles and SCUD missiles.

The Building of Baghdad

Today's troubled capital of Iraq got off to a glorious start when its construction was commissioned on July 30, 762, by the caliph Al Mansur. He wanted it to be the world's most perfect city. He envisioned the city as the capital of a global Islamic Empire, and wanted it to reflect the concept of paradise as described in the Qur'an. He assembled the world's best engineers and builders and more than 100,000 construction workers for the task. Because the basic footprint of the city incorporated two large semicircles that faced one another, early Baghdad was referred to as the "Round City." In the central square in the middle of the city Al Mansur erected his palace, the opulent Golden Gate Palace. The city was beautiful, indeed, and has been a leading center of Islamic cultural and religious life throughout its history.

Culture and Society. Approximately 95 percent of the Iraqi population is Muslim; about 65 percent comprises Shi'ites and 35 percent Sunnis. There is a small Christian minority with a long history in the country, and until they left in 1941, there was a sizable Jewish population with a long history as well. The main language is Arabic, although in the north of Iraq Kurdish is common too. In Baghdad, a distinctive dialect of Arabic is spoken that is thought to be related to the Arabic of nomadic tribes. The once rich culture life of the city has been disrupted by prolonged periods of warfare and terrorism.

Further Reading

Kennedy, Hugh. *When Baghdad Ruled the Muslim World: The Rise and Fall of Islam's Greatest Dynasty.* Cambridge, MA: Da Capo Press, 2005.
Refwan, Nissim. *The Last Jews of Baghdad: Remembering a Lost Homeland.* Austin: University of Texas Press, 2004.
Wiet, Gastron. *Baghdad: Metropolis of the Abbasid Caliphate.* Norman: University of Oklahoma Press, 1971.

BAKU

Baku is the capital and largest city in Azerbaijan, a small country on the western shores of the Caspian Sea in the Caucuses Mountains region of Eurasia. The city is on the shores of the Caspian, on the southern shore of the Absheron Peninsula that juts into the Caspian. The population is a little more than 2 million, making the city the largest urban center in the Caucuses. It is now a fast growing and dynamic city, rich in oil, and optimistic about development prospects for the future. It is one of the six finalist cities in the world for hosting the 2020 Summer Olympics. The most common explanation for the word Baku is that it comes from ancient Persian and means "place where the winds pound strongly."

Historical Overview. The settlement of what is now Baku dates back to the first century AD, but its rise as a city dates to the 12th century when it was designated as capital because the previous capital, Shamakhy, was destroyed by an earthquake. The city was periodically occupied by neighboring Persia, and in 1796 it fell to invading troops from expansionist czarist Russia. In a famous event on February 8, 1806, the Russian general who was the commandant of Baku was stabbed and killed at the gates of the city by the ruler of Persia's khanate of Baku. The Russians cemented their hold on Azerbaijan via the Treaty of Gulistan with Persia in 1813, by which they annexed most of the Caucuses region.

A reason for special interest in Baku was the discovery of oil in the region. The first well was drilled in 1846, with large-scale exploration and development commencing in the 1870s under licensing agreements with Russian authorities. A specialized oil-producing zone called Black City developed outside Baku. By the start of the 20th century, about one-half of the world production of oil was in the Baku region. The famous Russian writer Maxim Gorky described the oil fields and their flames from gas fires of Baku as the "grave of hell."

After the 1917 Russian Revolution, Azerbaijan enjoyed brief independence, albeit with war, but by 1920 the Red Army captured the city for Russia once again.

Azerbaijan was annexed to the Soviet Union soon thereafter as the Azerbaijan Soviet Socialist Republic, and Baku was made its capital. Much Soviet-style construction of government buildings and worker's apartment housing took place in Baku during the Soviet period. Oil from Baku was critical for supplying Red Army forces during World War II against Nazi invaders. In 1947, the first off-shore oil platform in the world was built in the Caspian Sea near Baku.

Azerbaijan gained its independence in 1991 after the collapse of the Soviet Union. Baku has changed greatly since then, replacing Soviet urban form with modern urban construction and a sleek new skyline. There has also been revival of religion in the landscape with many mosques being built after a period when religious expression was suppressed during the Soviet times. Baku has the look and feel of a boom city fueled by oil.

Major Landmarks. Old Baku is a UNESCO World Heritage Site. Major attractions are the Palace of the Shirvan Shachs, Maiden's Tower, old city walls, and old markets. On the Absheron Peninsula is the Atashgah Fire Temple and Yanar Dagh, a mountain that has been burning continuously for more than 1,000 years because of natural gas vents. The modern landscape of Baku includes the world's first high-rise office towers shaped like flames. The Bibi-Heybat Mosque was a historic place of Moslem worship from the 13th century. The Soviets destroyed it in 1936, but in 1990, just after the Soviet Union met its own demise, the Azeris rebuilt it in its original form. The National Library of Azerbaijan is another spectacular building.

Culture and Society. Baku offers a rich cultural life with many museums, concert halls, theaters, and other venues for great performances. The Baku International Jazz Festival is a popular annual event. The city also has a very rich nightlife, famous for activity all night long, and many opportunities for specialized shopping and bargain hunting. There are also beaches on the Caspian Sea, despite the oil industry. More than 90 percent of the population consists of ethnic Azerbaijanis. In the past, the city had large populations of Russians, Armenians, and Jews. Islam is now the predominant religion, but there is freedom of religion and other faiths are practiced as well. The official language is Azerbaijani, a language of the Turkic family.

Further Reading

Illis, Ben. *A Hedonist's Guide to Baku.* London: Filmer, Ltd., 2010.

LeVine, Steve. *The Oil and the Glory: The Pursuit of Empire and Fortune on the Caspian Sea.* New York: Random House, 2007.

Sattarov, Rufat. *Islam, State, and Society in Independent Azerbaijan: Between Historical Legacy and Post-Soviet Reality (with special reference to Baku and its environs).* Wiesbaden: Reichert Verlagm, 2009.

BAMAKO

Bamako is the capital and largest city in the Republic of Mali, a landlocked country in West Africa. The city is in the southern part of the country where the climate consists of a summer wet season and very dry winters, and where most of the nation's people live. Northern Mali, by contrast, is part of the arid Sahara Desert zone and is lightly settled. Bamako is on the Niger River, the longest river in West

Africa and the third longest on the African continent as a whole. It has developed on both banks of the river, although originally it was established on the north bank. The name Bamako comes from a word in the local Bambara language meaning "crocodile river." The population of Bamako is estimated to be about 1.8 million and perhaps as high as 2 million. The city is growing exceptionally quickly in comparison to other fast-growing African cities because of very high migration rates from the countryside and high birth rates.

Historical Overview. The site of Bamako has been settled for many centuries because of the fertile lands of the Niger River plain. In the 11th century Bamako was a thriving market town and center of Islamic learning in the Ghana Empire. The Ghana Empire was later succeeded by the Mali Empire and then by the Songhai Empire. The French colonized West Africa in the late 19th century and in 1883 what is today Mali became part of French Sudan. Bamako became capital of the French colony in 1899, taking over from the city of Kayes. From 1895 until 1960, French Sudan was part of a federation of French colonial territories in West Africa that was administered first from Saint-Louis (until 1902) and then from Dakar, both cities in Senegal. In 1960, Mali achieved independence and the Republic of Mali was established with Bamako as capital. Poor government and frequent droughts have hampered Mali's development. On March 22, 1991, some 300 Malian demonstrators were killed in the center of Bamako by government troops during a protest march against the government. The unpopular regime of General Moussa Traoré fell a few days later to a military coup with considerable popular support. In 2002, Amadou Toumani Touré, a leader of that coup, was elected president in a democratic election. He has since led the country to greater political stability and economic advancement. In 1988 Bamako was the site of an important World Health Organization conference about health conditions on the African continent.

Major Landmarks. The tallest building in Bamako is the 20-story BCEAO Tower, the headquarters for Mali of the Central Bank of West African States. Other important landmarks are the National Museum of Mali, the Bamako Grand Mosque, and the National Library of Mali. There are several notable monuments in Bamako as well, including Independence Monument and the Hamdallaye Obelisk. The Bridge of Martyrs across the Niger River was completed in 1960 and was renamed later in commemoration of the Mali citizens who were killed during the protest march in 1991. The other bridge across the river is the King Fahd Bridge named after the donor of funds for its construction, the ruler of Saudi Arabia. A third bridge is under construction with assistance from the Peoples Republic of China.

Culture and Society. The main language in Mali is Bambara, although French is the official national language. More than 90 percent of the population is Islamic. There is a small Roman Catholic minority in Bamako and a cathedral. Much of the population has migrated from the countryside in search of a better life or is descendant from such migrants. Bamako has a lively music scene with variety of local sounds. There are many street musicians. Football (soccer) is the most popular sport. In 2002, Bamako was one of five Mali cities to host the African Cup of Nations football tournament. The city is the subject of an award-winning 2006 film about African poverty and government corruption called *Bamako,* directed by Abderrahmane Sissako.

Further Reading

Duodo, Cameron. "A Stich-Up in Bamako?" *New African* 517 (May 2012): 24–27.
Lyons, Michael and Alkison Brown. "Has Mercantilism Reduced Urban Poverty in SSA? Perceptions of Boom, Bust, and the China-Africa Trade in Lomé and Bamako," *World Development* 38, no. 5 (2010): 771–82.
Williams, Stephen. "Bamako," *New African* 440 (May 2005): 62–65.

BANDAR SERI BEGAWAN

Bandar Seri Begawan is the capital and largest city of the Nation of Brunei, the Abode of Peace, which is the formal name for the small Southeast Asian country on the north coast of the island of Borneo that is called simply Brunei. The country is also called the Sultanate of Brunei. Bandar Seri Begawan is located on the northern bank of the Brunei River in the northern part of the country near Brunei Bay of the South China Sea. The population of the city is about 140,000 and that of the urban area is 277,000; the population of Brunei as a whole is about 402,000. Bandar Seri Begawan is sometimes called simply Bandar or BSB.

Historical Overview. Settlement of Brunei goes at least as far back as the 6th century, and that of the banks of the Brunei River to the 8th century. During the 15th–17th centuries, the Sultanate of Brunei ruled a much larger territory that encompassed all of northern Borneo and the southern Philippine Islands. Islam came to Brunei in the 16th century. From 1888 to 1984, Brunei was a British protectorate. It was occupied by Japan during World War II. The city, then called Brunei Town, was captured by the Japanese on December 22, 1941, and then liberated on June 10, 1945, under Operation Oboe Six. Oil was discovered in Brunei in 1929 and the first offshore well was drilled in 1957. Oil production now accounts for some 90 percent of Brunei's Gross Domestic Product. As the British colonial period drew to an end in Southeast Asia, Brunei opted out of joining the Malaysian Federation in 1962, and on January 1, 1984, became an independent country. Its government is a constitutional sultanate. Hassanal Bolkiah, the 29th Sultan of Brunei, has ruled Brunei since 1967, and is one of the world's longest-serving rulers still in power. Because of the oil wealth of his country, Bolkiah is also one of the richest people in the world, with a personal fortune estimated to be $20 billion. He has a personal collection of 1,932 automobiles, including 532 Mercedes-Benzes, 367 Ferraris, 185 BMWs, 177 Jaguars, 160 Porsches, 130 Rolls-Royces, and 20 Lamborghinis.

Major Landmarks. The Istana Nurul Iman is the royal palace and residence of Hassanal Bolkiah, the Sultan of Brunei. Access is private. The palace is also the working quarters for the Sultan and totals 1,888 rooms, 290 bathrooms, and a floor area of 2,152,782 square feet (200,000 sq m). The Sultan Omar Ali Saifuddin Mosque, built in 1958, is distinguished by a golden dome and interior walls of Italian marble and is said to be one of the most beautiful mosques in the world. The Jame'asr Hassanal Bolkiah Mosque is even larger. Major museums include the Royal Regalia Museum about the life of Sultan Hassanal Bolkiah, the Brunei History Center, the Malay Technology Museum, and the Brunei Stamp Gallery. Kampong Ayer (the water village) is a traditional Malay neighborhood in which houses stand on poles above the water and are reached by water taxi.

Culture and Society. Most residents of Bandar Seri Begawan are ethnically Malay, including citizens of Brunei and of neighboring Malaysia. The population also includes a Chinese minority and foreign expatriates in the diplomatic community and in oil production. Some of the service work is done by laborers from Malaysia, Indonesia, the Philippines, and South Asia. Income levels in Brunei are among the highest in the world on average, but that number is skewed because of the greatly disproportionate wealth of the royal family. Nevertheless, Brunei citizens enjoy a high standard of living and have access to free public health care and education. The official religion is Islam, although other religions are practiced as well. Malay is the official language, specifically *Bahasa Melayu,* but English and Chinese are widely spoken as well.

Further Reading

Leake, David. *Brunei: The Modern Southeast Asian Islamic Sultanate.* Jefferson, NC: McFarland, 1989.
Sidhu, Jatwan S. *Historical Dictionary of Brunei Darussalam.* Lanham, MD: Scarecrow Press, 2009.

BANGKOK

Bangkok is the capital and largest city of Thailand, a country in the heart of Southeast Asia that is bordered by Myanmar, Laos, Cambodia, and Malaysia. The city is near the center of the country where the Thailand's great river, the Chao Phraya, empties into the Gulf of Thailand. The river is the main geographical feature of the city, along with the low-lying, flat terrain of the lower Chao Phraya basin. In addition to the wide main channel of the river, there are many deltaic distributaries and canals in the city, resulting in a nickname for Bangkok as the "Venice of the East." In Thai, the name of the city is Krung Thep Maha Nakon or simply Krung Thep, meaning "City of Angels." The population of Bangkok is about 8.2 million and that of the metropolitan area is between 14.6 million and 20 million, depending on how the metropolitan area is defined. The city is by far the leading city of Thailand, as well as an influential economic center in Southeast Asia and in the global economy more generally. Many foreign companies have offices or factories in or near Bangkok, and its airport and seaport are among the world's busiest. The city is also a major destination for tourism from around the world.

Historical Overview. Bangkok dates to the early 15th century and was but a small village under the rule of the capital in Ayutthaya Kingdom until Ayutthaya's fall to the Burmese Kingdom in 1767. At that time, the newly crowned King Thaksin moved the capital to the village, being attracted by its strategic location near the mouth of the Chao Phraya. In 1782, the next king, Buddha Yodfa Chulaloke, moved the capital to the eastern bank of the river, where the heart of the modern city is now located. Thailand played a strategic role as a buffer between British colonies in Southeast Asia to the west and south and French colonies to the east, and maintained its independence. Bangkok prospered as a result and became an influential city in the region. It grew steadily in the 19th and 20th centuries as the center of Thailand's modernization, and then explosively in the second half of the 20th century as Thailand emerged as an important center of the global economy. In World War II, the city was occupied by Japanese forces and was bombed by the Allies. During the United States phase of the Vietnam War (1960s–1975), the city was a transportation hub for American troops and equipment, and a rest and recreation center for American soldiers. There has been considerable political turmoil in Thailand over the past two decades, and the city is clogged from time to time with mass demonstrations. In November 2008, antigovernment protestors occupied Suvarnabhumi Airport, the city's main link to the world, stranding many passengers and disrupting the economy.

Major Landmarks. The heart of Bangkok is the island of Rattanakosin, the "Old City." This is the site of the Grand Palace and its enormous complex of highly decorated religious buildings. Within the grounds of the Grand Palace is Wat Phra Kaew, the most sacred Buddhist temple of Thailand. One of its building houses the Emerald Buddha, Thailand's most sacred Buddha image. Nearby is Wat Po, another major temple and the site of the famous Reclining Buddha statue, while across the river is Wat Arun, the Temple of Dawn, a striking high structure that is covered with broken pieces of colorful Chinese porcelain. Wat Arun can be climbed for a dramatic view of the Chao Phraya and the heart of Bangkok. Other landmarks are Dusit Palace built by King Rama V and its Vimanmek Mansion, the largest golden teakwood house in the world; Jim Thompson's House, the National Museum, the Museum of Siam; the Victory Monument; the King Rama V Monument, and the Democracy Monument. In Siam Square is the newly opened Bangkok Art and Culture Center. Lumphini Park is the largest park in central Bangkok. Important districts include the Siam Square shopping district, Sukhumvit, and Chinatown.

Wat Arun, the Temple of Dawn, on the banks of the Chao Phraya River in Bangkok. (Photo courtesy of Roman Cybriwsky)

Culture and Society. Thailand is a devoutly Buddhist country and Bangkok reflects that aspect with thousands of temples and Buddhist shrines, and many religious festivals and traditions. The city is also a major business center, and has more than 1,000 skyscrapers and a great many busy shopping areas, including ultramodern malls and popular street-side night markets. Its roads are famously jammed, so residents and visitors alike often get to where they need to be by river bus or on the Bangkok rail transit system that rides above street level. The language of the country and that of the city is Thai, although the population of the city includes many migrants from mountain regions where local languages are spoken. There is also a Chinese minority in the city that has long been influential in local trade and commerce. At any given time there are tens of thousands of foreign tourists in Bangkok who visit the main landmarks of the city and enjoy shopping. There are also sex tourists from abroad and specific districts of the city that are known for prostitution and sexual exploitation of impoverished young girls and boys from Thailand's countryside and neighboring poor countries. There is a high prevalence of HIV/AIDS in Thailand, especially among sex workers.

Further Reading

Askew, Marc. *Bangkok: Place, Practice and Representation.* London: Routledge, 2002.
Kerr, Alex. *Bangkok Found: Reflections of the City.* Bangkok, Thailand: River Books, 2010.
Warren, William. *Bangkok.* Singapore: Talisman Publishing, 2002.

BANGUI

Bangui is the capital and largest city of the Central African Republic, a land-locked country located just north of the equator in the very center of the African continent. It is located in the southwest of the country on the northern bank of the Ubangi River, the largest right-bank tributary of the Congo River. Across the river from Bangui are the Democratic Republic of the Congo and the Congolese town of Zongo. There are rapids on the Ubangi just upstream from Bangui, so the city marks the limits of navigation on the river. The city is named after the rapids on the river. The population of Bangui is approximately 650,000.

Historical Overview. Bangui was founded in 1889 as a spin-off of a French military post on the Ubangi River during the height of European colonialism in Africa. The area was referred to as Haut-Oubangui (Upper Ubangi) at the time, and was then Oubangui-Chari, a part of French Equatorial Africa. Bangui was the French colonial administrative center. Independence came in 1960, but the history of the Central African Republic has continued to be plagued with violence and political instability. During the rule of Jean-Bédel Bokassa (1966–1979), the country was renamed the Central African Empire. The Central African Republic is among the poorest countries in the world and ranks as one of the worst in terms of investment potential and ease of doing business. Bangui itself is considered to be one of the world's worst cities in quality of life.

Major Landmarks. Bangui's most distinctive landmarks are a triumphal arch honoring deposed Emperor Bokassa and the presidential palace. There is also a busy central market. At the riverfront is a busy passenger ferry terminal with service to Zongo across the river and to Brazzaville, capital of the Republic of the Congo. Also of note are the University of Bangui founded in 1970 and Notre Dame Cathedral, Bangui's Roman Catholic cathedral.

Culture and Society. Before the period of colonialism, the native language of the Central African Republic and the Ubangi River area was Sango. It is still widely spoken in Bangui and elsewhere in the country, but French is the official language. The main religious groups are native religions, Protestants, Catholics, and Muslims. Many of Bangui's residents are migrants from the countryside who hope for a better life in the nation's capital. However, the Central African Republic is a very poor country and Bangui is not known as an attractive city. There are many squatter slums and there is a general lack of sanitation and potable water.

Further Reading

Kalck, Pierre. *Historical Dictionary of the Central African Republic.* Lanham, MD: Scarecrow Press, 2004.

Titley, Brian. *Dark Age: The Political Odyssey of Emperor Bokassa.* Montreal: McGill-Queen's University Press, 1997.

Woodfork, Jacqueline. *Cultures and Customs of the Central African Republic.* Westport, CT: Greenwood Press, 2006.

BANJUL

Banjul is the capital of The Gambia, a West Africa country that is the smallest nation on the African continent. Except a short coastline with the Atlantic Ocean, the country is bordered only by Senegal and consists of two narrow, more-or-less parallel strips of land on either side of the Gambia River that flows west to empty into the Atlantic. The capital city lies at the mouth of the river on the tip of a small island named Banjul Island or St. Mary's Island. It is connected by ferry service to points on the north side of the river and by bridge to the south. The population of Banjul is 357,238 (2003 census), making it the third-largest city in the Gambia after Serekunda and Brikama. The name Banjul comes from a corruption of the local Mandinka word for a specific fiber that was once gathered on the island for making rope.

Historical Overview. The valley of the Gambia River was settled many centuries ago by African tribesmen. Arab traders crossed the Sahara Desert and traded with communities on the Atlantic Coast. Conversions to Islam took place in the 11th and 12th centuries. In the 14th century the area of the Gambia River became part of the Mali Empire. Portuguese ships and then English ships traded in slaves from the mouth of the Gambia. From 1651 to 1656 there were attempts to establish colonial settlements near the mouth of the river by the Duke of Courland, from a vassal state of the Grand Duchy of Lithuania. In 1816, the British established a trading base on St. Mary's Island. It was named Bathurst after Henry Bathurst, the secretary of the British Colonial Office. Gambian independence from the United Kingdom was achieved in 1965 and the capital's name was changed to Banjul in 1973.

Major Landmarks. Arch 22 is a symbolic gateway to Banjul that was erected to commemorate a bloodless coup on July 22, 1994, that put the current president of the Gambia Yahya Jammeh in power. The arch includes a textile museum. The city's other two museums are the Gambian National Museum and the African Heritage Museum. Religious landmarks include the King Fahad Mosque and St. Mary's Anglican Cathedral.

Culture and Society. The Gambia is a diverse society with several important ethnic groups. Mandinka people are most numerous, but there are also Fula, Wolof, Jola, Serahule, and Serers. English is the official language, but tribal languages are commonly spoken. Islam is the dominant religion, embracing more than 90 percent of the population. The majority of the national population is rural, and many residents of Banjul have roots in the countryside. The Gambia has rich musical traditions. The popular American book and television series by Alex Haley, *Roots,* was set in the Gambia.

Further Reading

Hughes, Arnold and David Perfect. *Historical Dictionary of the Gambia.* Lanham, MD: Scarecrow Press, 2008.
Jagne, Hassan. *The River Gambia: Life of the Gambia.* Bloomington, IN: Xlibris Corporation, 2008.

BASSETERRE

Basseterre is the capital and largest settlement of the Federation of Saint Kitts and Nevis, also known as Saint Christopher and Nevis, a small, sovereign country consisting of two islands 2 miles (3 km) apart in the Windward Islands chain of the Caribbean Sea. The town is located on the southwestern coast of the larger island, Saint Kitts (also written as St. Kitts). In addition to administrative functions, the town is a banking center for islands in the region, a port, and center of international tourism. The population of Basseterre is about 15,500, while that of Saint Kitts and Nevis as a whole is about 51,300.

Historical Overview. The indigenous Amerindians on Saint Kitts were Kalinago people, but their population began to decline with European settlement. In 1493, Christopher Columbus and his crew were the first Europeans to arrive. French settlers came in 1538 but were later sent back by the Spanish. The first English settlement was established in 1623. This was followed by renewed French settlement and an agreement between France and England to share the island. In 1628, the last of the Kalinago were massacred. Although the Spanish reclaimed Saint Kitts in 1629, French and English claims won out. By 1713, France ceded its interests in the island to Great Britain. French and English planters imported slaves from Africa as labor on their plantations. With time, the descendants of slaves became the majority population group. Until the late 19th century, Saint Kitts and Nevis were viewed as different colonies and were administered separately. In 1967, the two islands were united with Anguilla, another of the Windward Islands to the north, into a single dependency called Saint Christopher-Nevis-Anguilla. Anguilla separated in 1971 and became a British colony on its own, which it still is, while Saint Kitts and Nevis became an independent state in 1983. Relations between the two islands are strained to the point that in 1998 the population of Nevis held a referendum (which failed) about breaking away from the neighboring island and becoming independent. The island's once-dominant sugar industry is now defunct and old cane fields are being converted into housing developments and resorts.

Basseterre itself was founded by French settlers in 1627. It developed into the largest port in the eastern West Indies and the administrative center for France's several island possessions. In 1727, the British made Basseterre the capital of the Saint Kitts. Several times in its history, the city has been devastated by hurricanes and earthquakes. Hurricane Georges that hit the city in September 1998 was one of the most recent natural disasters to strike the country.

Major Landmarks. Independence Square is the center of Basseterre. The Basseterre Roman Catholic Cathedral of the Immaculate Conception, the court house, and other prominent buildings face the square. St. George's Anglican Church is nearby, as is a statue of Britain's Queen Victoria. The Circus, a prominent traffic circle in the center of Basseterre, is home to the Berkeley Memorial, a four-faced clock that commemorates a political leader from the late 19th century. The Warner Park Sporting Complex was built in 2006 and was one of the host venues for the 2007 World Cricket Cup. Hosting this important event was a major achievement for a country as small as Saint Kitts and Nevis.

Culture and Society. About three-quarters of the population of Saint Kitts and Nevis is of African descent, while most of the rest are of mixed African and European descent or African-South Asian. About 3 percent of the population is South Asian. The population of Saint Kitts and Nevis peaked in about 1960, with considerable emigration afterward to the United Kingdom, the United States, and Canada. There are several popular festivals in Basseterre including Carnival at the end of the calendar year and into the New Year's holiday and the Saint Kitts Music Festival at the end of June. Cricket is the most popular sport, but rugby and football (soccer) are also played.

Further Reading

Dyde, Brian. *St. Kitts: Cradle of the Caribbean.* Oxford: Macmillan Education, 2008.

Mosimba. *The History of the St. Kitts & Nevis Carnival.* Bloomington, IN: Xlibris Corporation, 2008.

Richardson, Bonham C. *Caribbean Migrants: Environment and Human Survival on St. Kitts and Nevis.* Knoxville: University of Tennessee Press, 1983.

BEIJING

Beijing is the capital city of the People's Republic of China, also known as Communist China, a large Asian country that extends from the Pacific coast of the continent deep into the interior. The city is located in the northeastern part of the country (although not in what is known formally as China's Northeast), and is formally a municipality that is under the direct control of the national government. The province of Hebei almost surrounds Beijing. The population of Beijing is about 19.6 million, making the city one of the largest urban areas in the world and the second-largest city in China after Shanghai.

In addition to being the center of China's strong and highly centralized national government, Beijing is a significant cultural and educational center, the headquarters city for most of China's large state-owned companies, a major gateway to China for air passengers, and a hub of China's extensive road and rail networks. It is a city of rapid growth and continual urban construction, and is increasingly a landscape of new high-rise buildings and congested streets and highways. Traditional architecture and neighborhoods called *hutongs* are being lost to urban redevelopment. Air pollution is a major problem, both from within the urban environment caused by industry and vehicular traffic, and from outside the city in the form of blown dust from deserts in northern China.

Rebellious Ürümqi

Ürümqi is the capital and largest city (population 2.7 million) of the Xinjiang Uyghur Autonomous region in the far northwest of the People's Republic of China. It is a desert-oasis city and its name means "beautiful pasture." About three-quarters of the population is Han Chinese, as Chinese settlement has been heavy in this frontier area, and only 13 percent comprises ethnic Uyghurs, a Turkic ethnic group of Islamic religion. However, there is a strong sense of separatism among the Uyghurs and other minorities in the region, and a history of recent conflict with China about colonial control and minority rights. Ürümqi has recently been the scene of rioting and other unrest, especially in 2009, with the result that the Chinese military presence and Chinese government censorship of news about the region have been greatly increased.

The word Beijing means northern capital. In an earlier form of Romanization of Chinese, now outdated, the city was known as Peking. Beijing is the current of the "Four Great Ancient Capitals of China," the other three being Nanjing, Louyang, and Chang'an (now Xi'an).

Historical Overview. Beijing's site has been settled for millennia. It first gained prominence in the 11th-century BC as the walled city of Ji, and then as capital of the State of Yan during the Warring States Period (475–221 BC). It was one of five capitals the Liao Dynasty (907–1125 AD) and capital of the Jin Dynasty (1115–1234). After the Mongol conquest of China and establishment of the Yuan Dynasty (1279–1368), a new capital called Dadu in Chinese was built beside the ruins of the Jin capital. The Ming Dynasty (1368–1644) established its capital initially in Nanjing and renamed Dadu as Beiping (Northern Peace), but in 1403 the capital was moved back to Beiping and the city was renamed Beijing (Northern Capital). The Ming engaged in enormous new construction on the same symbolic north-south axis as Dadu, including four sets of walls, symbolic gates, and the Forbidden City, the Chinese imperial palace. The same complex served as capital of the Qing Dynasty (1644–1911) after the Manchus conquered the Ming. Beijing was briefly the Chinese capital after the fall of the imperial era, but the capital of the Republic of China shifted to Nanjing again in 1928 and Beijing was once more renamed Beiping. In 1937, the city was occupied by Japan and was made into the capital of the Provisional Government of the Republic of China, a puppet state that ruled northern China during the time of Japanese control.

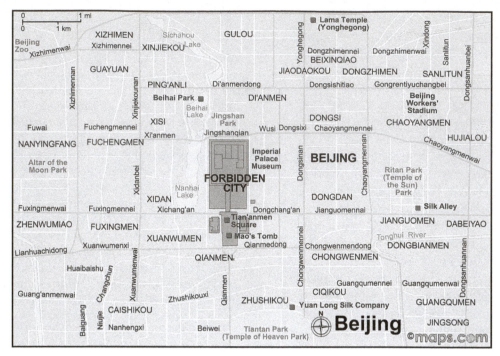

Beijing, China (Maps.com)

On October 1, 1949, after Japan's defeat in World War II and the end of the Chinese Civil War (1927–1949), Mao Zedong, leader of the Communist Party of China, announced the creation of the People's Republic of China. He renamed the city Beijing and made it capital once more. It has been capital of the People's Republic of China ever since. In 1966–1976 the city was a focus of the Chinese Cultural Revolution. In 1989, Beijing was the site of prodemocracy protests in Tiananmen Square that resulted in a bloody suppression by the People's Liberation Army. A lasting image of that event was the confrontation on June 5 of that year by a lone protestor facing a column of Chinese army tanks. In 2008, the city hosted the Summer Olympic Games, an event of enormous national pride in China.

Beijing's Forbidden City

Built between 1406 and 1420, the Forbidden City in Beijing was the residence of China's emperors for nearly 500 years from the Ming Dynasty through the Qing Dynasty, China's last, which ended in 1912. It was also the center of political and ceremonial functions of China's government. The complex is now the Palace Museum and covers 7.8 million square feet (720,000 sq m), includes 980 surviving buildings, and is a UNESCO World Heritage Site. The central north-south axis of the Forbidden City is the central axis of the modern city of Beijing. The "forbidden" aspect of the Forbidden City refers to the fact that people needed the emperor's permission to enter or leave. Beijing residents often refer to the complex as Gùgōng, meaning former palace.

Major Landmarks. The center of Beijing is Tiananmen Square, the largest public square in the world. It is surrounded by the Great Hall of the People, the Museum of Chinese History, the Museum of the Chinese Revolution, Qianmen Gate, and the Forbidden City. Tiananmen Square also houses the Chairman Mao Memorial Hall and the Monument to the People's Martyrs. The Forbidden City is a complex of 980 buildings covering 7.8 million square feet (720,000 sq m), and is a UNESCO World Heritage Site. The Temple of Heaven is a major landmark in the southeastern part of central Beijing. The Pagoda of Tianning Temple dates to the early 12th century and the Liao Dynasty. Beihai Park is a large imperial garden in the center of Beijing. More recent landmarks include the Beijing National Stadium, known colloquially as the Bird's Nest because of its shape, which was built for the 2008 Olympics, the National Center for the Performing Arts, and the uniquely shaped China Central Television (CCTV) Headquarters that was opened in 2008 in the Central Business District (CBD) of Beijing. The CBD itself is an ever-expanding collection of new landmark office towers such as China World Trade Center 3, Jianwai SOHO, and Yintai Center. The Badaling, Simatai, Mutianyu, and Jinshanling sections of the Great Wall of China are just to the north of Beijing and are extremely popular visitor attractions. Also located just outside the city are the elaborate Thirteen Tombs of the Ming Dynasty.

Culture and Society. The over whelming majority of Beijing residents are Han Chinese, with minority of other ethnic groups such as Manchu, Hui, and Mongols

Tiananmen Square and the mausoleum of Chairman Mao Zedong in the background. (Photo courtesy of Roman Cybriwsky)

Lhasa: The Dalai Lama's Capital

Lhasa is the capital of the Tibet Autonomous Region within the People's Republic of China (PRC) and, at an elevation of 11,450 ft (3,490 m), is one of the highest cities in the world. It had long been the center of Tibetan Buddhism and the capital of a Buddhist religious monarchy, but China incorporated Tibet into the PRC in 1951 and has held tight control since. The Dalai Lama, Tibet's traditional leader fled the country for India in 1959. Since then, the ethic Tibetan population has been greatly diluted by in-migration of Han Chinese, the presence of Chinese troops, and discrimination against Tibetans to speed up assimilation. Nevertheless, many Tibetans continue to aspire to nationhood, as do many supporters of the Tibetan cause around the world. The government of China exercises stern censorship with respect to issues in Tibet. The Dalai Lama's former palace, called the Potala, is now a state museum and is a popular tourist attraction along with the Jokhang Temple and other historic religious sites in the city.

also in the mix. The language of the city is Mandarin Chinese, with natives speaking a dialect that is unique to the city. Many of the city's residents are migrants from the Chinese countryside. Bicycle transportation has long been a feature of commuter life in Beijing, but recent years have seen increases in automobile usage and expansions to the city's mass transit system. In addition to being national capital Beijing is a major corporate center, with headquarters of 26 Fortune 500 companies (third-most in the world behind Tokyo and Paris) and more than 100 large Chinese firms. There are also many shopping districts and super shopping malls, as well as a thriving nightlife.

Further Reading

Chang, Sen-Dou. "Beijing: Perspectives on Preservation, Environment, and Development," *Cities: The International Journal of Urban Policy and Planning* 15, no. 1 (1998): 13–25.

Meyer, Michael. *The Last Days of Old Beijing: Life in the Vanishing Backstreets of a City Transformed.* New York: Walker Publishing Company, 2008.

Wang, Jun. *Beijing Record: A Physical and Political History of Planning Modern Beijing.* Singapore: World Scientific Publishing, Ltd., 2011.

BEIRUT

Beirut is the capital and largest city of the Republic of Lebanon, a small country in the Middle East on the eastern shores of the Mediterranean Sea that is bordered by Israel to the south and Syria to the north and east. The city is in the approximate midpoint of the country's coast on a peninsula that juts into the Mediterranean Sea. Behind the city are the Lebanon Mountains. In addition to being the national capital, Beirut is Lebanon's main port, a busy commercial and financial center, and a significant cultural center. It has reconverted from civil war in 1975–1990, and is now a popular destination for international tourists, shoppers, and night clubbers as well. There has been no census in Lebanon for some time, but the estimated population for the city of Beirut is about 750,000, while that of the metropolitan area totals between 2 and 3 million.

Historical Overview. Beirut is an ancient city with origins that date back at least 3,000 years, perhaps even 5,000 years. The first historical reference to the city is from letters written in the 14th-century BC that a local king sent to the pharaoh of Egypt. Recent archaeological work had revealed many layers of earlier settlements beneath the center of today's city, including sequentially those of Phoenicians, Greeks, the Roman Empire, Byzantium, Arabs, Crusaders, and the Ottoman Empire. From 635 to 1110, Beirut was ruled by the Yemeni-Arab Arslan family whose principality based in Beirut foreshadowed the modern Lebanese state. From 1001 to 1236 the city was ruled by Crusaders as part of their Latin Kingdom of Jerusalem. The powerful Crusader known as John of Ibelin, the Old Lord of Beirut, fortified the city and made it his capital. The city was later ruled by the Druze emirs during the Ottoman period, and competed with the nearby city of Acre (in today's Israel) for supremacy in regional maritime trade. As a result of conflict in the north of Lebanon in 1860 between Druze and Maronite Christian religious groups, Beirut took in many Christian refugees, significantly altering the religious make-up of the city. In 1894, a modern harbor was installed in Beirut, and in 1907 a rail link was completed to Damascus and Aleppo in Syria, greatly enhancing the city's advantage for trade between Europe and the Middle East. After the Ottoman Empire collapsed with the end of World War I, Beirut was part of French Mandate Lebanon. In 1943, Lebanon became independent and Beirut became a national capital.

Beirut prospered after independence as a popular tourist destination, cultural center, and center of banking and shopping in the Middle East as the oil industry of the region was expanding. During this heyday, Beirut was often referred to as the "Gateway to the Middle East" or the "Paris of the Middle East." In 1975, however, civil war broke out in Lebanon and raged until 1990. Beirut very heavy damaged in the fighting. The city was divided into a Muslim western part and a Christian east, with the downtown being a no-man's zone called the Green Line. Entire sections of the city were destroyed and abandoned. The worst battles occurred in 1978 when Syrian troops bombed the eastern section of Beirut, and in 1983 when West Beirut was under siege by Israeli forces during what came to be called the Hundred Days War. In 1983, a suicide bombing attack against military barracks took the lives of 241 American and 58 French soldiers, and 6 civilians. Beirut has been rebuilt since 1990, and is again an attractive and lively city. However, conflicts have continued. In 2005, Lebanese prime minister Rafic Hariri was assassinated in Beirut. That was followed by the so-called Cedar Revolution in which as many as one million citizens gathered at Beirut's Martyr's Square to protest Syrian interference in Lebanese affairs. In 2006, Israel bombed parts of South Beirut where there were strongholds of the militant Hezbollah movement.

Major Landmarks. Martyrs' Square in the heart of Beirut's downtown is a major landmark with a statue in the center that honors Lebanese patriots who were hanged on May 6, 1916, by Ottoman Turks because they opposed foreign occupation of their country. Another prominent downtown square is Place de l'Etoile (Nejmeh Square). It was built by the French in the early 20th century and now has a high clock tower at its center. The Corniche of Beirut is a lively area between the skyscrapers along the city's shoreline and the beaches and sunsets of the Mediterranean Sea. Pigeon Rocks is a natural arch rock formation in the waters off Beirut that is one of the city's most recognizable icons. The Garden of Forgiveness is a new monument under construction that commemorates the Lebanese Civil War. Important museums include the National Museum of Beirut, the Sursock Museum, and the Beirut Arts Center. The American University of Beirut was founded in 1866 and has a sizable campus on a promontory overlooking the Mediterranean Sea and approximately 6,900 students. A symbol of Beirut's religious diversity is the presence of Mohammad al-Amin mosque and St. George Maronite Cathedral as neighbor buildings in the downtown of the city.

Culture and Society. Beirut's population includes both Christians and Muslims. Its major religious communities include Maronite Catholics, Greek Orthodox, Armenian Apostolic, Armenian Catholics, Protestants, Sunni Moslems, Shia Moslems, and Druze. The east of the city is mostly Christian, the west is largely Sunni, and the south is largely Shia Moslem, although there is even more religious integration. The official language of the country is Arabic, but English, French, and Armenian are also widely used. Many Lebanese, especially Christians, live in diasporas abroad.

Further Reading

Kassir, Samir. *Beirut.* Berkeley: University of California Press, 2011.
Llewellyn, Tim. *Spirit of the Phoenix: Beirut and the Story of Lebanon.* London: I. B. Tauris, 2010.
Sawalha, Aseel. *Reconstructing Beirut: Memory and Space in a Postwar Arab City.* Austin: University of Texas Press, 2011.

BELFAST

Belfast is the capital and largest city in Northern Ireland, a part of the United Kingdom in the northeast of the island of Ireland that shares a border with the Republic of Ireland from which it was divided in 1921. The city is located at the western end of the Belfast Lough and at the mouth of the River Lagan, a combination that facilitated the city's development in the 18th and 19th centuries as a major shipbuilding center. It is the second largest city on the island of Ireland after Dublin, the capital of the Republic of Ireland. The population of Belfast is about 268,000, while that of the metropolitan area is about 642,000. During the industrial era, one of the city's nicknames was "Linenopolis" because of its role as a center of linen manufacturing.

Historical Overview. The site of Belfast has been occupied for millennia. There are remains of ancient Iron Age forts near the center of the city, as well as other archaeological evidence of a long history. A castle was built in the city in 1177 by John de Courcy, the Norman conqueror of Ulster. The modern history of the city began in 1611 with the construction of a new castle by Sir Arthur Chichester and the issuance of a charter of incorporation for Belfast in 1613. The Society of United Irishmen, an organization that led the Irish Rebellion in 1798, was founded in the city in 1791. In the 19th century, the city became a major industrial center with specialization in the manufacture of linen, tobacco products, and shipbuilding. With 35,000 workers, the city's Harland and Wolff shipyards became the largest shipbuilders in the world. The RMS *Titanic* was built at this shipyard in 1909–1911. There were major riots in the city in 1886 over the issue of British rule. The Irish War of Independence began in 1919 and resulted in partition of the Irish island in 1921 and designation of Belfast as capital of British Northern Ireland. There has been considerable conflict ever since between republican forces in favor of a united and independent Ireland and loyalists in Northern Ireland who prefer government under the British Crown. Sectarian violence between Irish Catholics and Protestants known as "The Troubles" raged in Northern Ireland from the late 1960s until the peace that came with the Belfast "Good Friday" Agreement of 1998. In the interim, hundreds of lives were lost. Although the situation is now much quieter, violence still reappears from time to time.

Major Landmarks. The Belfast City Hall is located in the center of Donegal Square and is perhaps the city's most recognizable landmark. It opened in 1909 and is an impressive domed building in a style that became a standard in the British Empire as a symbol of government authority. On the grounds of City Hall is a statue of British Queen Victoria and a monument to the victims of the Titanic. Another striking building is the 1904 Church of Ireland St. Anne's Cathedral. Other important landmarks are the Royal Courts of Justice, St. George's Market, the Albert Memorial Clock Tower, and the Grand Opera House. Waterfront Hall is the city's principal concert hall. Queen's College and the nearby Botanic Gardens are also noteworthy. The old industrial waterfront district is now being redeveloped into a trendy commercial and residential district called the Titanic Quarter. There are many murals throughout the city that depict various scenes from the conflicts of the "The Troubles."

Culture and Society. Despite the peace accord that was reached in 1998, much of Belfast is still segregated between opposing neighborhoods that are mostly Catholic and favor one Ireland ("Republican" areas) and those that are Protestant and favor Northern Ireland's union with Great Britain ("Loyalist" areas). This is reflected quite graphically in the city's famous murals. The various strongly segregated working-class neighborhoods in the city are known as "interface areas." The numbers of Catholics and Protestants in Belfast are almost equal. Minority groups include Irish Travelers and various immigrant populations from abroad, the most numerous of whom are Chinese.

Further Reading

Boal, Frederick W. *Enduring City: Belfast in the Twentieth Century.* Belfast: Blackstaff Press, 2007.

Maguire, William. *Belfast: A History.* Lancaster, UK: Carnegie Publishing, 2009.

Shirlow, Pater and Brendan Murtagh. *Belfast: Segregation, Violence, and the City.* London: Pluto Press, 2006.

BELGRADE

Belgrade is the capital and largest city in the Republic of Serbia, a landlocked country at the junction of central Europe and the Balkans. It lies in the north-center of the country at a natural crossroads of trade that has served it through history at the confluence of the Danube and Sava Rivers. The historic core of the city, Kalemegdan, is on the right bank of both rivers. The population of Belgrade is more than 1.1 million, while that of the metropolitan area exceeds 1.6 million. The name of the city in the local Serbian language is Beograd, meaning "White City."

Historical Overview. The area of Belgrade has been occupied for millennia. The rise of city is traced to a fortress built by the Celts in the third-century BC. It was conquered by the Romans near the start of the Christian era, and was then settled by Slavs in the sixth century. In the late 13th century it became the capital of Serbian monarch Stephen Dragutin. In 1402, Stephen Lazarevi'c made it the capital of Serbia. Ottoman Turks besieged the city from the 1440s and finally conquered it in 1521, designating it as capital of a local administrative district. During the Austro-Ottoman Wars, the city changed hands frequently between the Ottoman and Hapsburg Empires and saw much destruction from battles. A Serbian uprising against the Turks early in the 19th century resulted in a short-lived Serbian state, but the Ottomans regained control and held the city until Serbian rule was reestablished in 1867. From about 1918 until its final dissolution in 2006, Belgrade was the capital of Yugoslavia. The city was badly damaged in World War II and its Jewish population was massacred. After the war, Belgrade was made into an industrial city. After a long period of conflict among the various parts of Yugoslavia and considerable resettlement of ethnic and religious groups across emerging national borders, the independent Republic of Serbia was formed in 2006 with Belgrade as capital. All told in its history, Belgrade was the scene of 115 wars and was destroyed in battle 44 times.

Major Landmarks. The Kalemegdan-Belgrade Fortress is a major landmark from the city's history and the main attraction in large park along the riverside. The main square of the city is Republic Square. It is the site of the National Theater, the National Museum founded in 1844, and a landmark equestrian statue of Prince Mihailo Obrenovic that is a popular meeting place for people going downtown. Knez Mihailkova Street (Prince Mihailo's Street) is a busy pedestrian street that runs off the square with many shops, restaurants, and cafés. Nearby is Skadarlija, an historic old neighborhood with winding streets and vintage residential architecture. The Church of St. Sava, constructed beginning 1935, is the largest Eastern Orthodox church building in the world. The Belgrade Cathedral or St. Michael's Church, located near the historic fortress, is another major religious landmark. Other landmarks in the city include the White Palace, the Old Palace, and the National Assembly of Serbia.

Culture and Society. The main ethnic group of Belgrade comprises Serbians, accounting for over 90 percent of the population. The other ethnic groups include Croats, Montenegrins, and Roma, as well as new populations of Chinese and the Middle East. The official language is Serbian. The city attracts economic migrants from the countryside of Serbia and from other countries that were once part of the Yugoslavian federation. The Serbian Eastern Orthodox Church is the faith of the large majority of the population. There are minority faiths of Muslims and Roman Catholics. Until the tragic events of World War II, there was once a large Jewish community in Belgrade.

Further Reading

Hirt, Sonia. "Belgrade, Serbia," Cities: *The International Journal of Urban Policy and Planning* 26 (2009): 293–303.
Norris, David. *Belgrade: A Cultural History.* New York: Oxford University Press, 2009.

Waley, Paul. "From Modernist to Market Urbanism: The Transformation of New Belgrade," *Planning Perspectives* 26, no. 2 (2011): 209–35.

BELMOPAN

Belmopan is the new (since 1970) capital of Belize, the former British Honduras and the northernmost of the six countries of Central America. It is located near the nation's center, about 50 miles (80 km) inland from the old capital Belize City. The population of Belmopan is about 20,000, making it one of the smallest capital cities in the world. The name of the city comes from the Belize River and the Mopan River, two rivers that Belmopan is near to.

Historical Overview. Belmopan came into being after Hurricane Hattie destroyed low-lying, coastal Belize City in late October 1961. A decision was made to move the capital inland where it would be safer from Caribbean storms and closer to the center of the country. As Belize was at the time a colony of the United Kingdom called British Honduras, permission and funding for the project were obtained in London. Mr. Anthony Greenwood, the British Secretary of State for the Commonwealth and Colonies, visited British Honduras in 1965 and dedicated a monument on October 9 of that year at the side of a highway where the new city would be built. Construction work began in 1967 and the capital was officially opened in 1970. Construction still continues, as Belmopan is a work in progress. The United States Embassy in Belmopan opened in December 2006.

Major Landmarks. The center of Belmopan is Independence Plaza. The main landmark of Belmopan is the new National Assembly Building. Yim Sang is a 12-story hotel, the tallest building in Belmopan. There is a new bus terminal and central market, both of which date to 2003. Guanacaste Park is a beautiful nature preserve outside the city.

Culture and Society. Most residents of Belmopan are connected to national government work. Belize City, with a population of about 80,000, is still the largest and most cosmopolitan of the country's settlements. It is only about one hour drive away, so the two cities are closely interconnected. Commuters travel from residences in Belize City to work in Belmopan. There are plans for developing an industrial zone in Belmopan. The population of Belize and Belmopan specifically is a mix of races including people of European descent, Africans, indigenous Americans (particularly Mayans), and Asians. English is the official language, although Spanish and local languages are spoken as well. The use of Spanish has been increasing recently with stepped-up migration from neighboring countries in Central America. Approximately 80 percent of the population is Christian, with Roman Catholics being the largest denomination.

Further Reading

Shoman, Assad. *Belize's Independence & Decolonization in Latin America: Guatemala, Britain & the UN.* New York: Palgrave Macmillan, 2010.
Sutherland, Anne. *The Making of Belize: Globalization at the Margins.* Westport: Bergin & Garvey, 1998.
Thompson, Peter. *Belize: A Concise History.* Oxford: Macmillan Caribbean, 2004.

BERLIN

Berlin is the capital and largest city in Germany, the most populous country in the European Union and its largest economy. The city is located in the northeastern part of Germany about 37 miles (60 km) from the country's border with Poland. Its topographic region is the flat Northern European Plain that stretches from France well into Russia. It is one of the most dynamic urban centers of Europe and a city of global significance, both now and in recent history. The population of Berlin is 3.45 million, the second most populated city and seventh largest urban area in the European Union. The Berlin–Brandenburg Metropolitan Region has 4.4 million inhabitants. The city is unusually green in comparison to other large cities, with about one-third of the land area given to forests, parks, gardens, lakes, and other open spaces.

Historical Overview. The origins of Berlin date to the 13th century. In the 15th century the city was in the Margraviate of Brandenburg, a principality of the Holy Roman Empire that was ruled by the Hohenzollern family, and became capital after Frederick II Irontooth designated it to be his capital in 1440. The House of Hohenzollern ruled Berlin until 1918 as a royal dynasty of electors, kings, and emperors. In 1539, the city became Lutheran. The Thirty Years War of 1618–1648 destroyed much of the city and killed half the population. Afterward, Frederick William, the "Great Elector" invited immigration to repopulate Berlin and the principality of Brandenburg. Among those who settled the city were Huguenots who had been expelled from France, and settlers from Bohemia, Poland, and Salzburg. In 1701, Frederick I was crowned King of Prussia and designated Berlin to be the capital instead of Königsberg, giving the city new prominence. Under Frederick II, known as Frederick the Great, who reigned from 1740 to 1786, Berlin became one of the great centers of the European Enlightenment.

As a consequence of the Industrial Revolution in the 19th century, Berlin became a major center of manufacturing and an important railroad hub, and grew rapidly with German urbanization. In addition to being capital of Prussia from 1701 to 1918, Berlin was also capital of the German Empire from 1871 to 1918. After World War I in 1919, Germany replaced imperial rule with a liberal democratic parliamentary republic that is called the Weimar Republic. Between 1919 and 1933 when Adolf Hitler gained power, Berlin was the Weimar capital and thrived as a center of industry, art, intellectual life, nightlife, and cultural change. By virtue of the 1920 Greater Berlin Act, the city incorporated many of the neighboring municipalities and grew to about 4 million inhabitants. It was one of the world's largest and most vibrant urban centers.

Berlin was also the capital of Adolf Hitler's Nazi Party Third Reich (1933–1945), and saw enormous destruction and killing both before and during World War II. Most of the 170,000 Jews who lived in the city in 1933 did not survive, the greatest toll being taken by Hitler's death camps. There were hundreds of thousands of other deaths as well, most particularly during the Allied air raids of 1943–1945 and the advance to Berlin by the Soviet Union's Red Army as the war ended. The city was all but totally destroyed. After the war, Germany was divided among the victors, and Berlin, which found itself within a Soviet-controlled East Germany (eventually the Communist German Democratic Republic), was itself divided into four sectors, one administered by the Soviet Union (East Berlin) and three by the Western Allies, the United States, France, and Great Britain (collectively, West Berlin). As a result of Cold War tensions, the Allies had to operate an airlift of food and other supplies over East German air space from June 24, 1948, to May 1, 1949, to meet war survivors' needs in West Berlin. In 1961, East Germany constructed a wall between East and West Berlin (the Berlin Wall), as well as other barriers West Berlin and East German space, in order to isolate Berlin more and to keep its own citizens from fleeing to the West. During the Cold War (1946–1989), East Berlin was the capital of East Germany, while West Germany (the Federal Republic of Germany) moved its government administration to the city of Bonn.

The Berlin Wall was torn down by a popular uprising against Communism on November 9, 1989, and in 1990, East and West Germany were united as the Federal Republic of Germany. In 1991, the German Parliament voted to return the set of government to Berlin; the move was completed in 1999.

Major Landmarks. What little is left of the Berlin Wall is on display at the open-air East Side Gallery in the Berlin district of Friedrichshain. The Brandenburg Gate built between 1788 and 1791 is a former grand entrance to the city that is an icon of Berlin. It was badly damaged in World War II and was restored in 2000–2002. The Reichstag Building is the seat of Germany's parliament. It too was remodeled after the war and features a striking glass dome and skywalks. Potsdamer Platz is a public square where the Berlin Wall was located. The area is now the site of many redevelopment projects, including high-rise office towers and hotels, and shopping areas. The Memorial to the Murdered Jews of Europe, a Holocaust memorial, is nearby. Charlottenburg Palace is a beautiful, large palace from the end of the 17th century with formal gardens. Berlin Cathedral (Berliner Dom) is one of the many historic churches in the city. The Fernsehturm is a very high TV tower (1,207 ft; 368 m) that is a landmark of Berlin. Neue Nationalgalerie is a beautiful museum for modern art. Museum Island in the River Spree has several other museums and is on the UNESCO World Heritage Sites list.

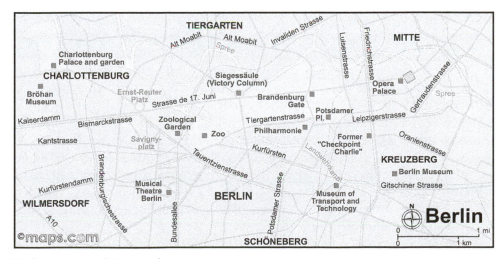

Berlin, Germany (Maps.com)

Culture and Society. Berlin is a fast-developing, dynamic city that is rapidly putting away the scars of nearly 50 years of division between East and West, and overcoming the economic and cultural differences that the division created. It is a mecca for young artists, musicians, writers, filmmakers, designers, and other creative people, and has once again come to be one of the intellectual capitals of Europe. The city has many theaters, concert halls, museums, and art galleries, as well as cafés, restaurants, and pubs. Its nightlife is among the liveliest anywhere. Berlin also boasts a great outdoors, with parks and forest, bicycle paths, and outdoor sports.

The population of Berlin is increasingly diverse with immigration from Turkey and various countries of southern and eastern Europe, among other countries. There are approximately 250,000 Turks in the city, making Berlin the largest concentration of Turks outside Turkey. There are also many Germans from the former Soviet Union, Poles, Serbs, Bosnians, and Russians. In comparison to other parts of German and northern and western Europe in general, Berlin has a youthful population. The main religious affiliation is Protestant, but the majority of Berliners identify with no religious body. Because of immigration, the Muslim population is increasing and now is 10 percent of the total.

Further Reading

Broadbent, Philip and Sabine Hake, eds. *Berlin Divided City: 1945–1989.* New York: Berghahn Books, 2010.

Hanf, Theodor. "Berlin or Bonn? The Dispute over Germany's Political Center," in John Taylor, Jean G. Lengellé, and Caroline Andrew, eds., *Capital Cities: International Perspectives/Les Capitales: Perspectives Internationales,* 295–323. Ottawa: Carleton University Press, 1993.

Ladd, Brian. *The Ghosts of Berlin: Confronting German History in the Urban Landscape.* Chicago: University of Chicago Press, 1998.

Stratigakos, Despina. *A Women's Berlin: Building the Modern City.* Minneapolis: University of Minnesota Press, 2008.

Till, Karen E. *The New Berlin: Memory, Politics, Place.* Minneapolis: University of Minnesota Press, 2005.

BERN

Bern is the capital of Switzerland, a mountainous and landlocked country in Western Europe. The city is located in the west-center of Switzerland in an area called the Swiss Plateau, and originally grew up on a hilly peninsula formed by a sharp bend of the Aare River. Bern outgrew its original site in the 19th century and has expanded since in all directions from the river's bend. The population of Bern is 133,920 (2010), the fourth-largest city in Switzerland. The 43 municipalities of the Bern urban agglomeration total 349,000 inhabitants, while the Bern metropolitan area had 600,000 inhabitants in 2000. The name Bern is also spelled Berne from its French equivalent. Its origin may stem from the word "bear," associated with the animal that was killed during a hunting expedition by the city's founder, or possibly from the Italian city of Verona. Since the 1220s, the coat of arms of Bern has had an illustration of a bear.

Historical Overview. Bern is said to have been founded in 1191 by Berthold V, Duke of Zähringen and member of a prominent ruling family that established many towns in what is today Switzerland and Germany. After the Duke's death in 1218, the city was ruled directly by Frederick II of the Holy Roman Empire. Bern joined the Swiss Confederacy in 1353, being one of the eight founding cantons. Bern conquered the cantons of Aargau in 1415 and Vaud in 1536, thereby becoming the largest city-state in the north of the Alps Mountains. Bern was occupied by French troops in 1798 during the French Revolutionary Wars. In 1848, it became the seat of government for the new Swiss Federation. In addition to government, the economy of Bern is based on education, cultural performances and entertainment, financial services, transportation, and industry. The prestigious University of Bern was founded in 1834, but its roots go back to the 16th century and the need to train Protestant ministers after the Reformation.

Major Landmarks. Bern's Old Town is a well-preserved medieval city that is on the UNESCO World Heritage Sites list. In the center of that district is Zytglogge, a clock tower from the 13th century that is an icon of Bern. Other landmarks are the Bundeshaus (the Swiss House of Parliaments), the Bern Historical Museum, the Bern Museum of Fine Arts, and the Paul Klee Center, a modern wave-like building that displays the largest collection of works by the famous Swiss painter. Einsteinhaus is a museum and is where Albert Einstein and his wife lived while he worked at the Swiss patent office, and where he did some of his own work and writing. It is a museum to Einstein's life and contributions to science. The Bear Pits are where live bears are kept on display as a reminder of the city's historical identification with bears these creatures.

Culture and Society. Switzerland is a prosperous country and Bern is a prosperous city with a population that is generally well educated, well employed, and comfortable. There is little glaring poverty. The main language is German, particularly a distinctive local dialect that is called Bernese German. French, Italian, and Romansh are also official languages of Switzerland in addition to German, and are heard as well in the country's capital city. The predominant religion is Christianity, with Roman Catholics and Protestants in approximately equal numbers. The number of Muslims is growing because of recent immigration. Nearly one-quarter of the population of Bern is foreign-born, mostly from European Union countries. Bern has a vibrant cultural live with a range of theaters, concerts, film festivals, and other events, as well as many fine restaurants and cafés. As a major university center, the city has been home to some of Europe's leading scholars, writers, and scientists.

Further Reading

Huber, Werner. *Building Berne: Contemporary Architecture Guide, 1990–2010.* Zurich: Scheidegger & Spiess, 2009.

Rérat, Patrick. "The New Demographic Growth of Cities: The Case of Reurbanisation in Switzerland," *Urban Studies* 49 (2012): 1107–25.

BISHKEK

Bishkek is the capital and most populous city of Kyrgyzstan, a landlocked Central Asian country that until 1991 was a part of the Soviet Union. The city is in the northern part of the country near the border with Kazakhstan. It is situated at about 2,625 ft elevation (800 m) in the valley of the Chui River, and has as a backdrop the high peaks of the Tien Shan mountain range. According to a census count in 2009, the population of Bishkek was 835,800. The population of the metropolitan region as a whole is more than 1,250,000.

Historical Overview. Bishkek may have originated as a caravan stop on the Silk Road some centuries ago, but it is essentially modern-era city that grew from an Uzbek mud fort that was constructed in about 1825. In 1862, the fort was destroyed by Russian troops who annexed the area to the Russian Empire and erected a fort of their own. In 1877, the Russians named the settlement Pishpek and began encouraging Russian frontier settlement of the rich farmlands of the Chui River valley. In 1926, Kyrgyzstan became a part of the Soviet Union as the Kirgiz Autonomous Soviet Socialist Republic, with Bishkek as capital. The Russians renamed the city Frunze in honor of Mikhail Frunze, a Russian revolutionary figure who was born in Bishkek. The city was developed by the Russians as a planned city with a grid street pattern, wide tree-lined streets, a network of irrigation canals, and orderly arrays of identical apartment block housing. When the Soviet Union broke up in 1991, Kyrgyzstan became an independent country and Bishkek a national capital. At that time, the city was renamed Bishkek.

Major Landmarks. The main square of the city is Ala-Too Square. It is the site of the Independence Monument and the National Historical Museum. The Parliament Building is nearby. It had previously been headquarters of the Communist Party. The Frunze Museum is fashioned from Mikhail Frunze's family home and displays information about the history of the city. An equestrian statue of Mikhail Frunze stands in front of the central rail station. The Osh Bazaar is a busy and colorful food market.

Culture and Society. Many Russians settled in Kyrgyzstan during the time of the Russian Empire and Soviet Union, and Bishkek had a majority of Russians as residents. After independence, the Russian population declined to about 20 percent in 2004 and to less than 10 percent by 2011. The decline is because of Russian outmigration and because of urbanization by ethnic Kyrgyz. Russian remains as the official language of the country, along with Kyrgyz. About 80 percent of the national population is Muslim, with most of the rest being Russian Orthodox. There has been considerable conflict in recent years between Kyrgyz ethnics and the country's Uzbek minority.

Further Reading

Flynn, Moya and Natalya Kosmarskaya. "Exploring 'North' and 'South' in Post-Soviet Bishkek: Discourses and Perceptions of Rural–Urban Migration," *Nationalities Papers* 40, no. 3 (2012): 453–71.

Holt, Blaine D. "Roads, Power, and Schools: A Brighter Future for Bishkek and the Region," *American Foreign Policy Interests* 32, no. 6 (2010): 386–96.

Schroeder, Philipp. "'Urbanizing' Bishkek: Interrelations of Boundaries, Migration, Group Size, and Opportunity Structure," *Central Asian Survey* 29, no. 4 (2010): 453–67.

BISSAU

Bissau is the capital and largest city of the Republic of Guinea-Bissau, a small and poor nation on the Atlantic Ocean coast of West Africa. The country is bordered by Senegal and the Republic of Guinea, as well as by the coastline. The city of Bissau is located on the north shore of the Geba River estuary off the Atlantic Ocean, and is also a port city as well as administrative center. There has not been a census of population in Guinea-Bissau since 1991. At that time, the city of Bissau counted 195,389 inhabitants. An estimate by the country's census bureau in 2007 gave the city's population as 407,424, approximately one-quarter of the national population total.

Historical Overview. Portuguese navigators first arrived in the area of Guinea-Bissau in the mid-15th century. By the 16th century, Portuguese were trading in slaves along the coast and on offshore islands. The town of Bissau was founded in approximately 1687, and was a base for Portuguese slave traders and Franciscan missionaries. By 1696, the town had a fort, a church, and a hospital, and was the main trading center of the Geba River valley. In 1775, the fort of São José de Bissau was constructed by the Portuguese in order to gain a stronger hold on the region and to keep larger numbers of African captives for shipment as slaves to Brazil. Bissau eventually became the capital of Portuguese Guinea, as Portugal's colony was called. In 1959, there was a bloody dockworkers' strike at the Pijiguiti dockyards that increased calls by Bissau-Guineans for an end to colonial rule. Independence for Guinea-Bissau was achieved in 1974. The main hero of the struggle for independence was Amílcar Lopes da Costa Cabral (1924–1973), who was assassinated shortly before the dream of national independence was realized. A civil war raged in the country in 1998–1999, causing great loss of life and damage to the Bissau and its economy. Many of the fighters on both sides were child soldiers.

Major Landmarks. The main landmarks in Bissau include the Fortaleza de São José de Bissau da Amura, the mausoleum of Amílcar Cabral, the Guinea-Bissau National Arts Institute, and a landmark monument honoring the workers who were killed in the Bissau Dockers' Strike. Much of the city was badly damaged in the 1998–1999 civil war, with many ruins still standing.

Culture and Society. Guinea-Bissau is an ethnically diverse country with a mix of African language groups and tribal affiliations. The official language is Portuguese but only 14 percent of the population can speak it. Most people speak a Portuguese-based creole language or native African languages. About one-half of the population is Islamic, and nearly one-half practice traditional indigenous religious beliefs. Christianity is the faith of about 10 percent of the population. Carnival is a major annual event in Bissau. Guinea-Bissau is one of the poorest countries on the planet, with some two-thirds of the population living below the poverty line.

Further Reading

Einarsdottir, Jónína. *Tired of Weeping: Mother Love, Child Death, and Poverty in Guinea-Bissau.* Madison: University of Wisconsin Press, 2004.

Lobban, Richard A. and Peter Karibe Mendy. *Historical Dictionary of Guinea-Bissau.* Lanham, MD: Scarecrow Press, 1995.

Lourenço-Lindelle, Ilda. "How Do the Urban Poor Stay Alive? Food Provision in a Squatter Settlement in Bissau, Guinea-Bissau," *African Urban Quarterly* 11, no. 2–3 (1996): 163–68.

Vigh, Henrik. *Navigating Terrains of War: Youth and Soldiering in Guinea-Bissau.* New York: Berghahn Books, 2006.

BLOEMFONTEIN

Bloemfontein is one of the three capital cities of the Republic of South Africa, the country at the southern tip of the African continent. It is the judicial capital, with Cape Town being the legislative capital and Pretoria being the executive or administrative capital. (See the separate entries for these two cities.) The city is located near the center of South Africa and is also the capital of Free State Province. The population is about 370,000, while that of Manguang Local Municipality of which Bloemfontein is a part is about 645,000. The nickname of Bloemfontein is "the city of roses" because roses are abundant. Manguang means "place of cheetahs" in the local Sesotho language.

Historical Overview. Bloemfontein was founded in 1846 as a British fort by Major Henry Douglas Warden who had purchased the site from a Boer farmer named Johannes Nicolaas Brits. The city was at various times in the Orange River Sovereignty, the Orange Free State Republic (1854–1902), and the Orange River Colony (1902–1910). In 1890, a rail line was built from Bloemfontein to Cape Town increasing the city's commercial advantage and improving transportation to the interior for British troops. In 1899, the city was the venue for the Bloemfontein Conference, a failed attempt to prevent the outbreak of the Second Boer War. During the Second Boer War in 1900, a camp was built at Bloemfontein by the British to house more than 20,000 Boer women and children against their will. Bloemfontein became the judicial capital of South Africa in 1910. The noted writer J.R.R. Tolkien was born in Bloemfontein in 1892.

Major Landmarks. The National Women's Memorial on the outskirts of Bloemfontein honors the Boer women and children who were imprisoned by the British in 1900 during the Second Boer War. Outside the city's old parliament building is an equestrian statue of Christiaan Rudolf de Wet, a Boer general, rebel leader, and prominent political figure. Other landmarks are the University of the Free State, the Choet Visser Rugby Museum, Free State Stadium, Bouden Observatory, and the Anglo Boer War Museum. There is an eclectic replica of the Eiffel Tower in one of the industrial suburbs of Bloemfontein.

Culture and Society. The population of Bloemfontein consists of black South Africans (72.6%), whites (22.2%), and people classified as colored (a mixed-race category; about 5%). The main languages of Bloemfontein are Afrikaans, Sesotho, and Setswana. The majority of the population is Christian, with the main denominations being Anglican, the Dutch reformed Church, the Afrikaans Baptist Church, and Roman Catholicism. The most popular sports are rugby, cricket, and football (soccer). The city consists of a mostly compact and walkable center surrounded by a sprawling ring of suburbs of various income levels and architectural types.

Further Reading

Marais, Lochner and Gustav Visser, eds. *Spatialities of Urban Change: Selected Themes from Bloemfontein at the Beginning of the 21st Century.* Stellenbosch: Sun Press, 2008.

Rex, Ralph and Gustav Visser. "Residential Desegregation Dynamics in the South African City of Bloemfontein," *Urban Forum* 20, no. 3 (August 2009): 335–61.

BOGOTÁ

Bogotá is the capital and largest city of the Republic of Colombia, a country in the northwest of South America that borders both the Caribbean Sea and the Pacific Ocean. The city is located near the center of the country to the west of the high plateau known as the Savannah of Bogotá, and is bordered on the east by the Eastern Cordillera of the Andes Mountains. The city's elevation is about 8,612 ft (2,625 m). The population of Bogotá is 6,840,116 (2011), while that of the metropolitan area as a whole is 7,881,156.

The city was known as Santa Fe de Bogotá during most of the Spanish colonial period, and had its named changed to simply Bogotá in 1819. In 1991, the name was again changed to Santa Fe de Bogotá, but in 2000, the city became simply Bogotá once more.

Historical Overview. Bogotá was founded in 1538 by Spanish conquistador Gonzalo Jiménez de Quesada during explorations of the interior of South America in search of gold and other riches, and received its charter from the Spanish crown of April 27, 1539. The indigenous Muisca inhabitants called the first settlement Bacatá, a word meaning "the end of the cultivated land." In 1549, the city was designated the capital of New Granada, the Spanish colony whose territory approximated today's Colombia. Construction of the main plaza and cathedral began in 1553. In 1739, the Viceroyalty of New Granada was established, which included present-day Venezuela, Ecuador and Panama, as well as Colombia, with Santa Fe de Bogotá as capital. Colombian independence was declared in 1810 and recognized in 1819. Afterward, Venezuela, Ecuador, and Panama each broke away separately to become independent countries themselves. On April 9, 1948, a popular former mayor of Bogotá, Jorge Eliecer Gaitán, was assassinated. This led to large-scale rioting in the city that is referred to as the *bogotazo,* and triggered a 10-year civil war, *La Violencia* (The Violence) that took some 300,000 lives across the country. The fighting was a precursor to the guerilla warfare led by the Revolutionary Armed Forces of Colombia (FARC) that still continues today in remote areas of Colombia. From the late 1970s, powerful drug cartels came to control much of Colombia, particularly the cities of Cali and Medellin, but Bogota too became a dangerous city with a very high crime rate. Kidnappings for ransom became a major problem. However, the situation in Bogota has improved somewhat since 2000 because of government successes in its war against crime due to improved policing, and because of improvements in living standards for Bogotá's poor.

Major Landmarks. Plaza de Bolívar (Bolívar Square) in the district of La Candelaria is the symbolic historic center of Bogotá. The Cathedral of Bogotá, built between 1807 and 1823, faces the square, as does the residence of the president of the Republic of Colombia, the Palace of Justice, the Colombian National Capitol, and Liévano Palace, the office of the Mayor of Bogotá. Other landmarks in La Candelaria include Iglesia del Carmen, the Biblioteca Luis A Arango, the Colonial Art Museum, and the Gold Museum. The residential streets of La Candelaria are lined with historic houses. Landmarks in other districts of the city include Torre Colpatria, which is Bogotá's tallest building, the National Museum, and the Museum of Modern Art. A ride on the funicular to the Cerro de Monserrate offers a spectacular panoramic view of Bogotá. Major parks include Simón Bolívar Metropolitan Park, the National Park (*Parque Nacional*), and the Bogotá Botanical Garden.

Culture and Society. Bogotá is a fast-growing city because of migration of peasants from the countryside. Many of the poorest arrivals live in squatter settlements at the edge of the city, mostly to the south and southeast. By contrast, the center of the city, the north, and the northwest are inhabited generally by the middle-class population. Inequality in income and opportunity is a major problem and an underlying cause for the city's high rate of crime. Middle-class and wealthy neighborhoods are heavily policed, as are the city's shopping centers, tourist attractions, and public transportation networks. The city's new mayor (as of 2012) has pledged to end the practice of people carrying guns on the city's streets. Despite the crime, Bogotá is a thriving city and is much loved by many of its residents. They point to a vibrant cultural life in the city with many museums, galleries, theaters, and other attractions, a strong sense of history and tradition in the life of the city, a lively restaurant and nightlife scene, and many modern shopping malls, among other pluses. Because of its many universities, Bogota has been called the Athens of South America.

Further Reading

Erlick, June Carolyn. *A Gringa in Bogotá: Living in Colombia's Invisible War.* Austin: University of Texas Press, 2010.

Mendieta, Eduardo. "Medellin and Bogota: The Global Cities of the Other Globalization," *City* 15, no. 2 (2011): 167–80.

Mockus, Antanas. "Building 'Citizenship Culture' in Bogota," *Journal of International Affairs* 65, no. 2 (2012): 143–46.

Skinner, Reinhard. "Bogota," *Cities: The International Journal of Urban Policy and Planning* 21, no. 1 (2004): 73–81.

BRASÍLIA

Brasília is the capital city of Brazil, the largest country of South America. It is a new, planned capital city that is strategically located near the center of the country between the vast and thinly settled Amazon region of the country and large coastal cities in the south. It has a population of about 2.5 million and is the largest city of Brazil's interior and the fourth-most populous in the country. It was the largest city in the world at the close of the 20th century that did not exist at the start of that century. The metropolitan area totals about 3.6 million inhabitants. The city is defined technically not as a city but as the Brazilian Federal District (Distrito Federal), a territorial unit somewhat analogous to a state. It was officially founded on April 21, 1960.

Historical Overview. As a new planned city, Brasília's history is recent. Brazil's constitution of 1891 called for a move of the country's capital from coastal Rio de Janeiro to the interior, a concept that had originated in 1827, not long after the 1822 declaration of independence of Brazil from Portugal. Realization of the goal for an interior capital came into being in the 1950s under the administration of President Juscelino Kubitschek. Lúcio Costa was the main planner for the new city, having been chosen for the task as the result of a competition with more than 5,000 entries, while Oscar Niemeyer was the chief architect of most nonpublic buildings. The chief landscape designer was Roberto Burle Marx. Initial construction before the city was officially opened took place between 1956 and 1960. At first there was resistance by Brazilian government workers to leave their beloved Rio de Janeiro for the interior, but the city is now also a popular place to live and work, and has grown considerably. In addition to the offices of Brazil's government, Brasília has offices and headquarters of many Brazilian companies and some 124 foreign embassies. Despite its recent origins, Brasília is a UNESCO World Heritage Site, mostly because of an iconic city plan and distinctive modernist architecture.

The story of Brasília's planning and development is discussed often in textbooks about city planning and architecture from opposing standpoints of the city's successes and its shortcomings. As a success, Brasília has indeed helped to open the Amazon as a frontier zone for Brazil's development, and relieved some of the overcrowding in the previous capital, Rio de Janeiro. Its center is highly organized, green, and efficient in terms of traffic flows. On the other hand, Brasília is said to be too monumental, too spread out, and too much at automobile scale rather than pedestrian scale. It has also been described as monotonous in appearance and aesthetically sterile. The main feature of the design is that when presented on paper, the city is shaped somewhat like an airplane seen from above where government buildings are lined up along the fuselage, the so-called Monumental Axis, and residential districts arranged along the two arched wings. The "nose" of the plane points at Lake Paranoá, an artificial lake that wraps partly around the city. Critics say that this is an example of urban planning from a drafting table rather than in consultation with the people who will live and work in the city.

Major Landmarks. Three Powers Square (Praça dos Três poderes) is at the lake end of the Monumental Axis and contains the Presidential Palace, the Congress, and the Supreme Court of Brazil. The Roman Catholic Brasilia Cathedral is an extraordinary and landmark modernist structure of concrete and stained glass that is midway along the axis. The Brasília National Museum and National Library (together the Cultural Complex of the Republic) are near the cathedral. The Juscelino Kubitschek Memorial is a museum dedicated to the life and work of the president of Brazil who brought Brasília to fruition. There are beautiful parks in the city, especially near the lake. The Juscelino Kubitschek Bridge across Lake Paranoá is also a major landmark.

The Capitals of Brazil

When Brasília was designated as the national capital in 1960, it was to stimulate the development of the interior of Brazil and to move the national frontier toward the riches of the Amazon. Prior to that, two coastal cities had served as national capital. The first was the city of São Salvador da Bahia de Todos os Santos, the "City of the Holy Savior of the Bay of all Saints," which was founded by Portuguese settlers on the northwest coast in 1549. It was the colonial capital until 1763 and is called simply Salvador. The center of the city is replete with history, is a focus of Afro-Brazilian culture, and is a UNESCO World Heritage Site. Rio de Janeiro became the capital in 1763 and led Brazil through the rest of the colonial period until independence in 1822, and then afterward until the move to Brasília. The city has a spectacular setting on the coast further south, is known for great beaches, the annual Brazilian *Carnival* (Mardi Gras) festival, and vibrant culture. Brazilians call it *Cidade Maravihosa*, "the marvelous city." In 2016, Rio de Janeiro will host the summer Olympics games, the first city on the South American continent to do so.

The National Congress Building of Brazil. (Corel)

Culture and Society. Brazil is a multiracial society and its capital city reflects that mix. The largest racial category is people of mixed race (48.2%), followed by whites (42.2%), and blacks (7.7%). In comparison to the country as a whole, Brasília has a much higher proportion of whites. About two-thirds of the population is Roman Catholic, with more than one-half of the rest being Protestant. The official language is Portuguese, an inheritance of Brazil's colonial history. Because of government work and the city's many banks and corporate offices, Brasília has a higher percentage of white-collar workers than does Brazil as a whole. As in the country as a whole, football (soccer) is a wildly popular sport. The main stadium is the National Stadium of Brasília.

Further Reading

Gautherot, Marcel. *Building Brasilia.* London: Thames & Hudson, 2010.
Holston, James. *The Modernist City: An Anthropological Critique of Brasília.* Chicago: University of Chicago Press, 1989.
Madaleno, Isabel Maria. "Brasilia: The Frontier Capital," *Cities: International Journal of Urban Policy and Planning* 13, no. 4 (1996): 273–80.

BRATISLAVA

Bratislava is the capital and largest city of Slovakia, a small landlocked country in central Europe. The city is located on the Danube River at the western edge of the country, and borders both Austria and Hungary. It was part of the Kingdom of Hungary, of which it was capital for a time, of the Austro-Hungarian Empire, and of Czechoslovakia from 1918 to 1993 when Slovakia and the Czech Republic became separate states and Bratislava became a national capital. For much of its history the city was known as Pressburg, a German name, and in Hungarian it was Pozsony. The population of Bratislava is about 431,000.

In addition to being a government center, the city is an important manufacturing city, most notably known for the assembly of German automobiles, and an emerging center of high-technology research, development and production. It is both an historic city with a traditional European historic urban form and a very modern city with upscale shopping malls and high-rise office towers and international hotels.

Historical Overview. The area of Bratislava has been settled for many centuries. From the first to the fourth centuries, it was a part of the Roman Empire. It was then settled by Slavs and in the 10th century became a part of Hungary. After the Ottomans conquered Hungary in 1526 but not Pressburg, the city became capital of Hungary in 1536. It flourished for centuries as a seat of government and center of learning and culture, particularly during the 18th-century reign of Queen Maria Theresa. Also during the 18th century, the city became a center of Slovak national consciousness. In the 19th century, the city became an important industrial center. After the fall of Austria-Hungary at the end of World War I, Pressburg became part of the newly formed country Czechoslovakia and was renamed Bratislava. The city became part of the first independent Slovak Republic in 1939 and was designated as capital, but was under Nazi control. Most of the city's Jews perished in the Holocaust. The Slovak republic was dissolved in the wake of World War II, and Czechoslovakia became a communist country and part of the Soviet Union-dominated Eastern Bloc. On March 25, 1988, the Bratislava Candle Demonstration against communist rule helped hasten the fall of the Eastern Bloc. Communist rule ended after the Velvet Revolution in 1989, and the split between Czech Republic and Slovakia took place in 1993 via the so-called Velvet Divorce.

Major Landmarks. Old Town in Bratislava has many historical buildings, including St. Martin's Cathedral, the Franciscan Church, and the old town hall which now houses an historical museum. There is also Michael's Gate, the last remnant of medieval fortifications. Bratislava Castle sits on a hill in the center of the city and is a dominant landmark with blocky, rectangular shape and four towers at each corner. It has been constructed and reconstructed periodically from the 10th century. The Grassalkovich Palace, constructed in about 1760, is now the office of the president of Slovakia. More recent landmarks include Nový Most (New Bridge) across the Danube River with its high tower atop which is a UFO-like restaurant, and the distinctively shaped Kamzik TV Tower with an observation tower and rotating restaurant.

Culture and Society. At different times in history, Bratislava had a majority German population, a majority Hungarian population, and now a majority Slovak population. Slovaks presently make up well over 90 percent of the population. The main language is Slovak and the religion of the majority of inhabitants is Roman Catholic.

Bratislava has many museums, theaters, and concert halls, and is home to the Slovak Philharmonic Orchestra. Many famous composers have lived and worked in the city.

Further Reading

Duin, Pieter van. *Central European Crossroads: Social Democracy and National Revolution in Bratislava (Pressburg), 1867–1921*. New York: Berghahn Books, 2009

Jacobs A. J. "The Bratislava Metropolitan Region," *Cities* 29, forthcoming, accepted on July 26, 2011, available on-line at www.elsevier.com/locate/cities.

Mihálíková, Silvia. "The Making of the Capital of Slovenia," *International Review of Sociology* 16, no. 2 (July 2006): 309–27.

Moravćiková, Henrieta. "Bratislava," in Emily Gunzburger Makaš and Tanja Damljanović Conley, eds., *Capital Cities in the Aftermath of Empires: Planning in Central and Southeastern Europe*. London: Routledge, 2010, 174–88.

Sloboda, Martin. *Bratislava*. Bratislava: MS Agency, 2006.

BRAZZAVILLE

Brazzaville is the capital and largest city in the Republic of the Congo, a country on the Congo River in central Africa. The country is sometimes called Congo-Brazzaville to better distinguish it from its neighboring country the Democratic Republic of the Congo, whose capital is Kinshasa. Brazzaville and Kinshasa are on opposite shores of the Congo River, Brazzaville to the north and Kinshasa to the south, the only place in the world where two national cities are within sight of each other. The population of Brazzaville is about 1 million, while that of its metropolitan area within the Republic of the Congo totals about 1.5 million. Kinshasa is a much larger city with about 10 million inhabitants, such that the two-country Brazzaville-Kinshasa urban area totals more than 12 million inhabitants.

Historical Overview. Brazzaville was named in honor of Pierre Savorgnan de Brazza, a French-Italian explorer and colonial administrator who founded the city on September 10, 1880, as a base for colonial expansion by France. The site had previously been a Bateke tribal village named Nkuna. Brazza's objective was to claim territory for France in competition with King Leopold II of Belgium, who was taking African lands in the Congo River drainage basin as his own personal possession. Hence, Brazzaville's location was on the river opposite Kinshasa, which was called Léopoldville in colonial times. In contrast to other Europeans who pillaged the continent and abused its people, most particularly the murderous Leopold, Brazza was much respected by Africans for his progressive administration of the French Congo and his love of Africa. He was referred to as the "peaceful conqueror." As a result, Brazzaville is one of the few African cities to have retained its colonial name.

France's claim to the Congo was recognized by the 1884 Treaty of Berlin. Brazzaville became the capital of, first, the French Congo and then of French Equatorial Africa, a federation of French colonies comprised of the Congo, Gabon, the Central African Republic, and Chad. In 1924, a railway was built to link Brazzaville with the ocean port Pointe-Noir. Congo-Brazzaville gained independence from France in 1960. In the decades since, the city has seen much political instability and violence. Civil wars in the 1990s took heavy casualties among civilians, and turned tens of thousands of Brazzaville residents into refugees.

Major Landmarks. Nameba Tower, named after a mountain in the Republic of the Congo, is a distinctive, cylinder-shaped office skyscraper that stands high above everything else in the city. Built during 1983–1986, the structure has 30 stories and is

348 ft high (106 m), and is the tallest building in all of central Africa. Its tenants are primarily government ministries and the offices of foreign aid and charity organizations. Other landmarks in Brazzaville are St. Anne's Basilica, the Cathedral of the Sacred Heart, the Congressional Palace, and the mausoleum of Pierre Savorgnan de Brazza.

Culture and Society. Brazzaville is generally a pleasant city and has recently become stable after the Republic of the Congo Civil War, but the city, like the country as a whole, is quite poor. The margins of Brazzaville have many squatter slums. Brazzaville's population is mostly ethnically diverse. The major ethnic group is Kongo, which comprises about half of the national population, with the Laari subgroup being prevalent in Brazzaville. There had been a community of European expatriates, mostly French, in the city, but most fled with the unrest in the 1990s. Most residents of the Republic of the Congo are Christian, with Roman Catholics and Protestants in approximately equal numbers.

Further Reading

Bernault, Florence. "The Political Shaping of Sacred Locality in Brazzaville: 1959–97," in David M. Anderson and Richard Rathbone, eds., *Africa's Urban Past*, 283–302. Oxford: James Currey, 2000.

Petringa, Maria. "Pierre Savorgnan de Brazza: Brief Life of a Lover of Africa, 1852–1905," *Harvard Magazine* 1997. http://harvardmagazine.com/1997/01/vita.html.

Tiepolo, Maurizio. "Brazzaville," *Cities: The International Journal of Urban Policy and Planning* 13, no. 2 (1996): 117–24.

BRIDGETOWN

Bridgetown is the capital and largest city of Barbados, an island country in the Lesser Antilles in the western part of the North Atlantic Ocean. The city is located on the southwest coast of the island on Carlisle Bay, and is centered on a sheltered harbor called the Careenage and the Constitution River. It was once called the Town of St. Michael and is located within the Parish of St. Michael. Barbadians commonly refer to Bridgetown simply as "The City" or "Town." The population of Bridgetown is about 97,000, roughly one-third of the total population of the island.

Historical Overview. Barbados was once occupied by Amerindians known as Arawaks and Tainos. The first Europeans were Spanish and Portuguese sailors who visited the island in the late 16th and early 17th centuries and enslaved the indigenous inhabitants, possibly causing the rest of the population to flee to other islands. The first English settlers arrived in 1627–1628 led by Charles Wolverstone. There were no indigenous inhabitants of Barbados when the English arrived, but an old bridge that was left over from the natives led to the naming of the Europeans' settlement as Bridgetown. The sugar industry began to develop in about 1640 and was the basis for importing slave labor from Africa. The descendants of these Africans would become the largest population group in Barbados. There is also an early Jewish settlement in Barbados, with the first synagogue dating to 1654, the oldest Jewish synagogue in the Americas. In 1824, St. Michael's Parish Church in Bridgetown was elevated to the status of a Cathedral, resulting in Bridgetown gaining the status of a city. In 1958, the Local Government Act was passed in Barbados, and on November 30, 1966, the island achieved independence from the United Kingdom.

Major Landmarks. The main landmarks of Bridgetown include the beautiful stone Parliament Building and its high clock tower, the Anglican Cathedral Church of St. Michael and All Angels, St. Patrick's Roman Catholic Cathedral, the Jewish synagogue (built in 1833 to replace the original structure that was destroyed in 1831 by a hurricane), Chamberlain Bridge across the Careenage, and Kensington Oval, the stadium that was the site of the 2007 Cricket World Cup final. The square in front of the Parliament Building was originally named Trafalgar Square and, like the same-named square in London, has a statue of Admiral Lord Nelson. The statue was erected in 1813 and, along with the square, is older than the parallel landmarks in London. In 1999, Bridgetown's square was renamed National Heroes Square. Another landmark is George Washington House, one of the many historic buildings in the city. It is singled out as a landmark because the future first president of the United States stayed there in 1751. Bridgetown is the only city outside the North American continent that Washington ever visited.

Culture and Society. Barbados is one of the most densely populated countries in the world, with 281,068 inhabitants on just 166 square miles (431 sq km) of territory. About 90 percent of the population is of African descent. Other ethnic groups include people from the United Kingdom, Ireland, the United States, Canada, as well as small numbers from China, India, Lebanon, and Syria. Collectively and colloquially, Barbadians call themselves Bajans. English is the official language of Barbados, but the language that is spoken on a daily basis by most residents is a distinctive dialect known as Bajan English or simply "Bajan." About 95 percent of the population is Christian, with Anglicans forming the largest single denomination. Among the Barbadians who have become globally famous is the pop music recording star Rihanna, born in the Parish of St. Michael in 1988.

The neo-Gothic style Parliament Building of Barbados. (Schoolgirl/Dreamstime.com)

Further Reading

Beckles, Hilary McD. *A History of Barbados: From Amerindian Settlement to Nation-State.* Cambridge, UK: Cambridge University Press, 1990.

Handler, Jerome. *Plantation Slavery in Barbados: An Archaeological and Historical Investigation.* Lincoln, NE: iUniverse, 2000.

Toy, Mike and Peter Laurie. *Barbados: An Island Portrait.* Oxford, UK: Macmillan Education, 2005.

BRUSSELS

Brussels is the capital and largest city of Belgium, officially the Kingdom of Belgium, a small country in western Europe that borders the Netherlands, German, Luxembourg, and France. In addition to being the Belgian capital, the city is considered to be the de facto capital of Europe because it is where much of the administration of the European Union is centered. It is also the headquarters city of the North Atlantic Treaty Organization (NATO). Brussels is located on the Senne River in the center of Belgium and has a population of about 1.1 million. The population of the metropolitan area of Brussels is about 1.8 million.

Historical Overview. The area where Brussels developed was settled as early as the sixth century, but the city itself is traced to 979 when Duke Charles of Lower Lotharingia transferred sacred relics to a chapel on an island in the Senne River that had been built some 400 years earlier by St. Gaugericus, and then constructed the first permanent fortification for the city on the same island. The river location was advantageous for trade and helped the city grow. In the late 12th century, the city got it first walls. A second set of walls was built in the mid to late 14th century, allowing the growing city to expand. The city manufactured luxury fabrics, which it exported to various important cities in Europe. In 1430, Brussels came under the control of the Duke of Burgundy until 1477. In the 16th century, the city was capital of the Low Countries of the Holy Roman Empire. In 1695, the city was bombarded by artillery by French troops sent by King Louis XIV, and suffered enormous damage, including destruction of the Grand Palace and some 4,000 other buildings. At various times in the 18th century, the city was under the rule of France and Austria. In 1815, it became part of the United Kingdom of the Netherlands.

In 1830, Brussels was the scene of the Belgian revolution leading to the formation of a state called Brabant, the precursor of Belgium, with Brussels as capital. Leopold I, the first King of the Belgians, began his rule in 1831. His son, Leopold II, who ruled from 1865 until his death in 1909, took possession of the interior of Africa after 1885 as his own possession and ruthlessly exploited the native population for personal gain in the rubber industry and exploited other resources. Somewhere between 5 and 10 million Africans lost their lives under the oppressive work regime of Leopold's Congo Free State (and later the Belgian Congo as the territory came to be known), and millions of others were maimed, often by cutting off the hands of those who did not seem to work hard enough. The profits went to Leopold personally and were used for palaces in Brussels and other locations, and to beautify Brussels to the standards of other European capitals with extraterritorial

The Capital of Europe

The European Union (EU) is an economic and political confederation of 27 European member countries that was organized under its present name by the Maastricht Treaty of 1993. The member states remain independent countries with their own governments, but many share a common currency, the Euro, and permit unrestricted border crossings from one country to another within the EU. Administration of the EU is divided among three cities near one another in three different countries near the center of Europe. Brussels, the capital of Belgium, is considered to be the de facto capital of the EU because it hosts the European Commission, the Council of the European Union, and the European Council, as well as a seat of the European Parliament. It also has the headquarters of NATO, a military alliance of which many EU member states are a part. A second capital is Luxembourg, the capital of the Grand Duchy of Luxembourg. It hosts the European Court of Justice, the European Court of Auditors, the Secretariat of the European Parliament, the European investment Bank, and the European Investment Fund. The third capital is Strasbourg, a French city near the border with Germany and the site of the European Parliament. There is no official capital of the EU and it has no plans to designate one; instead, the distribution of government functions among three cities in three countries underscores the multinational and cooperative character of the EU.

colonial possessions. World War I and World War II caused great damage in Brussels. The city is now a prosperous urban center known for European Union government and NATO headquarters, as well as the government of Belgium, and is home to many civil servants and international diplomats.

Major Landmarks. Much of medieval Brussels was destroyed in the 19th century during the course of urban modernization. The small historical district is centered on Grand Place, a beautiful city square that is especially nice when illuminated at night. Manneken Pis is a small bronze statue of a boy peeing that is a landmark of the city. There are many colorful stories as to the origin of this particular image. Other landmarks are the Palace of Justice, the Cathedral of the Sacred Heart, which is the fifth biggest church in the world, the Cinqauntenaire Triumphal Arch, the Royal Palace, and the Brussels Bourse. There are several heroic equestrian statues of Leopold II in the city despite the former ruler's guilt as a mass murderer on the scale of the other worst tyrants of the 20th century. Atomium in Heysel Park is an iconic remnant of a world's fair that was held in the city in 1958.

Culture and Society. Belgium is a country with three official languages, Dutch (Flemish), French, and German, all three of which are spoken in Brussels. English is also very widely understood. French is the most common language in the city. The two largest foreign groups in the city are from the French-speaking countries France and Morocco. Roman Catholicism is the primary religion, although most Belgians are nonpracticing. There is a growing Muslim population in the city, primarily from Morocco and Turkey.

Further Reading

Hochschild, Adam. *King Leopold's Ghost: A Story of Greed, Terror, and Heroism in Colonial Africa.* Boston, MA: Mariner Books, 1998.
Murphy, Alexander B. *Brussels.* New York: Wiley, 1994.

Romańczyk, Katarzyna. "Transforming Brussels into an International City: Reflections of 'Brusselization,'" *Cities: The International Journal of Urban Policy and Planning* 29, no. 2 (2012): 126–32.

BUCHAREST

Bucharest is the capital and largest city of Romania, a country at the junction of Southeastern and Central Europe. The city is in the southeast of the country on the banks of the Dâmbovița River in the southeast corner of the Romanian Plain. In addition to being the administrative center of Romania, the city is also the country's main cultural, industrial, and financial center, as well as its main airport hub. The population of Bucharest is about 1.7 million, while that of the metropolitan area totals about 2.0 million and as much as 3 million according to some estimates. With 1.7 million inhabitants, the city of Bucharest is the 10th largest in the European Union, although the population is decreasing because of emigration, low birth rates, and conversion of central residential precincts to commercial land uses.

Historical Overview. Bucharest dates back as a city to 1459 when it became the citadel of the Wallachian prince Vlad III, also known as Vlad Tepes (Vlad the Impaler) because of his habit of impaling his enemies. He was also known by his patronymic, Dracula, which means "son of the dragon." In the early 17th century, Bucharest was burned by Ottoman invaders. It was then rebuilt and became the permanent capital of the Wallachian court after 1698 beginning with the rule of Prince Constantin Brâncoveanu. In the 18th and 19th centuries, the city was controlled at various times by the Austrian Hapsburg Monarchy and by the Russian Empire. In 1813–1814, Bucharest suffered an outbreak of bubonic plague, and on March 23, 1847, it was devastated by an enormous fire that destroyed about one-third of the city. In 1862, the Principality of Romania was formed by the union of Wallachia and Moldavia and Bucharest became a national capital, and in 1881 it became capital of the newly proclaimed Kingdom of Romania. The city was modernized in the late 19th century and became an industrial center. Romania gained considerable territory after World War I and became "Greater Romania." Bucharest prospered as a result, and many new monuments and grand buildings were erected. Romania sided with the Germans in World War II and Bucharest suffered extensive damage. Most of the city's Jewish population was killed. After the war in 1947, Romania became a communist country within the orbit of the Soviet Union's eastern bloc of socialist states. From 1965 to until he was killed in the Romanian Revolution of 1989, Romania was governed by communist dictator Nicolae Ceaușescu. He tore down much of historic central Bucharest and erected enormous socialist-style government buildings in its place. With the overthrow of communism after 1989, Romania is making progress as a democratic country and Bucharest is once again evolving into a leading commercial and cultural center.

Major Landmarks. The most prominent building in Bucharest is the Palace of the Parliament built in the 1980s during the regime of Ceaușescu. It is the largest building in Europe and the second-largest in the world, and houses the Romanian Parliament, the National Museum of Contemporary Art, and a large convention facility. The Centrul Civic (Civic Center) is another gargantuan Ceaușescu-era construction. The Arcul de Triumf (The Arch of Triumph), built originally in

The enormous Palace of the Parliament in Bucharest is reportedly the world's largest and heaviest civilian building. It was built at an enormous cost to Romania between 1984 and 1997, and is a testament to the extravagance of the regime of former dictator Nicolae Ceaușescu. (Desislava Vasileva/Dreamstime.com)

1878 to celebrate national independence and subsequently rebuilt, most recently in 1936, is a national symbol patterned after the Arc de Triomphe in Paris. The 1888 Romanian Athenaeum is an opulent concert hall that is also a major symbol of Bucharest. Other landmarks are the National Museum of Art of Romania, the Museum of Natural History, the Museum of the Romanian Peasant, the National History Museum, the Military Museum, and the Central University Library. The Memorial of Rebirth opened in 2005 in Revolution Square honors those who gave their lives in the Romanian Revolution of 1989. The city also has a beautiful new National Sports Area for football (soccer), the nation's most popular sport.

Culture and Society. The main ethnic group of Bucharest is formed by Romanians, and Romanian is the principal language as well as the official language of the country. Nearly 90 percent of the population is Christian, mostly adherents of the Orthodox faith. The city is the seat of the Patriarch of the Romanian Orthodox Church. Bucharest has a sizable minority of Roma residents (formerly gypsies), a largely poor population that has experienced a history of discrimination and social exclusion. There is a lively cultural life in Bucharest, including theater, concerts, art galleries, and nightclubs with popular music.

Further Reading

Nae, Marina and David Turnock. "The New Bucharest: Two Decades of Restructuring," *Cities: The International Journal of Urban Policy and Planning* 28 (2011): 206–19.

Turnock, David. "Bucharest," *Cities: The International Journal of Urban Policy and Planning* 12 (May 1990): 107–18.

BUDAPEST

Budapest is the capital and largest city in Hungary, a landlocked country in central Europe. It is in the north-central part of the country on both banks of the Danube River, a central feature of the city's landscape. The geographical limits of the modern city were formed in 1873 with the amalgamation of hilly Buda and Óbuda on west bank of the river with Pest on a flat plain on the east bank. The population of Budapest was 1,733,685 in 2011, while that of the metropolitan area was 3,284,110. The metropolitan population is about one-third of that of Hungary as a whole. In addition to being the country's center of government, the city has been a major European political, cultural, and industrial center, as well as a transportation hub and popular tourist destination.

Historical Overview. Budapest was originally a Celtic settlement on the Danube River, but its growth as a city is traced to the early second-century AD when it became the main city of the province of Lower Pannonia in Roman Empire. The Romans named it Aquincum. After acquisition of Pannonia by Bulgaria in 829, Buda and Pest developed from two Bulgarian military outposts on either side of the Danube. The Kingdom of Hungary was founded at the end of the ninth century by the Árpád dynasty. Stephen I was the first Christian king, ruling from 1001 to 1038. In 1241, both Buda and Pest were destroyed in an invasion by Mongol Tatars. King Béla IV fortified both settlements with reinforced stone walls after that calamity, and had his own castle built high on a hill in Buda. In 1361, Buda became the capital of Hungary.

Buda was an especially important cultural and educational center in the European Renaissance under the leadership of King Matthias Corvinus. The city was pillaged by Ottoman Turks in 1526, attacked again in 1529, and finally occupied in 1541. It became a Muslim city with many mosques. Many of the residents converted to Islam. In 1686, Buda and Pest were recaptured by European Christian forces led by the Hapsburg Empire of Austria, but both Buda and Pest were completely destroyed in the decisive battle and Hungary was annexed by the Hapsburgs. The union of Austria–Hungary took place in 1867, with Vienna as one capital and Buda the other. The union of Buda and Pest into a single entity took place six years later. The new administrative center of Budapest developed on the east bank in the Pest district. Budapest developed again into a major cultural and educational center of Europe. At the end of the 19th century, about one-quarter of the population was Jewish and the city was referred to as the "Jewish Mecca." Budapest was greatly damaged in World War I and World War II, and many of its Jews were killed in the Holocaust.

After World War II, Hungary became a communist satellite state of the Soviet Union, and Budapest was its capital. A revolution against Soviet rule in 1956 was violently suppressed by Soviet troops. The Soviet invasion resulted in some 20,000 civilian deaths and sudden large-scale emigration to the West. Afterward, under the 1956–1988 rule of János Kádár referred to as "goulash Communism," there was increasing liberalization of the economy in Hungary and improvements to human rights. Castle Hill and the banks of the Danube River were included as UNESCO World Heritage Sites from 1987. Communism collapsed in 1989, and the country became a member of the European Union in 2004.

Major Landmarks. Budapest is filled with visitor attractions. The major landmark in Budapest is Castle Hill, the site of the Royal Palace, the National Gallery, the beautiful Matthias Church, and Fisherman's Bastion that offers a panoramic lookout of the Danube and the Pest district. Also on the Buda side of the river, in historic Óbuda are the remains of old Aquincum and its thermal baths. Statue Park is a collection of Communist-era statues that have been preserved for the sake of history. In Pest, the main landmarks are the spectacular neo-gothic Parliament Building, St. Stephen's Basilica, the Great Synagogue and Jewish Museum. Andrassy Street is lined with historic buildings and attractions such as the Hungarian State Opera House and the House of Terror, a museum in former secret police headquarters that spotlights the brutality of the Nazi and Communist regimes. City Park is at the far end of Andrassy Street.

Culture and Society. More than 90 percent of Budapest's population is Hungarian. The main minority group is Roma (gypsies), comprising about 4 percent of the total population. The major religion is Christian, the largest group of which is Roman Catholic. The Jewish population, once very large and prominent in the city, is now small in comparison. The official language of the country and the dominant language in Budapest is Hungarian, a language unrelated to most other languages of Europe.

Budapest is historic and picturesque, as well as greatly cosmopolitan, and rich with art, music, and other cultural opportunities. Opera, theater, and ballet are very popular. It is known as well for an exceptionally lively nightlife. Nicknames for the city include "Paris of Central Europe" and "Pearl on the Danube." Budapest is also known as "spa city" because of its many public baths and spa resorts. The history of spas in Budapest dates to Roman times.

The Hungarian Parliament Building on the Pest side of the Danube River. (PhotoDisc, Inc.)

Further Reading

Dent, Bob. *Budapest: A Cultural History*. New York: Oxford University Press, 2007.

Elter, Istvan and Pal Baross. "Budapest," *Cities: The International Journal of Urban Policy and Planning* 10, no. 3 (August, 1993): 189–97.

Enyedi, György. *Budapest: A Central European Capital*. London: Belhaven Press, 1992.

BUENOS AIRES

Buenos Aires is the capital and largest city in Argentina, a large country on the Atlantic Ocean side of southern South America. The city is on the western shore of the Rio de la Plata estuary, across which is the neighboring country Uruguay. The population of Buenos Aires is about 2.9 million, while that of the metropolitan area approaches 13 million. The Buenos Aires conurbation is, therefore, one of the two or three largest in South America and among the largest urban areas in the world. Much of the city's growth was based on the exceptional productive capacity of Argentina's farmland and industrialization. Residents of Buenos Aires are referred to as *porteños*, meaning "people of the port."

Historical Overview. Buenos Aires was founded initially in 1536 by Pedro de Mendoza, a Spanish explorer and was named Ciudad de Nuestra Señora Santa María del Buen Ayre (City of Our Lady Saint Mary of the Fair Winds). The site was abandoned in 1541 because of conflict with indigenous inhabitants. A second founding occurred in 1580 by Juan de Garay who arrived from the interior of the continent via the Paraná River and established a coastal base for Spanish trade. He named the port "Puerto de Santa María de los Buenos Aires." In 1776, the Viceroyalty of the Rio de la Plata was established with Buenos Aires as the capital. Its territory encompassed the land of modern-day Argentina, Bolivia, Paraguay, and Uruguay. The British attacked Buenos Aires in 1806 and 1807. The Argentine War of Independence commenced in 1810 after the May (1810) Revolution in Buenos Aires and lasted until 1818. A declaration of independence for Argentina was signed in 1816, and a constitution was adopted in 1853 after considerable political turmoil and more warfare.

Argentina began to develop economically after the mid-1870s with the arrival of waves of European immigrants and new investments in Argentina's land and resources. The port of Buenos Aires prospered and the city developed into a major business and industrial center. In architecture, modernity, and cultural life, the city was on par with the great capitals of Europe, and came to be referred to as the "Paris of South America." Elites dominated the country. In 1946, following a huge popular demonstration on October 17, 1945, in the city's Plaza de Mayo, Juan Perón was elected president. Together with his very popular wife Eva, he initiated a populist movement and economic reforms called Peronism that improved living conditions for industrial workers. María Eva Duarte de Perón died of cancer in 1952 at the age of 33. Her funeral in Buenos Aires was a landmark event and a time of unprecedented national mourning. On June 15, 1955, a splinter group from Argentina's navy bombed Plaza de Mayo from the air, killing 364 civilians. Perón was deposed shortly thereafter in a military coup.

Argentina has suffered for decades from economic chaos, the devaluation of currency, hyperinflation, corrupt and repressive government, and unemployment resulting from the closing of factories. During the "Dirty War" of 1976–1983, some 30,000 dissidents "disappeared" at the hands of the military dictatorship. Their mothers have been gathering regularly at the Plaza de Mayo in silent protest. Argentina is still healing from economic chaos and murderous political repression, and from a decades-long tumble from being among the most prosperous countries in the world to high rates of poverty and unemployment. In 1992 and 1994, terrorist bombs aimed at Jewish targets exploded in Buenos Aires, resulting in many deaths and injuries.

Major Landmarks. Buenos Aires is a city of many different districts, each with its own flavor and character. Plaza de Mayo is the main square. It contains the May Pyramid and an equestrian monument to General Manuel Belgrano, and is bounded by the Metropolitan Cathedral of Buenos Aires, Casa Rosada (home of the executive branch of Argentina's government), and the current city hall, among

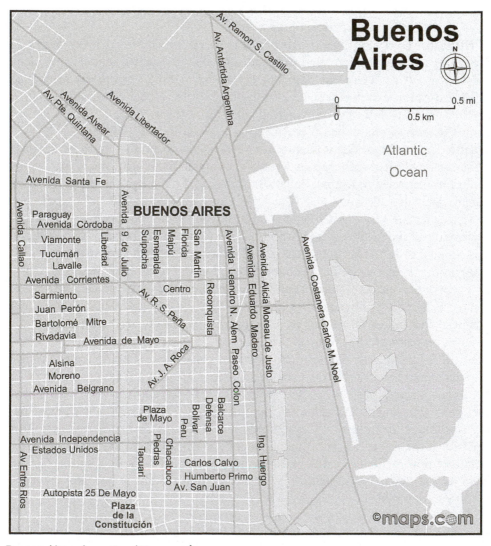

Buenos Aires, Argentina (Maps.com)

other notable buildings. The financial district is nearby. Florida Street is a popular downtown historic pedestrian street lined with shops. Teatro Colón is the beautiful opera house of Buenos Aires. Recoleta is an upscale district near the downtown. Its cemetery has the graves of many notables in Argentina's history, including that of Eva Perón. San Telmo is an area of colonial architecture, nice shops, and a bustling nightlife. La Boca is a working-class waterfront neighborhood with a popular zone for tourists. It features houses and shops painted in various bright colors, many street artists and clothesline art sales, and tango dancers performing on the streets. Puerto Madero is a revitalized old waterfront district that has made a transition from industrial economy and port uses to offices, shops, and upscale residences in renovated buildings and nee construction.

Culture and Society. The majority of *porteños* are of European origin, especially from Italian and Spanish descent. Other ethnicities are from various countries of Europe and from neighboring countries in South America. The city has the largest Jewish community in Latin America, estimated to number about 250,000. The most common religious affiliation is Roman Catholic. The Spanish that is spoken in Buenos Aires is a dialect known as Rio de la Plata Spanish. Because of the economic troubles, the population of Buenos Aires is greatly divided into those who live comfortably and those who are poor. Many new migrants from poor areas in Argentina, Paraguay, and Bolivia reside in large squatter communities at the edge of the city. In the center, one sees unemployed *porteños* and their family members begging for money or

The Argentine National Congress Building in Buenos Aires. (Photo courtesy of Roman Cybriwsky)

working as hawkers amidst lanes of jammed automobile traffic, and picking through the city's trash containers. There are an estimated 40,000 *cartoneros* (waste pickers) in the Buenos Aires area. The working-class port neighborhoods Buenos Aires are the birthplace of tango, a sensual dance genre that is closely identified with the city and the struggles of its poor people. There are many *milongas* throughout Buenos Aires where locals go to dance and relax, as well as theaters and dinner clubs where tango is performed by professional dancers for paying audiences.

Further Reading

Auyero, Javier. *Flammable: Environmental Suffering in an Argentine Shantytown.* Oxford: Oxford University Press, 2009.

Keeling, David. *Buenos Aires: Global Dreams, Local Crises.* Chichester: Wiley, 1996.

Sarlo, Beatriz. "Cultural Landscapes: Buenos Aires from Integration to Fracture," in Andreas Huyssen, ed., *Other Cities, Other Worlds: Urban Imaginaries in a Globalizing Age,* 27–50. Durham, NC: Duke University Press, 2008.

Wilson, Jason. *Buenos Aires: A Cultural History.* New York: Interlink Books, 2000

BUJUMBURA

Bujumbura is the capital and largest city of the Republic of Burundi, a small and very poor landlocked country in east-central Africa. Bujumbura is in the west of Bujumbura on the shore of Lake Tanganyika, one of the Great Lakes of East Africa. Its population is about 800,000. The city is referred to colloquially as "Buj."

Historical Overview. Burundi came under European colonial influence in the late 19th century. In 1899, it became part of German East Africa and then after World War I, it fell under Belgian administration following a League of Nations mandate as part of Ruanda-Urundi. The Belgians administered the territory indirectly through rule by elites in the local Tutsi population. After World War II, Ruanda-Urundi became a United Nations Trust territory under Belgian administration. Burundi separated from Ruanda in 1962 and both the countries achieved independence. As in Ruanda, there was longstanding conflict between the ruling Tutsi group and the majority Hutu population, with resultant warfare, mass killings, political assassinations, and political coups. The unrest continued from before independence until well into the 1980s. There were tens of thousands of deaths on both sides of the social divide, and mass flows of refugees into neighboring countries. As a result, Burundi remains one of the world's poorest and least developed countries.

Bujumbura was a small village named Usumbura until the German colonial period when it was made into a garrison town for the Europeans. Under Belgian rule it became the administrative center of Ruanda-Urundi. The name Bujumbura came into use following national independence in 1962.

Major Landmarks. Major landmarks in Bujumbura include the Roman Catholic cathedral, a large mosque, the Burundi Museum of Life, and the Burundi Geological Museum. The University of Burundi and Hope Africa University are located in Bujumbura. There is a nice sandy beach on Lake Tanganyika. Outside the town is a large stone where explorers David Livingstone and Henry Morton Stanley inscribed their names in 1871.

Culture and Society. Like neighboring Rwanda, Burundi has one major ethnic group, the Banyarwand, within which are three separate groups, the Hutu, Tutsi, and Twa. The Hutu group comprises about 85 percent of the population, while Tutsi comprises about 15 percent. The Twas form less than 1 percent of the total. The three groups share a common culture and language; therefore, they are not ethnic groups per se but social groups within Burundi. The language they speak is Kirundi, one of the official languages of Burundi along with French. Swahili is also spoken as a vernacular language. Christianity is the largest religious group, with Roman Catholics being most numerous. The poverty of Burundi is reflected in a low life expectancy (48.5 years in 2008), high rates of disease such as HIV/AIDS, and malnutrition. Bujumbura is a magnet for impoverished migrants from the countryside, but offers few opportunities. Many migrants live in unsanitary squatter communities at the city's edges.

Further Reading

Kidder, Tracey. *Strength in What Remains.* New York: Random House, 2010.
Uvin, Peter. *Life after Violence: A People's Story of Burundi.* London: Zed Books, 2009.
Watt, Nigel. *Burundi: The Biography of a Small African Country.* London: HURST Publishers, 2008.

C

CAIRO

Cairo is the capital and largest city of the Arab Republic of Egypt, commonly called Egypt, a country in the northeast of Africa between the Mediterranean Sea to the north and the Red Sea along much of its eastern boundary. The city is located in the northeast of Egypt, on both banks of the upper reaches of the Nile River just south of where the Nile fans out into a huge delta. The Suez Canal is approximately 60 miles (100 km) east of the city. The population of the city is about 7 million, while that of the metropolitan area is almost 20 million. It is the largest city in Africa and in the Arab world, and one of the largest and most crowded metropolitan areas in the world. In addition to its role in Egypt, the city is enormously important in the Arab and Islamic worlds as a cultural and educational center, center of media, business, trade, and finance. The word Cairo is al-Qāhira in Arabic and means "the conqueror." The city's nicknames include "City of a Thousand Minarets" and "Capital of the Arab World."

Historical Overview. The roots of Cairo trace back more than 5,000 years to the ancient Egyptian cities of Memphis, Giza, and Fustat that are popularly identified with the city because they or their ruins are now within the Cairo metropolitan area. The Great Pyramids of Giza and the Sphinx are some 4,500 years old and are also identified with Cairo although they are technically in Giza. Cairo itself was founded in 979 AD by the Fatimid General Gawar-al-Siqilli and was made into the new capital of the Fatimids replacing Mahdia in today's Tunisia. The al-Azhar Mosque was a major early building that became one of the world's oldest universities and major center for Islamic learning. In 1168, Cairo became the capital of Egypt, replacing Fustat that was set on fire to protect Cairo from the Crusaders. In 1171, the Fatimids were deposed by Saladin who established the Ayyubid dynasty and became the first Sultan of Egypt. Saladin fortified the city by constructing the Cairo Citadel on Mokattan Hill. In 1250, slave soldiers known as Mamluks seized control of Egypt and governed from the city through their own dynastic sultanate until 1517. The city grew during their rule and much infrastructure was constructed. In about 1340, Cairo reached about 500,000 in population and was the world's largest city outside China. The Black Death, however, took many lives in Cairo in the 14th and 15th centuries, with the result the city's population was reduced by at least 200,000.

The Ottomans seized control of Cairo in 1517 and ruled from Constantinople. Cairo was designated as a regional capital and was the second-largest city in the Ottoman Empire after Constantinople. French troops led by Napoleon occupied Cairo in 1798 and were driven out in 1801 by a combined force of British, Ottoman, and

Albanian troops. An Albanian named Muhammad Ali Pasha became the Ottoman ruler of Egypt from 1805 until his death in 1848. He instituted many social and economic reforms, as well as urban modernization projects, and is considered to be the father of modern Egypt. His grandson Isma'il Pasha who ruled from 1863 to 1879 was also a modernizer. He took cues from Paris and put in wide streets and boulevards, public squares and traffic circles, and public utilities. He also helped shape a modern downtown for Cairo. Direct rule by the Ottomans ended during his reign in 1867, after which Egypt became an autonomous tributary state of the Ottoman Empire until 1914. The British occupied Egypt in 1882 and then made the country a protectorate. Egypt achieved independence in 1922, but British troops remained in the country until after the Egyptian Revolution of 1952. On Black Saturday (January 26) in 1952 a devastating fire swept through downtown Cairo. President Gamal Abdel Nasser ruled Egypt with a strong hand from 1956 till his death in 1970. He was succeeded by President Anwar Sadat who was assassinated in Cairo in 1981. His successor, President Hosni Mubarak, presided over Egypt until huge crowds of demonstrators who gathered in Cairo's Tahrir Square chased him from office in the Egyptian Revolution of 2011. The country is now at a crossroad between a road toward democracy, military rule, or rule by fundamentalist Islamists.

Major Landmarks. The heart of contemporary Cairo is Tahrir Square, founded in the 19th century when the city was modernized and originally named Ismaila Square. The headquarters of the Arab League, the American University of Cairo,

"Arab Spring" protestors in Cairo's Tahrir Square during the 2011 Egyptian Revolution. (Khaled Desouki/AFP/Getty Images)

and the Museum of Egyptian Antiquities are nearby. The latter has more than 136,000 items on display. Khan el-Khalili is an old bazaar in the center of the city that dates back to 1382. A district known as Old Cairo retains much of the city's history, with ancient fortifications, historic mosques, Coptic churches, and many other structures from the past. Cairo University was founded in 1908. Cairo Tower is a free-standing concrete communications tower 614 ft (187 m) high built in 1956 that is an icon of modern Cairo. The upscale neighborhood of Heliopolis includes St. Mark's

Modernization of Cairo

Mohammed Ali Pasha (r. 1805–1848), the founder of modern Egypt, and his grandson Isma'il Pasha (r. 1863–1879), remembered as Isma'il the Magnificent, turned ancient Cairo into a modern city. They put in a modern road system and grand boulevards like those in Paris, added parks, introduced gas and lighting to the city, and built a theater and an opera house. Isma'il developed new European-style residential districts that contrasted with the crowded form of the Islamic medieval city. The two rulers also instituted reforms in city administration and finance, although their projects were very costly and put Cairo in debt. Financial crisis, then, became a pretext for British colonial incursion into Egypt.

Coptic Church with its distinctive domes and the highly ornate mansion of Baron Empain, a Belgian industrialist and Egyptologist from the early 20th century. The famous Pyramids of Egypt and the Sphinx are found across the Nile River from the main part of Cairo at the edge of the desert in Giza.

Culture and Society. Egyptians are the main ethnic group in Cairo, constituting more than 90 percent of the population. The official language of the country is Arabic. English is widely understood by the educated middle class. The predominant religion is Sunni Muslim, but there is a significant Christian minority comprised of members of the Coptic Orthodox Church of Alexandria and the Coptic Catholic Church. Copts have long lived peacefully in Egypt, but in recent years there have been victims of terrorist attacks by radical Islamists. The population of Cairo is strongly divided between rich and poor. The neighborhoods of the wealthy include Garden City, Zamalek, and Heliopolis. There are many gated housing developments in these districts. The poor live in very crowded conditions, as Cairo is extremely short of space and very densely built up, and often without basic amenities. One of the most famous squatter districts in Cairo is called the City of the Dead or the Cairo Necropolis, in which thousands of impoverished residents live amidst toms and mausoleums in an old cemetery.

Further Reading

Abu-Lughod, Janet. *Cairo: 1001 Years of the City Victorious.* Princeton, NJ: Princeton University Press, 1971.

Al Sayyad, Nezar. *Cairo: Histories of a City.* Cambridge: Belknap Press of Harvard University Press, 2011.

Ghannam, Farha. "Two Dreams in a Global City: Class and Space in Urban Egypt," in Andreas Huyssen, ed., *Other Cities, Other Worlds: Urban Imaginaries in a Globalizing Age,* 267–87. Durham, NC: Duke University Press, 2008.

Rodenbeck, Max. *Cairo: The City Victorious.* New York: Vintage Departures, 2000.

CANBERRA

Canberra is the capital of Australia, a large island country in the Southern Hemisphere between the Indian and South Pacific Oceans. The city is in the southeast of the main island of Australia in the northern part of the Australian Capital Territory (ACT). It is a planned city that was built specifically to be the national capital. The population of Canberra was 358,223 in 2011. It is the eighth largest city in Australia and the largest inland city.

Historical Overview. The Canberra region was inhabited originally by indigenous Australians and was first settled by Europeans in the 1820s as part of Australia's agricultural frontier. The area was a colony of England at the time, part of New South Wales. After the Commonwealth of Australia was formed in 1901, the two largest cities, Sydney on the coast in New South Wales and Melbourne on the coast in Victoria, competed to be capital. In 1903, a Royal Commission approved Dalgety, a town in New South Wales that now has a population of only 75, to be a compromise choice, but as supporters of Sydney and Melbourne could not agree on the choice, Canberra was selected as an alternative. It is 170 miles (280 km) southwest of Sydney and 410 miles (660 km) northeast of Melbourne. The selection of Canberra was ratified by the 1908 Seat of Government Act, and architects for the project were selected in 1911 after an international competition. They were Walter Burley Griffin and Marion Mahoney Griffin, a husband and wife team from Chicago in the United States. Construction of the planned city commenced in 1913. In the same year the city was given its name. Progress was hampered by World War I, the Great Depression, and World War II, as well as by political infighting in Australia. Walter Burley Griffin himself was fired from the project in 1920 after a dispute about funds for construction and pay. During the early years of construction, the nation's government was based in Melbourne, which was not allowed to refer to itself as a temporary "capital." The national legislature moved to Canberra on May 9, 1927, the official start date of Canberra as capital.

The construction of Canberra was greatly stepped up after World War II following a push by Prime Minister Robert Menzies who made the project a priority. The city's design, based on the Griffins' work, is a blend of geometric patterns, including circles, hexagons, and octagon from which streets radiate in various direction. Some of the streets are major axes that connect the sites of different branches of government. At the center is a large artificial lake that is now named Lake Burley Griffin. The city's residential zones that surround the planned center are not laid out geometrically. Many have the names of early Australian prime ministers or other figures in national history.

Major Landmarks. Canberra has many outstanding government buildings and museums. The Parliament House of Australia is a beautiful and remarkable building that was opened in 1988 and was, at the time, the most expensive building in the world. There is also the Old Parliament House (in use from the 1920s until 1988) and the High Court of Australia, as well as the National Library, the National Gallery, the National Portrait Gallery, and the National Museum of Australia. The Australian War Memorial is also one of the city's leading landmarks, and is a museum as well as a memorial building. Black Mountain Tower on a hill

A view across the government district of the planned city of Canberra, Australia. (Johncarnemolla/Dreamstime.com)

in a suburb of Canberra provides a panoramic view of the city. The Australian National Botanic Gardens are at the base of Black Mountain. Australian National University is also in the Canberra suburb of Acton and is regarded as one of the better universities in the world.

Culture and Society. Canberra is ranked as a beautiful and very livable city, with many fine neighborhoods, good roads, and many parks. There is also a rich cultural life, with a range of theaters, concerts, festivals, and art galleries, among other activities. Canberra's residents have the highest average education and incomes in Australia. Many people are highly mobile as well, and stay in Canberra only for relatively short periods of government work or university study. The rate of unemployment is lower in Canberra than in Australia's other large cities. More than one-fifth of the population was born overseas, a statistic that includes Canberra residents who are foreign diplomats as well as permanent immigrants. Principal sources of immigrants include New Zealand, the United Kingdom, China, Vietnam, and India.

Further Reading

Gordon, David A. "Ottawa-Hull and Canberra: Implementation of Capital City Plans," *Canadian Journal of Urban Research* 11, no. 2 (2002): 179–211.

Peirce, Sophie and Brent W. Ritchie. "National Capital Branding: A Comparative Case Study of Canberra, Australia and Wellington, New Zealand," *Journal of Travel and Tourism Marketing* 22, no. 3/4 (2007): 67–79.

McKenzie, Stuart, et al. *The Griffin Legacy: Canberra, the Nation's Capital in the 21st Century.* Canberra: National Capital Authority, 2004.

CAPE TOWN

Cape Town is one of the three capital cities of the Republic of South Africa, the country at the southern tip of the African continent. It is the legislative capital of the country, with Pretoria being the executive (administrative) capital and Bloem-fontein being the judicial capital. (See the separate entries for these two cities.) The city is on Table Bay in the far south of the country, and is also a port city, a major tourist attraction, and a leading financial and commercial center. It is known as one of the most beautiful cities in the world. The setting is called City Bowl, a natural amphitheater shaped by Table Bay and mountains named Signal Hill, Lion's Head, Table Mountain, and Devil's Peak. The population of the city is about 827,000, while that of the metropolitan area totals nearly 3.5 million.

Historical Overview. Cape Town occupies a strategic position at the southern tip of Africa connecting the Atlantic and Indian Oceans.The area was first mentioned by Portuguese explorer Batholemeu Dias in 1846, and then by Vasco da Gama in 1497. In 1652, Jan van Riebeeck established the first permanent European settlement at Table Bay as a supply station for ships of his employers, the Dutch East India Company (abbreviated VOC from the Dutch). England gained control of Cape Town from the Dutch first in 1795 until 1803, and then again in 1806. The Anglo-Dutch Treaty of 1814 permanently ceded the south of Africa to Great Britain, and Cape Town became capital of the newly designated British Cape Colony. When the Union of South Africa was formed in 1910, Cape Town became its legislative capital. It then remained legislative capital after the Republic of South

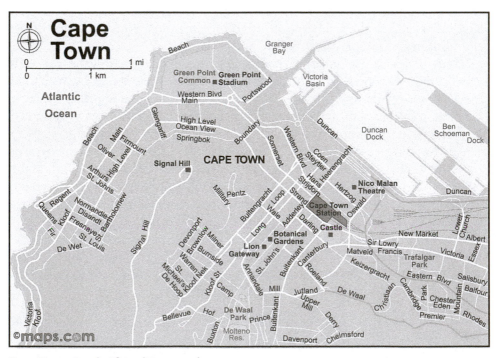

Cape Town, South Africa (Maps.com)

Africa was formed in 1961 and is the seat of the country's National Parliament. With the passage of the three successive Group Areas Acts by the apartheid government of South Africa, Cape Town's districts became strongly segregated by race. An area called District Six, which had a racially mixed population, was declared to be a whites-only district and all housing was demolished in order to forcibly remove all previous residents. Today post-1994, post-apartheid South Africa, Cape Town is becoming a socially diverse city with mixed neighborhoods.

Major Landmarks. The main landmark in Cape Town is Table Mountain, a distinctively shaped high mountain that forms a dramatic backdrop for the city as seen from the sea. The Victoria and Albert Waterfront in the heart of historic Cape Town offers shopping and entertainment for visitors, upscale office buildings and condominiums for residents, a beautiful marina, and spectacular views of both the sea and the city. Nobel Square at the waterfront honors South Africa's four awardees of the Nobel Peace Prize: the late Chief Albert Luthuli, Archbishop Desmond Tutu, and former presidents Nelson Mandela and F. W. de Klerk. Other landmarks are the University of Cape Town, the Victorian Cape Town City Hall, Cape Town Stadium built for the 2010 FIFA World Cup, the Cape Town International Convention Centre, Two Oceans Aquarium, Kirstenbosch National Botanical Garden, and the distinctive neighborhood called Cape Malay Bo-Kaap. The Castle of Good Hope built between 1666 and 1679 is South Africa's oldest surviving building. An equestrian statue of English businessman and mining magnate Cecil John Rhodes stands in Table Mountain National Park.

Culture and Society. About 48 percent of Capetonians are classified as colored, a diverse ethnic group comprised of people with mixed racial backgrounds. Black Africans make up about 31 percent of the total and whites about 19 percent. About 41 percent of the population speaks Afrikaans at home, 29 percent speaks Xhosa, and 28 percents speak English. English is a common language outside the home. More than three-quarters of the population is Christian.

Further Reading

Besteman, Catherine. *Transforming Cape Town.* Berkeley: University of California Press, 2008.

Samara, Tony Roshan. *Cape Town after Apartheid: Crime and Governance in the Divided City.* Minneapolis: University of Minnesota Press, 2011.

Western, John C. *Outcast Cape Town.* Minneapolis: University of Minnesota Press, 1997.

CARACAS

Caracas is the capital and largest city of Venezuela, a country officially named the Bolivarian Republic of Venezuela that is located in the northern part of South America on the coast of the Caribbean Sea. The official name of the city is Santiago de León de Caracas and its residents are called Caraquenians in English. The city is in the northern part of Venezuela in a narrow, elongated valley in the Venezuelan Coastal Range. A steep mountain range called Cerro Ávila (main peak 7,400 ft; 2,200 m) separates the Caracas Valley from the Caribbean Sea. The population of the District of Caracas is about 2.1 million, while that of the Metropolitan District of Caracas is about 4.2 million. The Caracas Valley has little room for urban expansion, so the city is very densely built and there are many high-rise buildings.

Historical Overview. The city of Santiago de León de Caracas was founded on July 25, 1567, by Spanish conquistador Diego de Losada. An earlier Spanish settlement near the site that was founded in 1562 by Francisco Fajardo did not survive, as it was destroyed in a rebellion by indigenous people who had occupied the valley earlier. In the 16th century, the city was frequently raided by pirates, even though it was not directly on the coast and was located behind a mountain barrier. In 1680, pirates sacked the city. The commercial production of cocoa in the surrounding farmlands caused the city to grow, and in 1777 it became the capital of the Captaincy of Venezuela. In the early 19th century, the city was a focal point of an independence movement for Venezuela. A Declaration of Independence for Venezuela was signed in Caracas on July 5, 1811, followed by the Venezuelan War of Independence, a rebellion against Spanish rule, led by Caracas-born South American Liberator Simón Bolívar. On June 24, 1821 Bolívar's troops defeated royalist forces in the decisive Battle of Carabobo. On March 26, 1812, much of Caracas was badly damaged by a massive earthquake.

Venezuela became a major exporter of oil after the discovery of the huge amounts of oil in the Lake Maracaibo basin in the early 20th century. The economy of Venezuela became pinned to oil as a result. The global oil crisis in 1973 increased prosperity in Venezuela because oil prices skyrocketed, but the drop of oil prices in the 1980s caused great economic hardship. On February 27, 1989, the dire economic situation sparked destructive riot in Caracas called the Caracazo ("the Caracas smash") and resulted in hundreds of deaths. There was considerable political instability as well, with attempted coups and the impeachment of the president, followed by a new constitution in 1999. Hugo Chávez has been president since about 1999, engineering what he calls a Bolivarian Revolution in the country toward socialism.

Major Landmarks. The symbolic center of Caracas is Plaza Bolívar, located at the site where the city was founded. A statue of Simón Bolívar stands in the center, while the surroundings include National Capitol, the government building for the city of Caracas, and the city's cathedral, Catedral Metropolitana. The National Pantheon is the final resting place for prominent figures in Venezuela's history. Other important sites are Simón Bolívar's birthplace, the Simón Bolívar Museum, the Museum of Colonial Art, and San Francisco Church. The Altamira neighborhood is a popular center for shopping, cultural life, and nightlife. Las Mercedes is said to be the largest shopping district in Latin America. Parque del Este (East Park) is a beautiful oasis of quiet in the bustling city. It has a small zoo. El Hatillo is a well-preserved colonial town in the suburbs of Caracas.

Culture and Society. Venezuela is a diverse society with many ethnic heritages. The majority of the population is of *mestizo*, or mixed Amerindian and European ethnic heritage, with Spanish surnames being most prevalent. Approximately 21 percent of the population is classified as white. Miss Universe 2009, Stefanía Fernández, is a Caraquenian of mixed Ukrainian and Spanish heritage. Spanish is the common language.

Caracas is divided between people who live comfortably and the poor, with separate neighborhoods for different economic classes. Crime rates are very high in comparison to most other cities in the world, especially murder rates. Some neighborhoods are controlled by criminal gangs.

The most popular sports are football (soccer) and baseball. The city has many theaters, libraries, and concert halls. There are national parks in the nearby mountains and beach resorts on the Caribbean Sea.

Further Reading

Hardy, Charles. *Cowboy in Caracas: A North American's Memoir of Venezuela's Democratic Revolution.* Willimantic, CT: Curbstone Press, 2007.

Lester, Jeremy. "Prometheus Unbound in Caracas," *Socialism and Democracy* 23, no. 3 (2009): 61–88.

Márquez, Patricia C. *The Street is My Home: Youth and Violence in Caracas.* Stanford: Stanford University Press, 1999.

Miller, Michael Eamonn. "Squatter City," *Virginia Quarterly Review* 87, no. 2 (2011): 188–210.

CARDIFF

Cardiff is the capital and largest city in Wales, a country on the west coast of the island of Great Britain, in the United Kingdom. In Welsh, the name of the city is Caerdydd. The city is also the county town (county seat) of the historical county of Glamorgan. Cardiff is located on the Bristol Channel at the mouth of the River Taff, and is about 150 miles (240 km) to the west of London. The population of the city is about 341,000, while that of its metropolitan area is about 861,000.

Historical Overview. The origins of Cardiff are traced to a Roman fort that was built around 75 AD. In 1081, King William I of England built a much larger fortification on the basis of the Roman fort, and strengthened it to face the sea against naval attacks. A walled town developed outside the castle and became a port and market center for the area inland. Cardiff developed into a larger city in the second half of the 18th century with the exploitation of coal to the north of the city. In 1794, a canal connected Cardiff and the coal fields, and in 1798 the first dock for coal exports was built at the terminus of the canal in Cardiff harbor. In the 19th century, the city was linked by rail to these and other coal deposits, expanded its port facilities, and developed an iron industry. By 1913, the port of Cardiff, known as Tiger Bay, had become the largest coal exporting port in the world. In 1922, the limits of Cardiff were expanded to include the town of Llandaff and its cathedral. The coal industry declined later in the century and the port went almost silent, but a revival has begun in the early 21st century based on general cargo. The city is the principal business and finance center of Wales.

Major Landmarks. The most prominent landmark in Cardiff is Cardiff Castle. It was given to the city in 1947 and stands prominently on a promontory in the historic center. Other landmarks are the old Cardiff Coal Exchange Building and Cardiff Central Market. Millennium Stadium, opened in 1999, is the Welsh national stadium for rugby. The Wales Millennium Centre is an arts venue that locals have nicknamed the Armadillo. Other landmarks are the Sendd, which is the National Assembly Building of Wales, the striking Pierhead Building with its clock tower known as " Baby Big Ben " nearby, and historic cathedral in Llandaff. The Red Castle (*Castell Coch*) is another famous castle in the Cardiff area. It is located to the immediate north of the city in the town of Tongwynlais.

Culture and Society. The population of Cardiff declined with weak economy in the 20th century but is now growing again with the city's restructuring to a service economy. The population is diverse because of immigration during the industrial period and recent immigration from the far corners of the former British Empire. The Butetown neighborhood has a high proportion of visible minorities from the Caribbean, West Africa, and South Asia. The main language of Cardiff is English, but Welsh is still spoken and is on the rise. About 11 percent of Cardiffians speaks Welsh.

Further Reading

Daunton, Martin J. *Coal Metropolis: Cardiff 1870–1914.* Leicester, UK: Leicester University Press, 1977.

Rees, William. *Cardiff: A History of the City.* Cardiff: The Corporation of the City of Cardiff, 1969.

CASTRIES

Castries is the capital and largest settlement in Saint Lucia, a small island country of about 174,000 inhabitants in the Lesser Antilles at the boundary of the Caribbean Sea and the North Atlantic Ocean. The town is located on reclaimed land at the end of a sheltered inlet in the northwest of the country. The bay is Saint Lucia's principal port for export of bananas and other commodities, as well as the harbor for passenger cruise ships and ferry boats. The population of the town of Castries is estimated at about 11,000 and that of Castries quarter (or parish) at 67,000 (2004).

Historical Overview. Castries was founded in 1650 by the French and was at first named Carenage (safe anchorage). It was renamed Castries in 1756 in honor of the Marquis of Castries, after a port city on the Mediterranean Sea coast of France. The marquis led a naval expedition to the island of Corsica in that year. The island changed often between French and British control. The indigenous Carib Amerindian population died off from diseases such as smallpox and measles, and African slaves were brought in to work in the European-owned plantations. A powerful hurricane destroyed the original settlement in 1780. On May 24, 1796, Castries was captured by the British. In the early 19th century, the British expanded the fortifications that the French had built on Morne Fortune, a high hill on the south side of the settlement, and rebuilt the center of Castries on a grid plan. They enlarged the port and made it into a fueling station for the British Royal Navy. A German U-boat sank two Allied ships in Castries harbor in World War II. A devastating fire destroyed part of Castries on June 19, 1948. In the latter half of the 20th century, tourism was developed as a major underpinning of the economy of Saint Lucia, with hotels and resorts built near Castries and along other parts of the island's coastline.

Major Landmarks. The center of Castries is Derek Wolcott Square named after the Saint Lucian poet and playwright who won the Nobel Prize for Literature in 1992. He still lives in Castries. Previously the square was named Columbus Square. The Cathedral of the Immaculate Conception faces the square on one side and the library on the other. Vigie Beach is just to the north of the George F. L. Charles Airport and stretches in parallel with the runway. The main airport for Saint Lucia is Hewanorra International Airport near the town of Vieux-Fort at the island's southern tip. The principal landmark in Saint Lucia is the Pitons, Gros Piton and Petit Piton, two volcanic spires that rise dramatically from the coast near the town of Soufrière in the southwest of Saint Lucia. The area has been designated as a UNESCO world heritage site.

Culture and Society. Saint Lucia's population is mostly of African descent or of mixed African European origin. English is the official language, although most residents speak a French-based Creole that is referred to as "Patwah" (*Patois*). About 70 percent of the population is Roman Catholic. Many Saint Lucians have emigrated to work abroad, particularly in the United Kingdom, the United States, and Canada. Remittances help support the economy of St. Lucia. Castries-born (in 1915) Sir Arthur Lewis won the Nobel Prize in Economics in 1979. He was the first black person to win a Nobel Prize in a category other than Peace. His prize in combination with that of Mr. Wolcott in 1992 gives Saint Lucia the highest rate of Nobel Prizes per capital of any country in the world.

Further Reading

Palmer, Jenny and Derek Wolcott. *Saint Lucia: Portrait of an Island.* Oxford, UK: Macmillan Education, 2008.

Tennyson, S. D. Joseph. *Decolonization in St. Lucia: Politics and Global Neoliberalism, 1945–2010.* Jackson: University Press of Mississippi, 2011.

CHIŞINĂU

Chişinău is the capital and largest city of Moldova, a small landlocked country in eastern Europe between Romania and Ukraine. It is located on the river Bîc in the center of the country. As of January 1, 2012, the population of the city proper was 723,500, while that of the municipality was 794,800. The city has also been known as Kishinev, its name from Russian pronunciation. The spelling "Chişinău" is based on the Romanian language and is the current preferred usage.

Historical Overview. The city was founded in 1436 as a monastery village in the Principality of Moldavia. In the 16th century, it fell under Ottoman control until 1812, after which it became part of the Russian Empire. The Russians made it the capital of their province of Bessarabia and the city began to grow. The 1830s and 1840s saw considerable planned urban development and construction of monuments in Chişinău designed to make the city an impressive regional capital. In 1862, the city had a population of 92,000, and in 1900 its population was 125,787, about 43 percent of whom were Jewish as Chişinău had become a major magnet for Jewish migrants from Eastern Europe from about the mid- to late-19th century. In the first decade of the 20th century, however, anti-Semitism resulted in murderous pogroms against the Jewish population, followed by the start of Jewish emigration. The late-19th and early-20th centuries also saw industrialization in Chişinău and the rise of the city as a regional railroad center. After the Russian

Revolution of 1917, Bessarabia declared its independence and joined the Kingdom of Romania. Chişinău became a local provincial capital.

The city was very heavily destroyed during World War II. Soviet Red Army occupation began on June 28, 1940, and then on November 10 that year a major earthquake devastated the city. On July 17, 1941, Chişinău fell after heavy fighting and bombardment to Nazi Germany, and was occupied by the Nazis until August 24, 1944, when it was retaken by the Red Army. During Nazi occupation, many thousands of Jewish residents were taken outside the city and shot. After World War II, most of the former Bessarabia became the Moldavian Soviet Socialist Republic in the Union of Soviet Socialist Republics (the Soviet Union), with Chişinău, then called Kishinev, as the capital. The Soviets rebuilt the city into a socialist style city with large prefabricated housing estate, broad streets, and monumental government buildings under the direction of architect Alexey Shchusev. Independence for Moldavia came in 1991 with the collapse of the Soviet Union, and the country was renamed Moldova.

Major Landmarks. The Cathedral of Christ's Nativity was originally built in the 1830s and was destroyed in World War II and again by Communist rule in 1962. It has since been rebuilt and stands as a major landmark in the center of Chişinău. Nearby is the Triumphal Arch that was built in 1841, then destroyed, and rebuilt in 1973. Other landmarks are the monument of St. Stephen the Great, the Chişinău Opera and Ballet Theater, the National Archaeology and History Museum of Moldova, the National Palace Concert Hall, the façade of the city's central rail station, and the historic Water Tower of Chişinău. One of the city's many Soviet-style high-rise residential areas has a distinctive architectural form that is referred to as the "Gates of Chişinău."

Culture and Society. Moldovans (Romanians) comprise more than 70 percent of the city's population. The main minority groups are Russians (14%) and Ukrainians (8%). There is also a visible Roma (Romani or gypsy) minority that has become vocal about ethnic discrimination against them and ethnic stereotyping. Nearly 90 percent of the population is Orthodox Christian. The official language of the country is Moldovan or Romanian (the same language), but Russian is still widely understood in the city, particularly by older residents. Although Moldova is making progress economically, it is one of the poorest countries of Europe. Many young people have gone abroad to work or have emigrated altogether. Because of emigration, the population of Chişinău has not grown in the 20 or so years after independence.

Further Reading

Brezianu, Andrei. *Historical Dictionary of Moldova*. Lanham, MD: Scarecrow Press, 2007.

Kosienkowski, Marcin and William Schreiber, eds. *Moldova: Arena of International Influences*. Plymouth, UK: Lexington Books, 2012.

COLOMBO

Colombo is the largest city in Sri Lanka, an island country in the Indian Ocean off the southern tip of the Indian subcontinent. It is commonly referred to as the national capital, although officially that designation belongs to Sri Jayawardenepura Kotte, also called Kotte, one of the suburbs of Colombo. Colombo had been the capital until 1982, but not all government functions have moved to Kotte, and Colombo retains many government buildings, including the President's House, the Presidential Secretariat, the Prime Minister's House, the Prime Minister's Office, the Supreme Court, the Central bank of Sri Lanka, the Ministry of Finance, Ministry of Foreign Affairs, and other offices of national government. Kotte is known mainly as the location of Sri Lanka's Parliament. In addition to sharing national administration duties with Kotte, Colombo is Sri Lanka's major port, commercial and financial center, and cultural center.

Colombo is located on the west coast of Sri Lanka at the mouth of the Kelani River. It is a city with many canals and a large lake in its center, the 160-acre (65 hectare) Beira Lake. The center of Sri Jayawardenepura Kotte is about 5 miles (8 km) from the center of Colombo. The population of Colombo is about 700,000, while that of the metropolitan area as a whole is about 5.6 million. The population of Sri Jayawardenepura Kotte alone is 116,000, while its metropolitan area, which is part of the Colombo metropolitan area, has about 2.2 million inhabitants.

Historical Overview. Colombo was not an important town for the Sinhalese and Tamil kingdoms that ruled various parts of the island of Sri Lanka (Ceylon) in ancient times, but its natural harbor was known to traders from China, Persia, and the Arabian Peninsula among others. Arab traders came to dominate export of spices, cinnamon, ivory, pearls, and other products from the island, and settled the early city. They came to be known as Sri Lankan Moors; their descendants still comprise an important minority in the city's population. The European colonial period commenced in 1505 when Portuguese were given permission by the king of Kotte to build a trading post at Colombo. After some conflict with the Moors, the Portuguese consolidated their hold on the area in 1551 and built a new fortified town near the harbor. The Dutch arrived on the island in 1638 and fought the Portuguese for control. In 1656, Colombo fell to the Dutch after a long siege that left few Portuguese survivors. The city then became a significant base for trade by the Dutch East India Company. The British captured the city in 1796, and in 1815 the island became the British Crown Colony of Ceylon with Colombo as capital. Much of the core of today's city was laid out under British rule, although the prominent district known as Fort is traced to the Portuguese fortifications. Ceylon became independent in 1948 and the words Sri Lanka became part of the official name of the country in 1972. Economic growth and stability in the country had been retarded by a prolonged civil war between 1983 and 2009 in which the government forces fought against secessionist rebels from the Tamil minority group that is concentrated in the north and east of the island. On July 24, 2001, the Liberation Tigers of Tamil Eelam carried out a terrorist attack on Colombo's Bandaranaike Airport, destroying many aircraft and taking seven lives. The danger of terrorism continues even after the 2009 cease fire.

Major Landmarks. Galle Face Green is an open area near along the sea near the Colombo financial district. It was laid out by the British in 1859 and was later used for golf and horse racing. Nowadays, it is a popular park and gathering place for recreation. The Central Business District is called Fort, after the old Portuguese fortifications. Its many landmarks include various government buildings of Sri Lanka, including the Old Parliament House which is now the Presidential Secretariat, and the office towers and international hotels of the city's financial center. There are two World Trade Center Towers that were constructed in 1997 and a tall Bank of Ceylon skyscraper. Other landmarks are Old Colombo Lighthouse, the Khan Clock Tower, Jami Ul Alfar Mosque, St. Paul's Church Miligiriya, St. Lucia's Cathedral, and the Murugan Hindu Temple. Colombo City Hall is also an impressive landmark. Major museums are the National Museum of Colombo, the Natural History Museum, and the Dutch Period Museum. In addition to government buildings, the Sri Lanka Civil War Memorial is a major landmark in Sri Jayawardenepura Kotte.

Culture and Society. The population of Colombo consists of a mix of ethnic groups: Sinhalese (41%); Tamils (29%); Moors (24%); and Indian Tamils (2%), plus smaller numbers of Malays, Chinese, Portuguese, Dutch, and others. In Sri Jayawardenepura Kotte, Sinhalese make up nearly 90 percent of the population and Tamils about 5 percent. Its main religious groups are Buddhists (Sinhalese), Hindus (Tamils), and Muslims (Moors). There are also Christians among the various ethnic populations. The official languages are Sinhala and Tamil. The conflict between the country's Sinhalese majority and secessionist Tamils still simmers in distant parts of the country, but it has little impact now on daily life in Colombo. English is widely understood by most educated people. By far, the most popular sport in Sri Lanka is cricket.

Further Reading

Arachchige-Don, Neville S. *Patterns of Community Structure in Colombo, Sri Lanka: An Investigation of Contemporary Urban Life in South Asia.* Lanham, MD: University Press of America, 1994.

Van Horen, Basil. "Colombo," *Cities: The International Journal of Urban Policy and Planning* 19, no. 3 (2002): 217–27.

CONAKRY

Conakry is the capital and largest city of the Republic of Guinea, formerly French Guinea, a country on the Atlantic Ocean coast of West Africa. The country is sometimes called Guinea-Conakry in order to distinguish it from the neighboring country Guinea-Bissau which has Bissau as the capital. The city of Conakry is on the coast of Guinea and is the country's principal port as well as administrative center. It was founded initially on a small island just off the coast called Tombo Island and has expanded from there along the long and narrow Kaloum Peninsula that extends from the mainland to near Tombo Island at its tip. The city, therefore, has a very confined geographical territory and faces great challenges for efficient flow of traffic, provision of fresh water and sanitation services, and other aspects of infrastructure. The population of Conakry has grown rapidly because of migration for the countryside and nearby countries, as well as from high birth rates. It is now estimated at approximately 2 million. Accurate counts for the city's population do not exist.

Historical Overview. Conakry was originally a small village on Tombo Island, and began growing into a city after the French acquired Guinea from the British in 1887, made it party of the Federation of French West Africa, and began to expand commercial interests. They made Conakry the capital of French Guinea in 1904. A railway track was built to the interior town of Kankan to bring groundnuts to Conakry for overseas export. Independence from France was achieved on October 2, 1958. Ahmed Sékou Touré became the first president and began a tradition of autocratic and corrupt rule that has marked Guinea ever since and kept it poor. In 1970, Conakry was invaded by Portugal and supporters from Guinean opposition forces in exile in neighboring Guinea-Bissau, then still a colony of Portugal, in an unsuccessful attempt to overthrow Touré because of his support to anti-Portuguese guerrilla fighters. Political instability continues in Guinea and the economy has hardly had a chance to develop normally. Infrastructure is so poorly developed that cuts of electric power and water supply are daily occurrences in Conakry.

Major Landmarks. The Conakry Grand Mosque is one of the largest mosques in West Africa. The Cathédrale Sainte-Marie, built in the colonial period in the 1930s, is the main place of worship for Conakry's Roman Catholic minority. Other places of note are the Presidential Palace, the Palace of the People (a large events hall), and the Conakry National Museum. The latter has displays about the ethnography and cultural history of Guinea's diverse population. The November 22, 1970 Monument honors those who defended President Touré from the Portuguese attack. As translated from the French, the inscription reads: "Monument of 22 November 1970. The Revolution is Imperative! Imperialism Finds its Grave in Guinea!"

Culture and Society. The main indigenous ethnic groups of Guinea are the Fulani, Mandingo, and Susu. Conakry also has migrants from neighboring West African nations and a minority population of non-Africans, mostly French, other Europeans, and Lebanese. About 85 percent of the population is Muslim, with most of the rest being Roman Catholic. French is the official language, but ethnic languages are commonly spoken. Literacy rates are quite low in Guinea, being under 50 percent for men and under 20 percent for women. Female genital mutilation is commonly practiced in the country.

Further Reading

O'Toole, Thomas and Ibrahima Bah-Layla. *Historical Dictionary of Guinea.* Lanham, MD: Scarecrow Press, 1995.

COPENHAGEN

Copenhagen is the capital and largest city of Denmark, one of the Scandinavian countries of northern Europe. The city is in the eastern part of the country, on the east shore of the island of Zealand, Denmark's largest island, as well as partly on the smaller island of Amager to which Zealand is connected by five bridges. It faces the Øresund to the east, the narrow strait that separates the Baltic and the North Seas and across which is Sweden and the city of Malmö. The population of

Copenhagen is about 1.2 million while that of metropolitan Copenhagen is about 1.9 million. In 2000, the Øresund Bridge linking Copenhagen and Malmö was completed, forming a binational metropolitan area with a population of nearly 2.6 million. The meaning of the word Copenhagen signifies an important economic role for the city: "merchants' harbor."

Historical Overview. The officially recognized founding date for Copenhagen is 1167, the year in which Absalon, the Bishop of Roskilde, constructed a castle on the small island of Slotsholmen in the harbor of Copenhagen. However, it is known from archeological evidence that a settlement existed on the site as many as 200 years earlier, albeit a small one. The year 1167 is thought of as the beginning of Copenhagen's rise to prominence. The city received its charters in 1254, and in the middle of the 15th century it became Denmark's capital. Copenhagen prospered during the long reign of King Christian IV (1588–1648), and many fine buildings were built. Sweden attacked Copenhagen in 1658–1659. In 1711, the plague killed about one-third of the city's residents. English naval vessels bombarded Copenhagen in 1801 and again in 1807, and did considerable damage to the city. Reconstruction afterward resulted in modernization of the city and expansion of its geographical limits. Nazi German troops occupied Copenhagen in World War II. In the decades afterward, the city managed to evolve into a major center of business and finance in the northern Europe-Baltic region. It is also a leading city for information technology (IT), and for research and development in the medical and biological sciences. Because of this, the Denmark-Sweden Øresund region has been referred to as "Medicon Valley." As the city expanded in the latter half of the 20th century, it has followed more or less an urban and regional development plan called the Finger Plan (for its shape like a palm and fingers) that was laid down in 1947.

Major Landmarks. Frederiksstaden is an historic district of Copenhagen that was largely built up during the 1746–1766 reign of Frederik V, and centers on the Danish royal residence Amalienborg Palace and the Marble Church with its enormous dome. An equestrian statue of Frederik V stands in a central plaza amidst the palace buildings. On the island of Slotsholmen is Christianborg Palace (construction started in 1733), today the seat of the Danish Parliament, the Prime Minister's Office, and the Danish Supreme Court. The building sits on the site of Absalon's 1167 castle. Other landmarks are the baroque, high-spired Church of Our Savior, the 1606 Rosenborg Castle and its gardens, and the Børsen, the old stock exchange building built in 1619 by Christian IV. It is distinguished by its "Dragon Spire," a 184-ft-high (56 m) spire that is shaped like the tails of four dragons twined together. The most visited landmark in Copenhagen is Tivoli Gardens, a greatly popular amusement park that opened in 1843. Dyrehavsbakken (Deer Park Hill) is the oldest operating amusement park in the world, dating back to 1583. The Little Mermaid statue is an icon of Copenhagen and another popular visitor attraction. It was unveiled in 1913 and is based on a fairy tale by Danish writer Hans Christian Andersen. More recent landmarks in Copenhagen include the Copenhagen Opera House, the Tycho Brahe Planetarium, and the Danish Royal Playhouse. Major museums include the National Museum, the National

Gallery, the Ny Carlsberg Glyptotek (the personal art collection of the scion of the Carlsberg brewery business), and the Louisiana Museum of Modern Art.

Culture and Society. Copenhagen has a deserved reputation as an open and tolerant city, as well as one that is ahead of most other cities in the world in terms of sustainability, green architecture and urban design, and alternative transportation such as bicycling. The city (as well as all of Denmark) provides excellent social services ranging from public transportation to public education to health care. The high quality of life, however, comes at the price of a high cost of living, as Copenhagen consistently ranks among the world's more expensive cities. About 90 percent of the population is Danish, with the rest being fairly recent immigrants and their offspring. The main sources of immigration are the neighboring countries of Europe, the Balkan Peninsula, Turkey, and South Asia. About 80 percent of the population identifies with the Lutheran Danish National Church. The city offers many annual festivals including the Roskilde Music Festival, the Copenhagen Jazz Festival, Copenhagen Distortion, Copenhagen Pride, Copenhagen Carnival, and Copenhagen Fashion Week.

Further Reading

Andersen, Hans Thor and John Jørgensen. "Copenhagen," *Cities: The International Journal of Urban Policy and Planning* 12, no. 1 (1995): 13–22.

Andersen, Hans Thor and Lars Winther. "Crisis in the Resurgent City? The Rise of Copenhagen," *International Journal of Urban and Regional Research* 34, no. 3 (2010): 693–700.

Kadushin, Raphael. "Copenhagen: Tidings of Good Cheer," *National Geographic Traveler* 26, no. 8 (2009): 38–43.

D

DAKAR

Dakar is the capital and largest city of Senegal, a small country on the Atlantic Ocean coast of West Africa. The city is located on the Cape Verde Peninsula and is the western-most point of the African continent, a geographical position that enabled the city to develop as an important port for maritime trade across the Atlantic and with Europe. The population of the city of Dakar is a little more than 1 million, while that of the metropolitan area is about 2.5 million.

Historical Overview. The Cape Verde Peninsula on which Dakar is located is the centuries-old home of the Lebou ethnic group and was ruled by the Jolof Empire of West Africa. In 1444, Portuguese navigators landed on the small island of Gorée about a little more than 1 mile (2 km) offshore and established a trading post and base for Portuguese ships. From 1536, Gorée also traded in slaves. The island reverted several times between Portuguese and Dutch rule before coming under the control of the English in 1664 and finally the French in 1677. The Lebou established a trading village on the mainland off Gorée in order to provide food and water supplies for European ships. They named that settlement Ndaakaru from which the word "Dakar" derives. In 1857, the French established a military base in Ndaakaru, which they began to call Dakar, and annexed the Lebou state. Soon, the French improved the port in Dakar, built a rail line to Saint-Louis on the coast in the north of Senegal, and in 1902 made Dakar the capital of French West Africa in place of the previous administrative center Saint-Louis. In 1906–1923, the French built a rail line from Dakar to Bamako in the interior, now the capital of Mali, which greatly enhanced the commercial potential of Dakar and facilitated administration of a large colonial possession. The city developed into a major industrial center, as well as French naval base and port. Many French and other European colonialists settled in Dakar in segregated European neighborhoods. On September 23–25, 1940, Dakar was the scene of a naval battle in which British vessels tried unsuccessfully to capture Dakar from Vichy French control. Senegal gained its independence from France in 1960 and Dakar became a national capital. Senegal's first president, the intellectual Léopold Sédar Senghor, was in office from 1960 to 1980. He promoted development of a distinctively African brand of socialism and a philosophy of *négritude* that espoused shared heritage among native African and rejection of colonialism. Also, he strove to shape Dakar into as sophisticated a city as any in Europe. The city is now one of the leading business centers of West Africa and a base for many foreign nongovernmental organizations (NGOs) with dealings in Africa.

Major Landmarks. Many foreign visitors come to Dakar to see Gorée Island, a UNESCO world heritage site that commemorates the evils of the African slave trade. A particular building there called the Slave House (*Maison des Esclaves*) that was built by Dutch slavers in 1776 has a famous "Door of No Return" through which shackled African slaves bound for the New World are said to have passed. Other landmarks are the Dakar Grand Mosque, the Dakar Cathedral, the IFAN Museum of African Arts, and the recently completed (2010) African Renaissance Monument. The latter is a striking, huge (161 ft; 49 m) bronze statue on a hill outside the city of a young woman, a man, and a child held aloft in the man's left arm, all looking out to sea. Les Almadies is the western-most point in Africa, a pleasant place on the shore with many seafood restaurants.

Culture and Society. Dakar is one of Africa's most diverse cities. The main ethnic groups of Senegal are Wolof, Pular, Sere, Jola, and Mandinka. The city also has a sizable Lebanese minority that is engaged in trade at the port, some 20,000 French residents, and residents from Morocco, Mauritania, Cape Verde, and other African countries. Many foreign tourists and business visitors come to Dakar. The official language is French, but African languages are commonly spoken. More than 90 percent of the population follows Islam. There is a proud traditional music heritage in Senegal, especially drumming rhythms, as well as a contemporary music scene based on traditional African music. There is also a deeply rooted tradition of African storytelling by *griots* who pass African history and values from generation to generation by words and music.

Further Reading

Arecchi, Alberto. "Dakar," *Cities: The International Journal of Urban Policy and Planning* (August, 1985): 198–211.

Shaw, Thomas M. *Irony and Illusion in the Architecture of Imperial Dakar.* Lewiston, NY: Edwin Mellen Press, 2006.

DAMASCUS

Damascus is the capital and second-largest city after Aleppo in Syria, officially the Syrian Arab Republic, a sovereign country in Western Asia. The country borders the eastern shores of the Mediterranean Sea, as well as Lebanon, Israel, Jordan, Iraq, and Turkey. The city is known in Arabic as al-Shām and also as the "City of Jasmine." It is one of the oldest continuously inhabited cities in the world. Damascus is located in the southwest of Syria in a hilly and semi-arid area about 50 miles (80 km) from the Mediterranean Sea. The Barada River is a trickle that runs through the center of the city. The population of Damascus is about 1.7 million, while that of the metropolitan area totals about 2.6 million.

Historical Overview. Damascus has been settled since the second-millennium BC, and was once capital of an Aramean kingdom that flourished from the late 12th-century BC to 734 BC. It was at various times later conquered by Assyrians, the armies of Alexander the Great, and then the Romans. From 661 to 750, the city was capital of the Umayyad Caliphate, which at its height was one of the largest empires in history. When the Abbasids overthrew the Umayyad rulers, the seat of Islamic power was moved to Baghdad, now the capital of Iraq, and Damascus declined. When power shifted to Cairo (now the capital of Egypt) and the Ayyubid dynasty in the 12th century, Damascus began to regain its status as a significant Islamic cultural and religious center. Ottoman rule came in the 14th century and lasted for about 400 years. During that time, the city was an important gateway for Hajj pilgrims on their way to Mecca. Christians and Jews also lived in the city during the Ottoman period, but became victims of violence in the 19th century. Most notable was the 1840 Damascus Affair against Jews and the massacre of Christians in 1860. Nationalist sentiment against the Ottomans grew in the late 19th and early 20th centuries, with a spike during the 1916–1918 Arab Revolt that sought to establish a large unified Arab state in place of Turkish rule. After the Ottoman Empire was dismantled in the wake of World War I, Syria became a League of Nations French Mandate, but in 1925–1927 the Syrians revolted against French presence in what is known as either the great Syrian Revolt or the Great Druze revolt. Syria eventually achieved independence in 1946 and Damascus remained the capital. Since 1971, Syria has been ruled by the al-Assad family, first by Hafez al-Assad until his death in 2000, and then by his son Bashar al-Assad. At the time of this writing, the 2011–2012 Syrian Uprising is raging in full with the intent of overthrowing the government.

Major Landmarks. The core of Damascus contains layers of history from various periods of the city's past. Portions of the old walls from the Roman period are still extant, as are seven old gates. Historic Christian sites include the Chapel of St. Paul, the House of St. Ananias, and the Mariamite Cathedral of Damascus, while the most significant Moslem sites include the Umayyad Mosque, the Sayyidah Ruqayya Mosque, and the Saladin Mausoleum. Another landmark is the ruins of the Citadel of Damascus. It dates back to 1076 and was designated as a UNESCO world heritage site in 1979. The largest khan in the city is the Khan As'ad Pasha. It dates to the mid-18th century. Other landmarks are the National Museum of Damascus, the Al-Fayhaa sports complex, Al-Hamidiyah Souk, and Al-Merjeh Square. Tishreen Park is the city's largest park.

Culture and Society. About 85 percent of the population of Damascus is Sunni Muslim and 10 percent is Christian. There are some 2,000 mosques in the city. There is also a small Jewish minority and a district known for historic Jewish settlement. Many residents of Damascus have rural roots, as the city has grown in recent times due to rural-to-urban migration. The common language of the city is Arabic. At present, the key fact about Damascus is the historic uprising that is underway against the authoritarian regime of Bashar al-Assad, and the kind of government, democratic or Islamic-fundamentalist, that would succeed Assad should he be toppled.

Further Reading

Burns, Ross. *Damascus: A History.* New York: Routledge, 2007.
Hudson, Leila. *Transforming Damascus: Space and Modernity in an Islamic City.* London: I. B. Tauris, 2008.

Keenan, Brigid. *Damascus: Hidden treasures of the Old City*. London: Thames and Hudson, 2001.

DAR ES SALAAM. *See* Dodoma.

DHAKA

Dhaka is the capital city of Bangladesh, a mostly low-lying, crowded country in South Asia that faces the Bay of Bengal and is bordered on three sides by India. There is also a border with Myanmar (Burma). The city is near the center of the country on the Buriganga River, one of the many distributaries that comprise the enormous delta of the Ganges River. Along with much of the rest of Bangladesh, Dhaka is prone to flooding during the summer monsoon season and to devastation from tropical cyclones. The climate is wet, hot, and humid. The city's name was once spelled Dacca.

The population of Dhaka is more than 15 million, making it not just the largest city in Bangladesh, but also one of the 10 largest cities in the world. The population continues to increase quickly because of migration from the Bangladesh countryside and because of high birth rates. Crowding and overpopulation are major problems, as are related concerns such as access to fresh water, sanitation, and traffic congestion.

Historical Overview. Dhaka's origins date back to the seventh century. It was at times ruled by an ancient Buddhist king, then by the Hindu Sena Empire in the 11th and 12th centuries, and then by the Sultanate of Bengal and the Delhi Sultanate before being taken over in 1608 by the Muslim Mughals. The Mughals designated Dhaka to be the administrative center of Bengal, a Bengali- (or Bangla) speaking region that comprised much of what is today Bangladesh and nearby parts of India. They called the city Jahangirnagar, "the city of Jahangir" after the Mughal emperor. The British East Indian Company gained rights for collecting taxation after defeating local rulers in the Battle of Plassey in 1757, and began ruling the region in 1793. British preference for Calcutta (now Kolkata) as their main business and administrative center was a setback for Dhaka, but the city grew nonetheless and became an important base for British and Bengali soldiers and center for local administration.

With Indian independence in 1947 and the creation of Pakistan as a separate country for Muslims, Dhaka became the capital of East Pakistan. Hindu residents of the area were forcibly resettled in India, while India's Muslims were sent across the border to the new Islamic state. There was enormous ethnic and religious violence during this time of Partition, and Dhaka absorbed huge numbers of Muslim refugees from adjoining parts of India. Tensions with West Pakistan led to a war of independence for East Pakistan and the establishment in 1971 of an independent Bangladesh with Dhaka as national capital.

Major Landmarks. The Bangladesh legislature is housed in a spectacular and monumental reinforced concrete structure called Jatiyo Sangshad Bhaban, the National Assembly of Bangladesh. It was designed by American architect Louis Kahn and was built in 1961–1982. It is one of the largest government buildings in the world. Other landmarks include Lalbagh Fort that was built by Mughal rulers

Jatiyo Sangshad Bhaban, the landmark National Assembly of Bangladesh in Dhaka. It was designed by American architect Louis Kahn. (Orhan Cam/Shutterstock.com)

in the late 17th century and where local patriots fought against British colonial troops, and the mid-19th century Ahsa Manzil, an opulent palace that has been remade into a museum. Karwan Bazar (also Kawran Bazar) is an important modern business center in Dhaka. Its origins trace to a customs check post during Mughal rule. The downtown of Dhaka has the offices of many Bangladeshi and foreign companies, as well as headquarters for two of the world's largest and best-known NGOs, the Nobel Prize-winning Grameen Bank and BRAC (Bangladesh Rural Advancement Committee). The Kamalapur Railway Station is a striking modern building that serves as the city's central passenger rail station. Shaheed Minar is a solemn monument honoring citizens who lost their lives in a massacre on February 21, 1952, while protesting in support of the Bengali language for East Pakistan as opposed to Urdu imposed from West Pakistan.

Culture and Society. The population of Dhaka is mostly Sunni Muslim and speaks Bengali. English is widely spoken as well. Many migrants from the countryside find work as streets vendors, in construction, and as drivers of cycle and auto rickshaws. The city has as many as 400,000 rickshaws and is sometimes referred to as the "rickshaw capital of the world." Islam is an important part of daily life. The city has many mosques and celebrates Islamic religious holidays as well as the national holidays of Bangladesh such as Language Martyrs' Day on February 21.

Further Reading

Khondker, Habibul Haque. "Dhaka and the Contestation over the Public Space," *City: Analysis of Urban Trends, Culture, Theory, Policy, Action* 13, no. 1 (March 2009): 129–36.

Paul, Bimal Kanti. "Fear of Eviction: The Case of Slum and Squatter Dwellers in Dhaka, Bangladesh," *Urban Geography* 27, no. 6 (2006): 567–74.

Rabbani, Golam. *Dhaka: From Mughal Outpost to Metropolis*, Dhaka: University Press, Ltd., 1997.

Siddiqui, Kamal, et al. *Social Formation in Dhaka, 1985–2005: A Longitudinal Study of Society in a Third World Megacity*. Surrey, England: Ashgate Publishing Company, 2010.

DILI

Dili is the capital and largest city of the Democratic Republic of Timor-Leste, commonly called East Timor, a small country on the eastern half of the island of Timor, the easternmost of the Lesser Sunda Islands in Southeast Asia. The city is on the northern coast of Timor Island on Obmai Strait, and is a port city. The population of Dili is 193,563.

Historical Overview. Dili was founded in about 1520 by Portuguese settlers and missionaries. In 1702, the eastern half of Timor Island was declared to be a Portuguese colony, Portuguese Timor. Dili became the capital in 1769. The western half of Timor Island was part of the Dutch East Indies, the large Southeast Asian archipelago that became Indonesia. In World War II in 1942, Japanese troops captured Timor Island and occupied Dili until September 26, 1945, when the territory was returned to Portugal. An estimated 60,000 Timorese, some 13 percent of the total population, were killed during three years of resistance to Japanese rule. In 1974, Portugal announced that it would end its colonial hold on the eastern part of Timor. East Timor declared independence from Portugal on November 28, 1975, but nine days later Indonesian troops invaded the territory and annexed it as the 27th province of Indonesia, Timor Timur, Indonesian for East Timor. An estimated 60,000–100,000 Timorese were killed as a result of the Indonesian invasion. On November 12, 1991, pro-independence demonstrators in the Santa Cruz cemetery were fired upon by Indonesia troops resulting in some 2509 deaths. The incident is called the Dili Massacre. In 1999, East Timor was put under the supervision of United Nations peacekeepers, and on May 20, 2002, East Timor declared independence from Indonesia and the first new sovereign state of the new millennium was born.

Major Landmarks. The Cristo Rei of Dili (Christ the King) is an 88.6-ft-tall (27 m) statue of Jesus standing on a globe, is perhaps the most distinctive landmark in Dili. It is on a peninsula just outside the city and was erected in 1996 as a gift from the government of Indonesia in commemoration of the 20th anniversary of the annexation by Indonesia of East Timor. Other landmarks are the Roman Catholic church at Moteal which was a center of East Timor's independence movement against Indonesia, and the Church of the Immaculate Conception, the largest Roman Catholic Cathedral in Southeast Asia.

Culture and Society. About 97 percent of the population of East Timor is Roman Catholic, making East Timor one of the most heavily Roman Catholic countries in the world and the second Asian nation after the Philippines to have a Catholic majority. The country's two official languages are Portuguese and Tetum, an Austronesian language that is spoken widely on Timor Island. A dialect of Tetum called Tetun-Dili is spoken in Dili and combines the local language with Portuguese. East Timor is mostly a poor country. It is still recovering from years of war. There are hopeful prospects for joint development with Australia of petroleum and natural gas reserves in the waters southeast of Timor Island.

Further Reading

Cristalis, Irena. *East Timor: A Nation's Bitter Dawn*. London: Zed Books, 2009.

Fernandes, Clinton. *The Independence of East Timor: Multi-Dimensional Perspectives—Occupation, Resistance, and International Political Activism*. Eastbourne, UK: Sussex Academic Press, 2011.

Molnar, Andrea Katalin. *Timor Leste: Politics, History, and Culture*. New York: Routledge, 2010.

DJIBOUTI

Djibouti is the capital and largest city of the Republic of Djibouti, a small C-shaped country on the Horn of Africa that is bounded by Somalia, Ethiopia, and Eritrea, and by the Red Sea and the Gulf of Aden. The country is named after its principal city. The city of Djibouti is on a peninsula on the southern shores of the Gulf of Tadjoura, which is an inlet of the Gulf of Aden. It has the nickname "Pearl of the Gulf of Tadjoura," as well as "French Hong Kong on the Red Sea." The population of Djibouti city is 567,000, more than one-half that of the country as a whole.

Historical Overview. Djibouti city was founded in 1888 on uninhibited land by French colonialists. In 1896, the French made it the capital of their colony French Somaliland. In 1946, the colony became an overseas territory of France, and in 1967 it was renamed the French Territory of the Afars and the Issas in recognition of local ethnic populations. In 1977, the territory became an independent country and changed its name to the Republic of Djibouti. From 1977 to until he stepped aside in 1999, the country was governed by a dictatorial president, Hassam Gouled Aptidon, and from 1991 to 2001, it was locked in civil war between the Djiboutian government, supported by France, and the opposition Front for the Restoration of Unity and Democracy (FRUD). More recently there has been conflict with neighboring Eritrea over control of the Ras Doumeria Peninsula which is claimed by both countries.

Major Landmarks. The city has a large Central Market, a multiuse national stadium called the Stade National Gouled, Djibouti's Presidential palace, and the Hamoudi Mosque. Other features of Djibouti city are beaches along the Gulf of Tadjoura and the Aden Bay Casino in the Sheraton Hotel. The Ethio-Djibouti Railways links the Port of Djibouti with Addis Ababa, the capital city of Ethiopia some 487 miles (784 km) to the southwest, and gives landlocked Ethiopia a maritime port.

Culture and Society. The population of Djibouti is made up of two major ethnic groups, Somalis and the Afar, with the main Somali clan being Issas. French and Arabic are the official languages of Djibouti, but Somali and Afar are widely spoken as well. More than 94 percent of the population is Muslim, with most of the remainder of the population being Christian. The majority of Djibouti's citizens are poor, with unemployment being about 40–50 percent.

The country maintains close cultural and economic ties to France, and depends on France and other international aid for economic support.

Further Reading

Alwan, Daoud A. and Yohanis Mibrathu. *Historical Dictionary of Djibouti*. Lanham, MD: Scarecrow Press, 2000.

Thompson, Virginia and Richard Adloff. *Djibouti and the Horn of Africa*. Stanford, CA: Stanford University Press, 1968.

DODOMA

Dodoma is the administrative capital city of Tanzania (formally the United Republic of Tanzania), a country in East Africa on the coast of the Indian Ocean. The city is located near the very center of the country, and was selected as capital for that reason. Its population is approximately 325,000. The previous capital of Tanzania was Dar es Salaam, an Indian Ocean port about 302 miles (486 km) to the east and the country's largest city with more than 2.5 million inhabitants. Plans to make Dodoma the new capital were announced in 1973 and the new National Assembly Building was opened in 1996. However, many Tanzanian government offices remain in Dar es Salaam, as do most foreign embassies and consulates, and Dodoma is busy mainly when the legislature is actually in session. Many Tanzanians are doubtful as to whether the transfer of capital from Dar es Salaam to Dodoma will ever be finally completed. Dar es Salaam is also the main business center of Tanzania.

Historical Overview. Dodoma was founded in about 1905 at the time of construction of the Tanzanian central railway from Dar es Salaam to Kigoma on Lake Tanganyika in the interior. The region was then called German East Africa and the rail line was called both *Tanganjikabahn* and *Mittellandbahn* in German. The town served as a rail station and resupply depot for the rail line. After World War I, Great Britain took over the region where Dodoma is situated and administered it as Tanganyika. Independence came in 1961. In 1964, Tanganyika joined with Zanzibar, a nearby island in the Indian Ocean, to form Tanzania. The first President Julius Nyerere took the country on a socialist path. In 1986, a master plan for the new capital in Dodoma was drawn up by American architect James Rossant with support from the United Nations.

Dar es Salaam, which means "port of peace," was formerly called Mzizima ("healthy town") and began as a fishing village. The constriction of the railroad to the interior by the German East Africa Company stimulated the city's growth. On August 7, 1998, in an act of terrorism by radical Islamists directed against the United States, a truck bomb exploded simultaneously outside U.S. embassies in Dar es Salaam and Nairobi, Kenya, killing a total of 223 and injuring more than 4,000 individuals.

Major Landmarks. The new parliament building (the Bunge) is the most significant landmark in this still-developing city. Other landmarks are Ismaili Mosque, two universities in the city that were opened in 2007, St. John's University of Tanzania, an Anglican institution, and the University of Dodoma. A statue of Julius Nyerere stands in the center of Dodoma. The Dodoma Railway Station is one of the city's oldest structures and is run-down.

Culture and Society. Dodoma's population reflects the main indigenous ethnic groups of Tanzania: Gogo, Tangi, and Sandawe. There are also small minorities of Arabs and Indians. About two-thirds of the population is Christian and one-third Muslim. English and Swahili are the two official languages. Football is a popular sport in Tanzania and Dodoma is represented by two teams in the Tanzanian Premier League.

Further Reading

Ngware, Suleiman and J. M. Lusugga Kironde, eds. *Urbanising Tanzania: Issues, Initiatives, and Priorities.* Dar es Salaam, Tanzania: University of Dar es Salaam, 2000.
Yeager, Rodger. *Tanzania: An African Experiment.* Boulder, CO: Westview Press, 1989.

DOHA

Doha is the capital and essentially only city in Qatar, formally the State of Qatar, a small sovereign country in the Middle East on a peninsula in the Persian Gulf. The country's only land border is with the much larger Saudi Arabia. The city is a port city on the Persian Gulf and is on the eastern side of the Qatar peninsula. Its population is 998,651 (2008). Together with adjacent suburbs, the population of Doha comprises about 80 percent of the total population of Qatar.

Doha is a fast-growing city due to Qatar's major industries exports of oil, liquefied natural gas, and petrochemicals. The skyline of the city sports many new high-rise buildings, plus many buildings under construction. Many of the skyscrapers are by prominent designer architects and have distinctive postmodern form. The city is mostly new and modern, with a high standard of living for Qatar citizens. The city is a frequent venue for large international conferences and sporting events. Doha will be the prime venue for the 2022 FIFA World Cup and is currently one of the six candidate cities to host the 2020 Summer Olympics. Education City in Doha is an international center for research and higher education.

Since mid-19th century, Qatar has been an absolute monarchy ruled by the Al Thani family. The present Emir of Qatar is Hamad bin Khalifa Al Thani.

Historical Overview. Doha was founded in 1825 and was originally a fishing village named Al-Bida. The economy of the city was based mostly on fishing and pearling until the 1930s. The pearling industry was crippled at that time by competition from cultured pearls from Japan, but the discovery of oil in the late 1930s open the way for development of the current economy. Large-scale exploitation of Qatar's petroleum reserved began in the late 1940s.

The Municipality of Doha was formally established in 1850. The Al Wajbah fortress was built in the west of Doha in 1882, and was the location of a landmark battle in 1893 when Sheikh Jassim bin Mohammed At Thani defeated the invading Ottoman army. In 1916, Doha was made the capital of the British Protectorate of Qatar. In 1917, Sheikh Abdullah bin Qassim Al Thannibuilt constructed the Al Out fortress in the center of Doha. The city became a national capital in 1971 following independence for Qatar.

Major Landmarks. The Diwan Al Amiri is the architecturally spectacular governmental headquarters of the State of Qatar. It was opened in 1989. Al-Corniche is an enjoyable seaside promenade around Doha Bay that affords views of the dynamic skyline of the city and a park setting. The Museum of Islamic Arts designed by I. M. Pei is a major new landmark on the Corniche. Other attractions along the Corniche include a giant statue of Orry, the oryx mascot of the 15th Asian Games that were held in Qatar in 2006, and Doha Heritage Village, an outdoor museum dedicated to traditional ways of living and economy in Qatar. There are many super shopping malls in Qatar, including City-Centre-Doha and Villaggio, a mall with canals that is designed to resemble Venice. Qatar University is in the northern suburbs of Doha.

Culture and Society. Qatari nationals are a minority in Doha, as most residents are foreign expatriates. Some are foreigners employed at the top of the economy in finance, trade, and in the petroleum industry as executives, engineers, and other specialists, while many other foreigners do the hard work of construction in Qatar or work in the service sector. The largest numbers of the city's foreigners are from South Asia: Pakistan, India, Bangladesh, and Sri Lanka. Islam is the predominant religion in Qatar, embraced by more than three-quarters of the total population and a much larger fraction of Qatari nationals. Sharia law is the basis for governance in Qatar. The official language of the country is Arabic.

Further Reading

Chaddock, David. *Qatar*. London: Stacey International, 2006.
Damluji, Aalma Samar. *The Diwan Al Amiri, Doha, Qatar*. London: Laurence King Publishers, 2012
Rahman, Habibur. *The Emergence of Qatar: The Turbulent Years 1627–1916*. London: Kegan Paul, 2006.

DUBLIN

Dublin is the capital and largest city of the Republic of Ireland, a European country on the island of Ireland in the North Atlantic Ocean off the European continent. The city is at the midpoint of the east coast of Ireland, where the River Liffey empties into the Irish Sea. The river bisects Dublin into districts called Northside and Southside. The population of the city of Dublin is about 525,000, while that of the urban area is about 1.0 million and that of the wider metropolitan area about 1.8 million. The word Dublin comes from Irish and means "black pool."

Historical Overview. The first known settlement at Dublin's site was established by Vikings in the ninth century. The Normans invaded in 1169 and in 1171 King Henry II of England became sovereign. Dublin Castle was built in 1204 by England's King John. In 1348, the Black Death decimated the city, as it did much of the rest of Europe. In the 16th century, Dublin became the center for English administration of Ireland. Trinity College was founded in 1592 in a campaign by the English throne to make Ireland Protestant. The city prospered because of wool trade with England and with processing of agricultural exports, and became one of the largest urban centers in the British Empire in the 17th century. In 1759, the Guinness

Brewery was founded and grew to become the city's largest employer. After the 1800 Act of Union, Ireland was governed from London, which hurt both Irish pride and the economy of Dublin. There was a revolt against English rule in Dublin in 1916 that is called the Easter Rising, which was followed by the Irish War of Independence (1919–1921). Irish independence was declared in 1916, ratified in 1919, and recognized in 1922. Dublin became the nation's capital, first of the Irish Republic (1919–1922), then of the Irish Free State (1922–1949), and eventually of the Republic of Ireland (from 1949). From 1995 to 2007, the economy of Ireland expanded greatly and Dublin prospered with growth in high-technology industry, research and development, financial services, and other branches of the economy. This is called the Celtic Tiger period. Tourism has also evolved to be a strong part of Dublin's economic base.

Major Landmarks. Dublin had many landmarks from history and recent times. Dublin Castle is a well-known city icon, as is Christchurch Cathedral, a construction from the 11th century. The Old Library at Trinity College along with its famous Long Hall is another significant attraction, as is the original copy of the beautifully illustrated Book of Kells which is on display in the library. Leinster House of Kildare Street houses the Oireachtas, the Irish Parliament. Kilmainham Gaol is the prison where the Irish patriots who led the Easter Uprising were executed, while the General Post Office on O'Connell Street was the site of their rebellion. Glasnevin Cemetery features the graves of many notables in Irish history. Henrietta Street, which dates to 1720, is an authentic repository of Georgian residential architecture. Among the many museums in the city are the Dublin Writers Museum, the Jeanie Johnson Famine Ship Museum about the transportation of emigrants from Ireland during the Great Famine in 1845–1852, the National Museum of Ireland, the National Gallery of Ireland, and the Irish Museum of Modern Art. Other attractions are Abbey Theatre, the National Theatre of Ireland, the Guinness Storehouse at the famous brewery, and the Old Jameson Distillery of Irish whiskey.

A newer landmark and one of a different sort is the Spire of Dublin on O'Connell Street. It has an interesting story. Built in 2002–2003, it is officially named the Monument of Light and is a 398-ft- high (121 m) stainless steel, pin-like spire. It was built as part of an upgrade of the McConnell Street area, which was seen as being somewhat rundown, as well as a replacement for an earlier monument at the site, Nelson's Pillar. That monument honored Horatio Nelson, an English naval hero of Napoleonic Wars, and had a statue of Nelson atop a high pillar. It had been built in 1808, three years after Nelson was killed at the Battle of Trafalgar. The monument was blown up on March 8, 1966, by a bomb set off by the Irish Republic Army as an anti-England act on the 50th anniversary of the Easter Uprising.

Culture and Society. Approximately 80 percent of the population of Dublin is Irish, most of whom are Roman Catholic. The official languages are Irish and English. In recent years, in response to the Celtic Tiger economy, there has been increased immigration to Ireland, particularly to Dublin, from central and Eastern Europe, especially from Poland and Lithuania, as well as from China and Nigeria.

Dublin is noted for an outstanding nightlife and attracts many additional tourists as a result. The best-known nightlife zone is Temple Bar on the Southside.

There are many street musicians and other entertainers, and a great variety of pubs. The old industrial waterfront district of Dublin is being remade into a pleasant zone of upscale residence, shopping, and recreation.

Further Reading

Kincaid, Andrew. *Postcolonial Dublin: Imperial Legacies and the Built Environment.* Minneapolis: University of Minnesota Press, 2006.
MacLaran, Andrew. *Dublin: The Shaping of a Capital.* London: Belhaven, 1993.
Whelan, Yvonne. *Reinventing Modern Dublin: Streetscape, Iconography, and the Politics of Identity.* Dublin: University College Dublin Press, 2003.

DUSHANBE

Dushanbe is the capital and largest city of Tajikistan, a landlocked mountainous country in Central Asia. The city is situated in the west of the country at the confluence of two rivers, the Varzob and the Kofarnihon. Its population is about 679,000. The name Dushanbe means "Monday" in the native Tajik language and comes from the fact that the site was once a periodic market that functioned on Mondays.

Historical Overview. Although archaeological evidence indicates that the site of Dushanbe may have been settled as far back as the fifth-century BC, the settlement was never more than a small village until the 20th century. The Tajik Autonomous Soviet Socialist Republic was created within Uzbekistan in 1924, and in 1925 the city was designated as capital. In 1929, when the Tajik Soviet Socialist Republic was created within the Union of Soviet Socialist Republics in place of the "autonomous" republic within the Uzbek republic, Dushanbe was renamed Stalinabad in honor of Soviet leader Joseph Stalin. In 1961, in the process of de-Stalinization, the city's name reverted to Dushanbe. During the Soviet period (1921–1991), the city was made into a silk and cotton production center, and many Russians and other ethnic groups from around the Soviet Union were resettled into the city. In 1990, there was rioting in the city over rumors that the Soviet government planned to resettle large numbers of ethnic Armenians into the city. In 1991, Tajikistan gained its independence when the Soviet Union fell apart. The statue of Soviet Union founder Vladimir Lenin that had stood in front of the parliament building was toppled almost immediately. There was a period of civil war from 1992 to 1997 about issues of ethnic representation in the new government that resulted in considerable damage to the city. Many Russian residents fled Tajikistan during this time, as some factions in the conflict called for establishment of an Islamic state. Tajikistan's economy has grown steadily since the end of the conflict, and Dushanbe has evolved into a prosperous commercial, industrial, and cultural center.

Major Landmarks. Major landmarks in Dushanbe include the Palace of Unity (Vahdat Palace), the Tajik State National University, the Dushanbe Government Building, and the Tajik National Museum. The city also has a towering monument to Amir Ismail Samami, a leader of the Samanid Empire that flourished in the 9th and 10th centuries and that is regarded as the first Tajik state.

Culture and Society. The population of Dushanbe is mostly ethnic Tajik, with minority populations of Uzbeks and Russians. The large majority of residents are

Muslims. Since the end of the Soviet period, there has been a revival of Islamic religious worship and education in Dushanbe. The main language of Dushanbe is Tajik, a Persian tongue, although Russian is still spoken, especially for interethnic communication. There had been a minority of Bukharan Jews in Tajikistan since the second century but very few are left. In 2008, the government of Tajikistan permitted the demolition of the historic Bukharan Jewish Synagogue in Dushanbe to make way for an urban modernization project.

Further Reading

Bergne, Paul. *The Birth of Tajikistan: National identity and the Origins of the Republic.* London: I. B. Tauris, 2007.

Manja, Stephan. "Education, Youth and Islam: The Growing Popularity of Private Religious Lessons in Dushanbe, Tajikistan," *Central Asian Survey* 29, no. 4 (2010): 469–83.

Middleton, Robert and Huw Thomas. *Tajikistan and the High Pamirs.* Hong Kong: Odyssey Books, 2012.

EDINBURGH

Edinburgh is the capital and second-largest city (after Glasgow) of Scotland, a country to the north of England that is part of the United Kingdom. The city is in the eastern portion of the Central Lowlands of Scotland and is bounded by the Firth of Forth, the estuary of the River Forth, to the north. In addition to being the main administrative center for Scotland, the city is an important cultural and educational center, and in history was one of the leading centers of the European Enlightenment. It has been known as the "Athens of the North." The population of Edinburg is about 486,000, while the city-region has about 1 million inhabitants.

Historical Overview. Humans have occupied strategic Castle Rock, a high-volcanic rock formation that dominates the center of city, since at least the ninth-century BC, but little is known about the first settlement. The site was part of Roman fortifications in the northern reaches of the Roman Empire. The city itself dates back to the 11th century and the construction of a castle atop the volcanic crag by David I. In 1329, King Robert I (Robert the Bruce) granted Edinburg a charter that confirmed the city's privileges as a royal burgh. There were frequent wars with England. The city became a prosperous mercantile center and center of learning. The University of Edinburgh was founded in 1583. In 1603 King James VI of Scotland succeeded to the English throne, uniting the two kingdoms in what is known as the Union of the Crowns. The 1707 Act of Union bound Scotland with England into the Kingdom of Great Britain. The two parliaments were merged and Edinburgh lost its status as a sovereign capital. In the 18th century, Edinburgh was a city of poor migrants from the Scottish countryside and also produced considerable intellectuals. Among the famous scholars, writers, and scientists that the city produced were David Hume, Adam Smith, James Boswell, Oliver Goldsmith, and Benjamin Rush. The *Encyclopedia Britannica* was founded in Edinburgh in 1768. From about 1830, Edinburgh developed into an industrial center. Its industries included baking, brewing, coach making, paper, book making, pharmaceuticals, and rubber. In the 1890s, there was urban renewal in Edinburgh led by noted city planner Sir Patrick Geddes. In the 1920s, the city was the focus of a Scottish scholarly and literary renaissance. The historical precincts of the city are designated as UNESCO world heritage sites.

Major Landmarks. Edinburgh Castle sits high above the city and is Edinburgh's principal landmark. It has been in use continuously for about 1,000 years. The Abbey and Palace of Holyroodhouse are the main landmarks of Old Town. Also in Old Town are St. Giles' Cathedral, Mary King's Close, the Scottish Parliament, the Museum of Scotland, and the Royal Museum. The Scott Monument in New Town

Edinburgh Castle atop volcanic Castle Rock in the center of the historic city. (Robert Zehet-mayer/Dreamstime.com)

was built in 1846 to honor Sir Walter Scott. The National Gallery of Scotland and the Scottish Gallery of National Modern Art are also in New Town. The campus of the University of Edinburg is also a major landmark.

Culture and Society. The majority of Edinburgh's population is of Scottish ancestry, with English ancestry ranking second. Some 22 percent of the population was born outside Scotland, with the largest number among this group being English-born. There are immigrants from China and South Asia, as well as service industry workers from Eastern Europe. The Church of Scotland is the largest religious denomination. There are also many Roman Catholics. English is the official language, but Scottish Gaelic and Scots are also in use. At any given time, there are many foreign tourists in the city as well as university students from abroad. The city is known for many arts and music festivals, and other cultural events.

Further Reading

Campbell, Donald and Alan Massie. *Edinburgh: A Cultural and Literary History*. Northamp-ton, MA: Interlink Books, 2004.
Fry, Michael. *Edinburgh: A History of the City*. London: Pan, 2010.
Knox, Paul L. "Edinburgh," *Cities: The International Journal of Urban Policy and Planning* (May, 1984): 328–34.

F

FREETOWN

Freetown is the capital and largest city of Sierra Leone, a small country on the Atlantic Ocean coast of West Africa. The city is in the west of the country at the estuary of the Sierra Leone River and is a major port as well as an administrative center. It is also one of the leading financial, educational, and cultural centers of West Africa. The population of Freetown was counted at 772,872 in 2004 and is now estimated to be more than 1 million.

Historical Overview. The history of Freetown is closely tied to resettlement in Africa of freed slaves from England, North America, and the West Indies. The first settlers were 400 former slaves from London who established the African "Province of Freedom" in 1787 near the present site of Freetown under the auspices of an English abolitionist organization, the Committee for the Relief of the Black Poor. They were followed by the founders of Freetown itself who arrived in March 1792 and began the settlement with a prayer service under a large tree (the "Cotton Tree") that is still a symbol of the city. There were approximately 1,500 first settlers. They were former slaves from the United States and the Caribbean, and arrived in 15 ships from Nova Scotia in Canada. By 1798, the town had some 300–400 houses. Indigenous inhabitants attacked the town in 1801, causing the British to take control of the town and making it a Crown Colony in 1808. From 1808 to 1874, Freetown served as capital of British West Africa. The British navy used its port as a base for combating the slave trade. The city continued to grow rapidly over the 19th century with the arrival of freed slaves from the United States and the Caribbean. Sierra Leone achieved independence from Great Britain in 1961 and the Republic of Sierra Leone was declared in 1971. The country was torn by civil war in the late 1990s and early 2000s, and Freetown was twice occupied by rebels. The conflict took tens of thousands of lives and exploited child soldiers. The situation is quieter now, but tensions still simmer and the city is swollen with refugees.

Major Landmarks. The large Cotton Tree in the center of Freetown is a national treasure that represents the birth of the city in March 1792. Fourah Bay College, founded in 1827, is the oldest institution of higher learning in West Africa. Major places of worship include St. John's Maroon Church built around 1820, St. George's Cathedral (1828), and Foulah Town Mosque built in the 1830s. There are many beaches near Freetown; Lumley Beach is one of the most popular.

Culture and Society. The population of Freetown descends from many indigenous African ethnic groups and from the Creole descendants of Africa slaves in North America and the Caribbean who had returned to the African continent after being freed. The Creoles comprise about 5 percent of the national population but are disproportionately represented in Freetown. Krio, the vernacular language of the transplants, evolved into the common language among the diverse ethnic groups of Sierra Leone, although English is also spoken, especially in the capital, and is the official national language. About 60 percent of the population of Sierra Leone is Muslim and 20 percent Christian. The most popular sport in Sierra Leone is football (soccer), with the most important matches being played in Freetown's National Stadium.

Further Reading

Clifford, Mary Louise. *From Slavery to Freetown: Black Loyalists after the American Revolution.* Jefferson, NC: McFarland and Company, 2006.
Gberie, Lansana. *The Dirty War in West Africa: The RUF and the Destruction of Sierra Leone.* Bloomington: Indiana University Press, 2005.

FUNAFUTI

Funafuti is an atoll (a ring of coral reef islands) that is the official capital of Tuvalu, a Polynesian island nation in the Pacific Ocean between 6° and 10° south of the equator that is comprised of four reef islands and five atolls. Funafuti is in the south-center of the island chain at 8°31″ S. It is the country's largest island group and, with a population of 4,492, the most populated. The population of Tuvalu as a whole is only 10,544, making the nation one of the least populated sovereign states in the world.

The largest island of Funafuti is Fongafale. One of its four villages is Vaiaku, the seat of Tuvalu government. For that reason, Fongafale or Vaiaku are sometimes mentioned as the capital of Tuvalu. However, the official capital is the Funafuti, the atoll as a whole.

Historical Overview. The original settlers of Funafuti were transplant from Samoa. The first Western visitor was Arent Schuyler de Peyster in 1819, who commanded the privateer *Rebecca* under the British flag and named Funafuti Ellice's Island after the English politician Edward Ellice who owned his ship's cargo. An American exploration expedition under the command of Charles Wilkes visited the island in 1841. The United States claimed the island under the 1856 Guano Islands Act. In the 1860s, Peruvian slave raiders visited the islands and carried off people to work on guano islands off South America. The same decade saw Christian missionaries arrive on the island and the Christianization of the population. The famous English author Robert Louis Stevenson visited Funafuti in 1890. In 1892, the island group that is now Tuvalu became a British protectorate, and in 1916 it became part of the Gilbert and Ellice Islands Crown Colony. In 1974, the Gilbert Islands and Ellice Islands separated, with the Gilberts becoming the independent nation Kiribati and the Ellice Islands independent Tuvalu on October 1, 1978. Funafuti and the other islands of Tuvalu are imperiled by rising sea levels triggered by global climate change.

Major Landmarks. The most prominent building in Funafuti is the local church Fetu Ao Lima. There is also Tausoalima Falekaupule, the traditional island meeting house. Tausoalima means "hand of friendship" and Falekaupule means "traditional island meeting hall." Another landmark is the remains of a U.S. military aircraft that crashed on Funafuti during World War II.

Culture and Society. Most Tuvaluans are of Polynesian background. They speak the Tuvaluan language, although English, which is not commonly used, is also an official language of the country. About 97 percent of the population belongs to the Church of Tuvalu, a Protestant Christian church. Birth rates are high in the country and population has increased rapidly. Emigration, mostly to Australia and New Zealand, has relieved some of the population pressure. Most Tuvaluans are engaged in fishing and subsistence farming, but other sources of income include men's jobs as seamen for foreign ships, sale of stamps, coins, and fishing licenses, and commercialization of Tuvalu's '.tv' Internet domain name. Tuvaluans have maintained the traditional dance and music culture of their islands, as well as traditional cuisine and many other aspects of traditional community life, but may well lose their land to rising tides.

Further Reading

Farbotko, Carol. "'The Global Warming Clock Is Ticking So See These Places While You Can': Voyeuristic Tourism and Model Environmental Citizens on Tuvalu's Disappearing Islands," *Singapore Journal of Tropical Geography* 31, no. 2 (July 2010): 224–38.

Warne, Kennedy. "Tuvalu: Drowning or Waving," *Geographical Magazine* 80, no. 10 (October 2008): 54–62.

GABORONE

Gaborone is the capital and most populous city of Botswana, a large, land-locked country in southern Africa. It is in far south of the country on the Ngotwane River only 9 miles (15 km) from Botswana's border with the Republic of South Africa. The population of Gaborone is conservatively estimated at about 192,000, and there are perhaps 50,000 or more additional inhabitants in the nearby towns of Mogadit-shane, Tiokweng, and Gabane. This too is a conservative estimate, as the population of the metropolitan area has been increasing rapidly because of migration from the countryside. Gaborone has referred to itself as "Africa's fastest-growing city."

Historical Overview. The name Gaborone comes from Kgosi Gaborone, the name of a popular Botswana chief who settled in the area in 1881. He called his settlement Moshaweng, but local European settlers referred to it as Gaborones, as in "Gaborone's Village." In the 1880s, the British arrived in the area and created a Protectorate over Botswana that was called the Bechuanaland Protectorate. Its boundaries more or less corresponded to the boundaries of modern-day Botswana. They built a fort near Gaborones in order to protect railway and telegraph construction, and a local administrative headquarters that was called Government Camp. In 1965, the capital of the Bechuanaland Protectorate was transferred to Gaborones from Mafeking, a city that today is just inside the border of South Africa. Botswana gained its independence from the United Kingdom in 1966 and Gaborones was confirmed as its capital. The city's name was changed to Gaborone in 1969.

Most of Gaborone's growth took place after national independence, so the city is quite new and modern. It was planned in line with "garden city" principles and includes wide streets, ample green space, and walkways for pedestrians. There is little of the historical settlement left, and the old fort is in disrepair. Rapid recent growth has produced automobile-scale development in much of the metropolitan area, including shopping centers with large parking lots, bust suburban commercial streets, and edge-of-the-city office parks.

Major Landmarks. The center of Gaborone is a commercial area and large public square called The Mall. Nearby is the National Museum and Art Gallery. Also in the center are modern office buildings, including headquarters for the Southern African Development Community and Debsawana, the huge company that operates Botswana's lucrative diamond mines. The Three Dikgosi Monument (Three Kings Monument) is also near the Central Business District. It was erected in 2005 and honors three African chiefs who negotiated with the British when the Bechuanaland Protectorate was established. At the edge of Gaborone, Kgale Hill, also known as the Sleeping Giant, offers a magnificent view of the entire city. Outside the city are the Gaborone Game Reserve and the Mokolodi Nature Reserve.

Culture and Society. The main languages of Botswana are English and Setswana (also called Tswana, after the main tribe of Botswana's citizens, for whom the country is named). Both are official languages of the country. Most citizens are Christians, especially Anglicans, Methodists, and members of the United Congregational Church of South Africa. In comparison to many other countries of Africa, the standard of living is quite high and there is political stability. Hence, the population of Botswana (and Gaborone in particular) includes many refugees and immigrants from abroad, especially from strife-torn neighboring Zimbabwe. The rate of HIV/AIDS infection in Botswana is high, estimated to be 24 percent for adults in 2006. The popular book series "The No. 1 Ladies' Detective Agency" written by Alexander McCall Smith is set in Gaborone.

Further Reading

Kent, Anthony and Horatius Ikgopoleng. "Gaborone," *Cities: The International Journal of Urban Policy and Planning* 28 (2011): 478–94.
Mosha, A. C. "The City of Gaborone, Botswana," *Ambio* 25, no. 2 (1996): 118–25. http://www.botswanatourism.co.bw/gaborone.php

GEORGETOWN

There are many places named Georgetown across the world, but this one is the capital and the largest city of the Co-operative Republic of Guyana, generally called Guyana, and formerly named British Guiana, a country on the Caribbean Sea coast of South America. The city is on the coast in the north of the country, just east of the wide mouth of the Demerara River. It lies below high-tide level, and is therefore protected from the sea by a seawall and a network of canals that drain water from the city. In addition to being the capital of the country, Georgetown is also a financial center for the region, a port, and the headquarters city for CARICOM, the Caribbean Regional Integration Organization. The population of the city is about 239,000, while that of the metropolitan area was about 355,000 in 2009.

Historical Overview. The indigenous inhabitants at the mouth of the Demerara River were Carib and Arawak Amerindians. Christopher Columbus may have sighted the area in his third trans-Atlantic voyage in 1498. The Dutch were the first to colonize the region, establishing three colonies. The British gained control of the area in 1781 and the Dutch ceded them their claims in 1814. The French entered the picture in 1782 and established Georgetown as their colonial capital, naming it La Nouvelle Ville ("the new city"). The Dutch gained the area back in 1784 and renamed the city Stabroek. The British renamed it Georgetown on April 19, 1812, in honor of English King George III. The Europeans imported African slaves to work on sugar plantations along the coastal plain of Guyana. The descendants of these slaves eventually became a major population group of the country.

Georgetown was accorded official status as a city by the British crown in 1842. There was a devastating fire in 1945 that destroyed large parts of the city. In 1966,

Guyana gained its independence from the United Kingdom and in 1970 it became a republic. For some years afterward, the U.S. Central Intelligence Agency (CIA) meddled in Guyana politics to make sure that its government would not align with socialist countries. In 1978, the Jonestown tragedy occurred in another part of Guyana in which 918 people, mostly Americans, died in a mass murder/suicide that stemmed from their membership in a religious cult led by American Jim Jones. Georgetown gained notice at that time as the way for journalists, investigators, and others to get to and from the site of the tragedy, and as the site of Guyana's own investigations and judicial proceedings. In 1998, Guyana put forward the historic district of Georgetown for recognition by UNESCO as a World Heritage Site, but the application has not yet been successful.

Major Landmarks. The main government landmarks in Georgetown are the Parliament Building, the State House (the residence of the president; built in 1852), and the Court House. These structures represent the three branches of national government. Independence Square and Promenade Gardens are in the center of the city. Other landmarks are the Cenotaph (the national War Memorial), Independence Arch, the Walter Roth Museum of Anthropology, the National Museum of Guyana, the Roman Catholic Brickdam Cathedral, and St. George's Anglican Cathedral. The Stabroek Market dates to 1792 and is marked by an iconic cast iron clock tower on its roof. The Georgetown Lighthouse built in 1817 by the Dutch is another prominent landmark. Umana Yana is a traditional conical thatched building that was built in 1972 by Wai-Wai Amerindians in preparation for Guyana's hosting of an international conference of non-aligned nations. The Guyana International Conference Center was built as a gift from the Peoples Republic of China. The Demerara Harbour Bridge (6,074 ft long; 1,851 m) is a floating bridge that crosses the Demerara River 4 miles (6.4 km) south of Georgetown.

Culture and Society. The population of Guyana is heavily concentrated in the Georgetown area and along a narrow coastal strip, while the interior and back country, the territory of Amerindian groups, is lightly populated. Georgetown's population is ethnically and racially heterogeneous. Most residents are descendants from slaves brought from Africa (Afro-Guyanese) or indentured servants that the British had imported from South Asia (Indo-Guyanese), or reflect a mix of these groups with one another, and/or with Europeans and Amerindians. For the country as a whole, 43.5 percent of the population is classified as Indo-Guyanese and 30.2 percent as Afro-Guyanese. There are unresolved ethnic tensions between these two groups. The official language of the country is English, although Guyanese Creole is very widely used. Languages from various regions of India are also spoken, as are those of Amerindian populations. The religious mix of Guyana is 38 percent Christian, 28 percent Hindu, and 21 percent Muslim.

Further Reading

De Barros, J. *Order and Place in a Colonial City: Patterns of Struggle and Resistance in George-town, British Guiana, 1889–1924.* Montreal: McGill-Queen's University Press, 2002.

Edwards, René, Suk Ching Wu, and Joseph Mensah. "Georgetown, Guyana," *Cities: The International Journal of Urban Policy and Planning* 22, no. 6 (2005): 446–54.

Williams, Brackette F. *Stains on My Name, War in My Veins: Guyana and the Politics of Cultural Struggle.* Durham, NC: Duke University Press, 1991.

GUATEMALA CITY

Guatemala City is the capital of the Republic of Guatemala, one of the six countries of Central America. The city's official name is La Nueva Guatemala de la Asunción, but it is known locally as simply Guatemala or Guate. The city is at 14°37′ N latitude, in the south-central part of the country in a mountain valley called Valle de la Ermita and is surrounded by volcanic peaks. The nearest peak is Mount Pacaya, located some 19 miles (30 km) to the southwest. It is an active volcano 8,373 ft high (2,552 m), and has been erupting regularly since 1965. The elevation of Guatemala City itself is approximately 4,900 ft (1,500 m), high enough to moderate tropical-latitude temperatures.

Guatemala City's population is reported to be about 1.1 million for the city proper and 4.1 million for the metropolitan area as a whole, but both the numbers are likely to be serious undercounts. There is heavy migration from the countryside and from the neighboring countries. Many residents, especially those in burgeoning squatter settlements at the city's edges, are not reported in official statistics. Guatemala City is the largest urban area in Central America.

Historical Overview. Guatemala City is located at the site of an ancient Mayan city called Kaminaljuyu that was founded some 3,000 years ago and was continuously inhabited for about 2,000 years. Archaeological evidence shows that Kaminaljuyu was an advanced and splendid city with many great ceremonial buildings and a sophisticated hydraulics system. The city was abandoned in approximately 1200 AD, and much of Guatemala City is built atop its ruins. However, a portion of the ancient site is preserved as a park in the center of Guatemala City.. The origins of Guatemala City itself date to the beginning of the Spanish colonial period in Central America in the 16th century. A monastery known as El Carmen was founded here in 1629. Guatemala City rem.ained a small town until after 1773, when the capital of the Captaincy General of Central America, a Spanish colonial territory that covered much of today's Central America, was moved there from Antigua Guatemala following the destructive Santa Marta earthquake. Independence from Spain was achieved in 1821, and Guatemala City became the capital of the Federal Republic of Central America, also known as the United Provinces of Central America. The federation was dissolved in a civil war in 1838–1840, after which Guatemala emerged as an independent country in its own right. During the 1898–1920 rule of dictator Manuel Estrada Cabrera, Guatemala was more-or-less controlled by a large and powerful American company, the United Fruit Company which exported bananas. As a result, Guatemala came to be referred to as a "banana republic."

Guatemala City has been devastated by periodic earthquakes. The worst tragedies occurred in 1917–1918 and 1976. Another large quake took place in September 2011.

Major Landmarks. The heart of Guatemala City is the Parque Central and Placio National, the national palace and the city's grand central plaza. The city's Roman

Catholic cathedral, Catedral Metropolitana, also faces the plaza. The National Palace of Culture, too, is in this area. Nearby is Parque la Aurora with a museum of archeology and ethnology, the Museum of Modern Art, the city's zoo, and other attractions. Ruins of Kaminaljuyu are also in the center of the city. The Minerva Park features a huge relief map of the Guatemala City region that was made in 1904 and that shows local topography in great detail. Zone Ten is Guatemala City's most modern and prosperous neighborhood, and the site of the city's financial district, many popular shopping centers and restaurant streets, and Zona Viva, the city's premier zone for nightlife.

Culture and Society. Guatemala City's population is a mix of Spanish, indigenous American, and *mestizo* (mixed native and European population) people, as well as several other ethnicities. The Roman Catholic faith is most numerous, encompassing more than half of the population, while Protestants account for about 40 percent. The native arts and musical styles of Guatemala thrive in Guatemala City and can be seen and heard at many venues. In recent years, Guatemala City has developed a garment and textile manufacturing industry that makes use of cheap local labor for foreign companies and an export market. There has also been conflict between poor migrants and the government about issues of economic opportunity and citizens' rights to a place in the city.

Further Reading

O'Neill, Kevin Lewis and Kedron Thomas, eds. *Securing the City: Neoliberalism, Space, and Insecurity in Postwar Guatemala.* Durham, NC: Duke University Press, 2011.

HANOI

Hanoi is the capital and second-largest city (after Ho Chi Minh City) of Vietnam, officially the Socialist Republic of Vietnam, a country on the Indochina Peninsula in Southeast Asia. The city is in the northern part of the country in the fertile Red River delta and on the right bank of the main channel of the river. Its port city is Haiphong, Vietnam's third-most populous city. The population of Hanoi is about 2.6 million, while that of the metropolitan area is about 6.5 million. The city is growing rapidly with migration from rural areas and natural population increase, and is a booming industrial center with successive years of double-digit annual growth rates of industrial growth rates since the 1990s. According to predictions by PricewaterhouseCoopers, Hanoi will be the world's fast growing city in GDP at least until 2025.

Historical Overview. The area of Hanoi has been inhabited for at least 5,000 years. Co Loa Citadel, about 12 miles (20 km) north of the present center of Hanoi, dates to about 200 BC and may have been the first permanent settlement in the region. In 1010, Ly Thai To chose the site of Hanoi to be his capital because he claimed to have seen a dragon rising there from the Red River. He named the city Thăng Long, meaning "Rising Dragon." The city remained capital of the Lý Dynasty until 1397, at which time the city was renamed Đông Đô (Eastern Capital). After the Ming Chinese invasion in 1408, the city was renamed Đông Quan (Eastern Gateway). Subsequent names for this frequently renamed city included Đông Kinh (also "Eastern Capital") and Bắc Thành (Northern Citadel). In 1831, the Nguyen Dynasty ruler Minh Mạng named it Hạ Nội, meaning "Between Rivers." The city was occupied by the French in 1873, and in 1887, as Hanoï, it became capital of French Indochina. Its name is now written Hanoi. Japanese troops occupied the city during World War II. After the war, Ho Chi Minh proclaimed Vietnam independence in 1945 and the city briefly became the capital of the Viet Minh government until the French returned in 1946. A war of independence followed, and in 1954 Hanoi became capital of an independent North Vietnam. The Vietnam War raged from 1955 to 1975, ending with the defeat of American forces and fall of Saigon (renamed Ho Chi Minh City) in the south of Vietnam, and the unification of the country on July 2, 1976, with Hanoi as capital.

Major Landmarks. The historic center of Hanoi has remains of the old citadel that was constructed in 1010 by Lý Dynasty. In 2010, the site was designated as a UNESCO world heritage site. The Flag Tower of Hanoi, built in 1812, is part of this site and is an iconic symbol of the city. Other landmarks in the old quarter of Hanoi are the Temple of Literature, site of the first university in Vietnam, and the One Pillar Pagoda, part of an ancient temple dating to the 11th century. Landmarks of the French colonial period include the Hanoi Opera House built between 1901 and 1911, the State Bank of Vietnam (formerly the Bank of Indochina), the Presidential Palace built between 1900 and 1906 as a residence for the French Governor-General, and the Roman Catholic Saint Joseph Cathedral built in 1886. The Hotel Metropole was built in 1901 and is now operated by Sofitel as a 5-star international hotel. Among the city's many museums are the National Museum of Vietnamese History, the Vietnam National Museum of Fine Arts, the Vietnam Museum of Ethnology, the Vietnam Museum of Revolution, and the Ho Chi Minh Museum. Hỏa Lò Prison was built in the late 1880s by the French to house Vietnamese political prisoners. During the Vietnam War, it housed American prisoners of war and was sarcastically called the Hanoi Hilton. The Ho Chi Minh Mausoleum is another major landmark of the city.

Culture and Society. The main ethnic group of Hanoi is the Viet population, officially known as the Kinh to distinguish it from minority groups in Vietnam's population. Vietnamese is the official language. Most Vietnamese identify with the Buddhist religion, although not all practice it. There is also a Roman Catholic minority in Hanoi. The "áo dài" is traditional Vietnamese dress that is worn for special occasions such as weddings and religious festivals. It was once worm by both women and men, but now men wear it only rarely. High school girls in Vietnam commonly wear white *áo dài* as the required school uniform.

Further Reading

Boudarel, Georges and Nguyen Nav Ky. *Hanoi: City of the Rising Dragon.* London: Rowman & Littlefield Publishers, 2002.

Logan, William S. *Hanoi: Biography of a City.* Sydney: New South Wales University Press, 2000.

Logan, William S. "Hanoi, Vietnam: Representing Power in and of the Nation," *City: Analysis of Urban Trends, Culture, Theory, Policy, Action* 13, no. 1 (March 2009): 87–94.

van Horen, Basil. "Hanoi," *Cities: The International Journal of Urban Policy and Planning* 22, no. 2 (2005): 161–73.

HARARE

Harare is the capital and largest city of Zimbabwe, a landlocked country in southern Africa. The city is located in the north-center of the country in an agriculturally fertile region with access to fresh water from the Mukuvisi River. In addition to being the capital, Harare is Zimbabwe's main communications and commercial center, and the site of the University of Zimbabwe, the nation's largest institution of higher education. The population of Harare was about 1,606,000 in 2009, while the metropolitan area had about 2.8 million inhabitants in 2006.

Historical Overview. Harare was founded by British settlers of Africa led by Cecil Rhodes in 1890. They called their colony British Rhodesia and the city Salisbury (initially Fort Salisbury). Independence from the United Kingdom was achieved in 1965, at which time the country was renamed Rhodesia. In 1979, it was named Zimbabwe Rhodesia, and then simply Zimbabwe. The name Harare came into use in 1982. The name comes from the name of the Shona chieftain Neharawa. As a colonial city, Harare (Salisbury) was segregated by race. Whites occupied the center of the city where there were planned streets and squares, while blacks lived in spontaneous settlements in the outskirts that were not recognized as legitimate parts of the city until after 1907. Segregation patterns broke down with independence and with subsequent emigration from Zimbabwe of much of the white population. Emigration of whites was spurred largely by an aggressive land reform program that was introduced by President Robert Mugabe. On March 26, 2010, Joina City, a large shopping mall and business complex, was opened in the center of downtown Harare after 13 years of start-and-stop construction. It is referred to as "Harare's New Pride."

Major Landmarks. The National Botanical Gardens are near the center of Harare and have more than 900 species of trees and shrubs from all parts of Zimbabwe. Other landmarks are the National Gallery, the National Archives, the Queen Victoria Museum, the Museum of Human Sciences, and the Harare City Library (formerly the Queen Victoria Library). The National Heroes Acre is a burial ground and national monument that honors Zimbabweans who fought for national independence and other recognized heroes of the country. Eastgate Centre, opened in 1996, is another modern retail and office complex in central Harare in addition to Joina City.

Culture and Society. Harare continues to struggle with the enormous problems of poverty and income inequality that were inherited from the time of British colonialism. The situation has worsened with economic decline since 2000, including declines in exports, a shortage of foreign exchange, extremely high rates of inflation, and shortages of food and consumer goods. The government of President Mugabe has also been accused of widespread human rights violations, suppression of political opposition, and suppression of freedoms of the press and freedoms of speech. Operation Murambatsvina ("Operation Drive Out Rubbish") is a campaign that was started in 2005 as a crackdown against urban slums and illegal housing, but that provided little in the way of replacement housing for people who were displaced. As a result, many poor Zimbabweans have crossed into neighboring countries as refugees. White Zimbabweans, who numbered about 278,000 in 1975, have emigrated en masse because of economic policies directed against their wealth and now number fewer than 50,000. Zimbabwe has one of the highest rates of HIV/AIDS infections in the world and extremely low life average expectancies (44 for men and 43 for women).

Further Reading

Rakodi, Carole. *Harare: Inheriting a Settler-Colonial City—Change or Continuity?* Chichester: John Wiley, 1995.

Scarnecchia, Timothy. *The Urban Roots of Democracy and Political Violence in Zimbabwe: Harare and Highfield, 1940–1964.* Rochester: University of Rochester Press, 2008.

Yoshikuni, Tsuneo. *African Urban Experiences in Colonial Zimbabwe: A Social History of Harare before 1925.* Harare: Weaver Press, 2007.

HAVANA

Havana is the capital and largest city in the Republic of Cuba, the largest country in the West Indies and Caribbean Sea. The country consists of one main island, Cuba, a second sizable but much smaller island nearby, and various small offshore islets. It is one of the last remaining socialist countries in the world. Havana is on the north coast of the western part of the main island and faces the Strait of Florida. Key West, Florida, the nearest U.S. territory, is about 106 miles (170 km) across the strait to the north. Havana's setting is at Havana Bay, which is entered from open water via a narrow inlet and opens into three main harbors: Marimelena, Guanabacoa, and Atarés. The Almendares River passes through the city from south to north and enters the Straits of Florida to the west of Havana Bay. The population of Havana is about 2.1 million. In addition to being the center of government for Cuba, the city is also the country's main port and commercial center, main cultural and education center, and a growing destination for foreign tourism, particularly the city's historical center (Old Havana; *La Habana Vieja*) and beaches east of the city.

Historical Overview. Spain first colonial port on Cuba was founded in 1515 on the island's south coast, but local conditions were poor and Havana was founded again in 1519 at its excellent natural harbor location on the island's north coast. The city prospered immediately and became Spain's base for exploration and conquest in the New World as well as a key port for galleons laden with gold and silver in transport from the Americas to Spain. In 1589, the city was designated the capital of Cuba, replacing Santiago de Cuba on the other end of the island. In 1634, a Spanish royal decree honored Havana as "Key to the New World and Rampart of the West Indies." The city's harbor and inlet were heavily fortified to protect against raids by English, French, and Dutch pirates. In 1762, the city was captured by British naval forces in the Seven Years War and held for six months until being returned to Spain by the peace treaty. The sugar industry became a major part of Cuba's economy and exploited slaves from Africa as labor. Cuba remained loyal to Spain in the early 19th century when other colonies in the Americas rebelled and became independent nations, but by the end of the century Cubans, led by José Martí, also sought independence. In 1898, the battleship USS *Maine* exploded in Havana Harbor, killing more than 242 crew members. Spain was blamed, although it is still not certain exactly what happened, and the United States and Spain began a short war that ended with transfer of several Spanish colonial possessions including Cuba to U.S. control. Cuban independence became official in 1902, but close economic, cultural, and political ties with the United States remained. Over the first half of the 20th century, U.S. business interests dominated the Cuban economy, and dictators friendly to American business ran the country and prospered as well. Havana became a busy foreign, mostly American, business center and a playground for foreign tourists, also mostly Americans. The city became known for grand hotels, casinos, a bustling nightlife, and American mobsters.

On January 1, 1959, a revolution led by Fidel Castro and referred to as the 26th of July Movement succeeded in deposing the corrupt government of dictator Fulgencio Batista, and took control of the whole country. Castor's troops arrived in the city on January 8, with Castor taking up residence and setting up provisional government headquarters in the Hilton Hotel in the city's Vedado district. The hotel, which had been a symbol of foreign control of Cuba and social inequality, was renamed Hotel Habana Libre, "Hotel Free Havana." American interests were chased from Cuba and businesses were nationalized, and the country turned to the Soviet Union for strategic and economic assistance. Cuba became a socialist state with both the economy and public services in government hands. The United States initiated an economic boycott of Cuba, but Soviet aid and ample harvests of sugar helped to sustain the country. When the Soviet Union fell apart in 1991, Cuba entered a time of great economic difficulty and material shortages. It has since liberalized parts of the economy in order to stimulate growth, and opened doors more widely for tourism. Fidel Castro retired from public office in 2011 because of ill health. His brother Raúl Castro has succeeded him.

Major Landmarks. Old Havana is a UNESCO World Heritage Site and is a neighborhood of historic architecture, narrow streets, pleasant squares, and vintage American automobiles from before the revolution. The baroque Catedral de San Cristobal is the most iconic historical landmark in the district. Other landmarks in central Havana are the Capitol Building, the Museum of the Revolution, the Hotel Habana Libre, the Plaza de Armas, and the city's historic fortifications, led by Castillo de la Real Fuerza and the Castillo del Morro. The Plaza de la Revolutión centers on a prominent statue of José Martí and a large image of the revolutionary leader Che Guevara on the wall of the Ministry of the Interior building. The Malecón is a popular walkway along the city's seawall. A park in the Verdado district has a much-loved statue of musician John Lennon. Many tourists also make it a point to visit a cigar factory and a rum distillery.

Culture and Society. Most of the population of Havana is of European-Spanish ancestry. There is also a large minority of blacks and mulattoes who are descendants of Cuba's slave population. Before the revolution, Cuba was a highly stratified society with whites at the top of the economy, but the policy of the government since has been one of equal opportunity regardless of ethnicity or racial identity and greater economic equality. Many of Cuba's wealthiest and best-educated people fled Cuba with the revolution, resettling primarily in the United States, making it necessary for socialist Cuba to train a new generation of specialists in all sorts of professions. The language of Cuba is, of course, Spanish, and historically, the population was heavily Roman Catholic. Catholicism is still practiced, but by fewer people as the Castro government discouraged church attendance. However, a visit to Havana by Pope Benedict XVI in March 2012 showed that there are still strong religious feelings in the country and more openness to religion on the part of the government. The culture of Cuba includes great love for song and dance, including musical styles such as salsa, rumba, and mambo, a delicious cuisine, and a passion for baseball.

Further Reading

Birkenmaier, Anke and Esther Whitfield, eds. *Havana Beyond the Ruins: Cultural Mappings after 1989*. Durham, NC: Duke University Press, 2011.

Estrada, José Alfredo. *Havana: Autobiography of a City*. New York: Palgrave Macmillan, 2007.

Scarpaci, Joseph, Roberto Segre, and Mario Coyula. *Havana: Two Faces of the Antillean Metropolis*. Chapel Hill: University of North Carolina Press, 2001.

Williams, A. R. "The Rebirth of Old Havana," *National Geographic* 195 (June 1999): 36–45.

HELSINKI

Helsinki is the capital and largest city in Finland, a country on the Baltic Sea in northern Europe adjacent to a northern extension of European Russia to the east and Sweden and northern Norway to the west. The city is in southern Finland on an extension of the Baltic Sea called the Gulf of Finland, and occupies a series of peninsulas and islands on an irregular coastline. It is at a latitude of 60°10′ N, making Helsinki the second northern-most capital city in the world after Reykjavik in Iceland. Its population is about 602,200, while that of the metropolitan area, which encompasses the nearby towns of Espoo, Vantaa, and Kauniainen in addition to Helsinki proper, totals nearly 1.4 million (2012). The metropolitan area is the world's northern-most urban center with 1 million or more inhabitants. Helsinki's nearest neighbor capital city is Tallinn, Estonia, located about 50 miles (80 km) south across the Gulf of Finland.

Historical Overview. The history of Helsinki and Finland in general is closely tied to neighboring Sweden and Russia. It was founded as a trading port in 1550 by King Gustav I of Sweden, but failed to prosper because of epidemics and warfare. It was called Helsingfors by the Swedes. In 1809, after Russia defeated Sweden in the Finnish War, Finland was annexed into the Russian Empire and Helsinki was designated as the capital of the Grand Duchy of Finland in place of Turku, a city seen as too close to Sweden. In 1827, the Royal Academy of Turku was relocated to Helsinki and became Helsinki University. As the city developed, it took on architectural aspects of the beautiful Russian capital, St. Petersburg. Finland gained its independence from Russia in 1917 in the wake of the Russian Revolution and the fall of czarist rule, and Helsinki became a national capital. In 1952, Helsinki hosted the summer Olympic Games.

Major Landmarks. The most famous landmark in Helsinki is the striking white Helsinki Cathedral (or Lutheran Cathedral) that was completed in 1852 and stands as a symbol of the city. It was designed by Carl Ludvig Engel as part of a grand plan for Helsinki's central square, Senate Square. A statue of Russian Czar Alexander II stands in Senate Square near the cathedral. Aleksanterinkatu, named after Alexander I of Russia, is an important commercial street in the center of Helsinki that was designed by Engel. The Church in the Rock, which was dug out of solid rock and has a roof made of copper strips, is another major attraction because of its unique design. It was completed in 1969. Other attractions in Helsinki are the University of Helsinki's main building, the Olympic Stadium, the National Museum of Finland, the Ateneum Art Museum, and Finlandia Hall, a concert venue. The Sibelius Monument honors world-famous Finnish composer Jean Sibelius. Soumenlinna is an historic fortress that was built by the Swedes in the mid-1700s on an island near Helsinki. It is registered on the UNESCO list of World Heritage sites for its distinctive military architecture and its role in history in the conflict between Sweden and Russia.

Culture and Society. Finnish and Swedish are both official languages in Helsinki, although the vast majority speak Finnish as a native language. About three-quarters of the population is Lutheran. Helsinki has a very high standard of living and is a clean, safe, and well-organized city. Forest products and high-technology companies such as the mobile telephone manufacturer Nokia underpin the economy. The population enjoys a rich cultural life and outdoor sports in every season. Almost every household in Finland has access to its own sauna. Spending time in a sauna is seen as relaxing and good for health, as well as an occasion for family and friends to be together.

Further Reading

Kent, Neil. *Helsinki: A Cultural and Literary History.* New York: Interlink Books, 2005.
Moorhouse, Jonathan. *Helsinki: Birth of the Classic Capital, 1550–1850.* Helsinki: SKS, Finnish Literature Society, 2003.

HONIARA

Honiara is the capital and largest city of the Solomon Islands, a country of many islands in Oceania to the east of Papua New Guinea. Honiara is on the north shore of Guadalcanal Island, one of the larger and more central islands in the Solomon Islands chain. The city's population was counted at 49,107 in the 1999 census and estimated at 78,190 in 2009. The name "Honiara" is translated from one of the local languages as meaning "facing the southeast wind."

Historical Overview. The Solomon Islands have been inhabited by Melanesia people for many centuries. The first European to visit was the Spanish navigator Álvaro de Mendañain in 1568 who gave the islands their name (Islas Salomón). Christian missionaries began visiting the islands in the 19th century. In 1893, the islands became a protectorate of Great Britain. Coconut plantations were the basis of the European economy. During World War II, the islands were the scene of fierce battles between Japanese and Allied forces, with great loss of life on both the sides. The Battle of Guadalcanal, also known as the Guadalcanal campaign, was fought between August 7, 1942 and February 9, 1943 and resulted in an important strategic Allied victory.

Honiara was made the capital of the Solomon Islands in 1952, replacing Tulagi on Florida Island that had been the British administrative center. Independence from the United Kingdom was achieved on July 7, 1978. In the 1990s and 2000s the islands were torn by civil unrest and ethnic violence, with much of Honiara's Chinatown neighborhood destroyed in April 2006 by rioters angry at the alleged overly large influence of Chinese businessmen on national government. Australian troops are still in place to help keep peace. The Solomon Islands ranks as one of the world's poorest countries, with more than three-quarters of the population engaged in subsistence farming or fishing.

Major Landmarks. The Guadalcanal American Memorial and the Japanese War Memorial on Mt. Austin are perhaps the most important landmarks in the Honiara area. There are also other monuments related to World War II battles in the city and elsewhere on Guadalcanal. Other landmarks are Government House, the National Museum, botanical gardens, the yacht club, and the Solomon Islands Visitors Bureau.

Culture and Society. About 95 percent of the people of the Solomon Islands are Melanesian and 3 percent Polynesian. There is a minority of Chinese people as well, most prominently in Honiara. Some 97 percent of the population is Christian, reflecting more than 100 years of active foreign missionary work on the islands. The official language of the Solomon Islands is English, although there are 70 local languages spoken as well. The people of the Solomon Islands are torn awkwardly between traditional society and influences from abroad, and strive for a healthy balance between the two. Honiara has a high crime rate. Foreign tourists are especially targeted by muggers.

Further Reading

Dinnen, Sinclair and Stewart Firth, eds. *Politics and State Building in Solomon Islands.* Canberra: Australian National University, 2008.
McDonald, Ross. *Money Makes You Crazy: Custom and Change in the Contemporary Solomon Islands.* Dunedin, NZ: University of Otago Press, 2003.

ISLAMABAD

Islamabad is the capital of Pakistan, a populous, mostly Islamic country in South Asia that ranks sixth in the world in population size (177,000,000 in 2011). Islamabad has about 1.7 million people, the 10th largest total in Pakistan. The Islamabad-Rawalpindi Metropolitan Area has 4.5 million people. Rawalpindi (population 3.3 million) is an ancient city next to Islamabad; the two cities together are referred to as Twin Cities. Islamabad is located in Islamabad Capital Territory (ICT) on the Potohar Plateau in the north of Pakistan. The name Islamabad means "City of Islam," reflecting the dominant religion of the Islamic Republic of Pakistan and the importance of Islamic traditions and requirements in daily life.

Historical Overview. Although the area of Islamabad has been inhabited since prehistoric times and has been a crossroad of trade and armies for centuries, the city is quite recent. Pakistan became an independent country in 1947 with the dissolution of British India, and at the time Karachi, the country's main seaport and largest city, was chosen as the capital. The site of Islamabad was selected for a new capital in 1958 after a detailed study, and construction took place in 1960. The first move of the government from Karachi was to Rawalpindi until construction in Islamabad was far enough along. The designer of the city's master plan was a Greek firm headed by Konstantinos Apostolos Doxiadis. He produced a grid plan for the city within a triangular shape, and monumental views and avenues. The site of Islamabad was chosen because it was nearer to the center of Pakistan's population than the giant coastal city. Also, Islamabad was near the national military headquarters in Rawalpindi and could therefore be more effectively defended. Another consideration was that the site was nearer to the disputed territory in Kashmir in the north, thereby helping to strengthen Pakistan's territorial claims on the regions.

Islamabad has grown rapidly since being opened and has attracted migrants from all parts of Pakistan. It is occasionally a flashpoint for the country's political conflicts, and has at times been the locus of terrorist attacks against non-Pakistani targets. In addition to being the government center, Islamabad has also developed into a major business center and center of higher education and related research. In contrast to the crowding and chaos that is typical of other large cities in Pakistan, Islamabad is more spread out, greener, quieter, and more orderly.

Major Landmarks. The most iconic landmark of Islamabad is the Faisal Mosque, the national mosque of Pakistan. It was constructed in 1986 and is the largest mosque in South Asia and one of the largest in the world. It is named after King Faisal of Saudi Arabia who financed the project. Other landmarks are

Parliament House, the Pakistan Monument, the Red Mosque, and the ruins of historic Rawat Fort from the 16th century. Major museums include the Lok Virsa Museum and the National Art Gallery. The Margalla Hills to the north of the city have beautiful parks and trails.

Culture and Society. The population of Islamabad is overwhelmingly Islamic, like that of Pakistan as a whole. The official language of Pakistan is Urdu and is widely spoken in the city because it is the common language for the country's diverse ethnic groups. English is also commonly used. The mother tongue of most residents (72%) of Islamabad is Punjabi, followed by 10 percent native Urdu speakers and 10 percent Pushto speakers. The adult literacy rate in Islamabad is 87 percent, the highest in Pakistan.

Further Reading

Botka, Dusan. "Islamabad after 33 Years," *Ekistiks* 62, no. 373–375 (1995): 209–36.

Doxiadis, C.A. "Islamabad: The Creation of a New Capital," *Ekistiks* 72, no. 430–435 (2005): 113–30.

Yakas, Orestes. *Islamabad: The Birth of a Capital.* Karachi, Pakistan: Oxford University Press, 2001.

J

JAKARTA

Jakarta is the capital and largest city in the Republic of Indonesia, a large and populous country of more than 17,500 islands in Southeast Asia and Oceania. The city is officially called the Special Capital Territory of Jakarta. The word Jakarta comes from the old Javanese word Jayakarta, which means "complete victory."

Jakarta is located near the center of the archipelago on the northwest coast of Java, Indonesia's most populous and most densely settled island. It is at the mouth of the Ciliwung River and faces Jakarta Bay in the Java Sea. The city is generally low lying and is prone to flooding. It is at 6°12′ S latitude and has a wet, tropical climate. The volcanic peaks of the Parahyangan Highlands are in the interior of Java behind the urban area. The population of Jakarta proper is nearly 10 million, while that of the wider metropolitan area, which includes the nearby cities of Bogor, Depok, Tangerang, and Bekasi, has as many as 28 million inhabitants and is the world's second-largest metropolitan area after Tokyo, Japan. The Jakarta metropolitan area is sometimes referred to as Jabodetabek, an invented word coming from the first letters of each of the five constituent central cities.

The Jakarta metropolis is an enormous urban region that grew up initially behind the old inner harbor, and then in the second half of the 20th century sprawled landward in all directions across farmland and villages because of rapid population growth spurred by migration of peasants from the countryside. There are hundreds of *kampung* districts (village-like neighborhoods) in the city, as well as many street markets and other vendors, various industrial zones with labor-intensive assembly plants, and poverty-stricken squatter settlements. The city suffers from some of the worst traffic congestion and air pollution problems compared to any other city in the world. Jakarta also has many upscale residential districts, and countless glistening shopping malls and high-rise office towers, hotels, and condominium buildings.

Historical Overview. The origins of Jakarta date to the fourth century and the Kingdom of Sunda, when it was a trading port named Sunda Kelapa, or "coconut harbor." The city exported coconuts, peppers, and other products. The first Europeans to arrive were Portuguese traders from their base in Malacca, now Malaysia, in 1513. The Portuguese were expelled in 1527 by an attack by Fatahilla, a general from a different part of Java. The city then became a part of the Sultanate of Banten and was renamed Jayakarta. Dutch traders arrived in 1596, followed by traders from England in 1602. The Dutch, led by Jan Pieterszoon Coen, soon defeated their British rivals and took sole command of the region on behalf of the Dutch East India Company. In 1619, the Dutch renamed the city Batavia. They constructed their buildings in the Dutch architectural style, planted shade trees along the city's streets, and dug drainage canals that showed hints of their capital, Amsterdam. The city was much admired by Europeans who referred to it as "Queen City of the East." Chinese immigrated to Batavia and also entered trade, resulting in a massacre of 5,000 Chinese by Dutch residents in 1740 and restrictions as to where in Batavia they could live and work. In 1800, the islands of Indonesia became a national colony of the Netherlands, the Dutch East Indies.

The name Jakarta was applied to the city by Indonesian nationalists during World War II who fought the Dutch for national independence. Indonesian independence was declared in 1945, making the country one of the first of Europe's colonial possessions to break away in the wake of World War II. Indonesia's founding president Sukarno was engaged in aggressive nation building, modernization, decolonization of the landscape, and construction of heroic monuments such as the 433-ft-high (132 m) National Monument (Monas). In 1965, he fought off a coup attempt and undertook a violent purge of Indonesia's Communists, resulting in some 500,000 deaths. Under Indonesia's next president, Suharto, whose "New Order" administration was in power from 1967 to 1988, the modernization of Jakarta and monumental construction continued, and Indonesia became more closely linked to a global economy that exploited the nation's natural resources. There was considerable economic growth and a rising middle class, but the East Asian Economic Crisis of the late 1990s brought on new poverty and civil unrest. Rioters targeted Chinese-Indonesians because the ethnic group was seen to be disproportionately prosperous. Since 2000, the center of Jakarta has been the target several times of deadly terrorist bombs set off by radical Indonesian Islamists.

Major Landmarks. The symbolic center of Jakarta is Merdeka Square, Indonesian for Freedom or Independence Square. The National Monument stands in its center. During the Dutch era, the square was called Koningsplein, "King's Square." The square also has an equestrian statue of Javanese Prince Diponegoro, a 19th-century opponent of Dutch rule. On a busy traffic circle along Jalal Thamrin, the main avenue of Jakarta, is the Salamat Detang (meaning "welcome") Monument. It shows a happy male and female couple waving greetings. The West Irian Liberation statue in another part of central Jakarta shows a male figure breaking the chains that had cuffed him. It celebrates the belated liberation of Indonesia's eastern-most province from Dutch rule, but as it stands near the Ministry of Finance headquarters, people sometimes joke that it is the "We are Broke" monument.

Other important landmarks include the city's central mosque, Istiqlal Mosque, and the Roman Catholic Jakarta Cathedral. Taman Mini Indonesia Indah (Beautiful Indonesia Miniature Park) is a park constructed in 1975 that displays daily life in all 26 provinces of Indonesia. Ancol Dreamland is a popular resort complex and amusement park along Jakarta's waterfront. There is a Chinatown neighborhood in the Glodok district of Jakarta, and there are many examples of Dutch architecture and historic museums near the city's original port, notably the Jakarta History Museum in the former city hall of Batavia. Three of the largest shopping malls in the city are Mall Taman Anggrek, Pacific Place Jakarta, and Grand Indonesia Mall.

Culture and Society. Indonesia is the world's most populous Islamic country, and Jakarta has more Muslims than any metropolitan area in the world. As the national capital of a country comprised of many regional cultures and languages, Jakarta is a diverse city, although Javanese culture and the national language Bahasa Indonesia, a form of Malay, predominate. Chinese-Indonesians are an influential minority, but have often been victims of ethnic violence. There are also religious tensions in Indonesia between Muslims and Christians, mostly in outlying islands, but sometimes evident in Jakarta as well.

Further Reading

Abeyasekere, S. *Jakarta: A History*, Singapore: Oxford University Press, 1989 (revised edition).
Cybriwsky, Roman and Larry R. Ford. "Jakarta," *Cities: The International Journal of Urban Policy and Planning* 18, no. 3 (2001): 199–210.
Silver, Christopher. *Planning the Megacity: Jakarta in the Twentieth Century*. London: Routledge, 2008.
Wilmar, Aalim and Benedictus Kombaitan. "Jakarta: The Rise and Challenge of a Capital," *City: Analysis of Urban Trends, Culture, Theory, Policy, Action* 13, no. 1 (March 2009): 120–28.
Ziv, Daniel. *Jakarta Inside Out*. Jakarta, Indonesia: Equinox Publishing, 2002.

JERUSALEM

Jerusalem is the capital and largest city of Israel, officially the State of Israel, a small parliamentary republic in the Middle East on the eastern shore of the Mediterranean Sea. Not all countries in the world recognize the existence of Israel, and even most of those that do, do not officially recognize Jerusalem as the country's capital. As a result, the city has no foreign embassies. The Embassy of the United States is in Tel Aviv, a city on the Mediterranean about 37 miles (60 km) to the west. Most other foreign embassies are in Tel Aviv as well. Jerusalem is one of the most contested cities in the history of humankind, as it is held holy by the three major Abrahamic religious traditions: Judaism, which is the main religion of Israel, Christianity, and Islam. The city is adjacent to the West Bank, a contested region between the Jordan River and the eastern part of Jerusalem that is claimed by both Israel and Palestinians. Israel refers to as Judea and Samaria and has encouraged Israeli-Jewish settlements to be built in the area despite international pressure to the contrary, while Palestinians see the area as the basis from which to recover a Palestinian state. The population of Jerusalem is about 801,000, while that of the metropolitan area is a little more than 1 million.

Historical Overview. There is perhaps more written about the history of Jerusalem than any city on earth. The city is one of the oldest in the world. Among the earliest inhabitants were Canaanites. Archaeological evidence shows a walled city under a part of today's Jerusalem that goes back some 4,000 years. Hebrews migrated from Egypt in about 1250 BC and conquered Jerusalem about 200 years later. King David ruled from about 1003 to 970 BC and made the city a political and religious capital. His son Solomon built the Holy Temple (later known as the First Temple) in about 957 BC on Mount Moriah and became the repository of the Ark of the Covenant. The Babylonians destroyed the temple and the city in 587 BC. Babylonian captivity lasted for 50 years and was replaced with Persian rule. The

Second Temple was completed in 516 BC and the walls of Jerusalem were rebuilt in about 445 BC. Greek rule followed Persian rule and lasted from 332 BC to 167 BC when Jewish dominion was restored. In 63 BC the Romans took control of the city. King Herod expanded the city and built walls and palaces, and greatly enlarged the temple. Jesus was born during the rule of Herod and, according to central tenets of Christian belief, was later crucified by Romans in Jerusalem and then three days later resurrected from the dead. The Second Temple was destroyed in 70 BC during the First Jewish-Roman War. The city was rebuilt as a Roman city by the emperor Hadrian in 135. From 324 to about 638 the city was under Byzantine Christian rule. During that time Emperor Constantine I constructed Christian sites in Jerusalem, most notably the Church of the Holy Sepulcher. Arabs conquered the city in 634. Crusaders also took Jerusalem from 1099 to 1187, after which the city reverted to Islamic Muslim rule. Among the major Islamic structures that were built in the city are the Dome of the Rock and the Al-Aqsa Mosque. The Dome of the Rock was constructed in 691 on the site of the Second Jewish Temple and commemorates the Night Journey in 621 by the prophet Mohammad, while the Al-Aqsa Mosque is at the point from which Muslims believe that Mohammad ascended into heaven. The heart of the Dome of the Rock is the Foundation Stone, seen by Jewish tradition as the spiritual junction between heaven and earth and the holiest site of Judaism.

In 1517, Jerusalem fell to the Ottoman Turks who maintained control until 1917. The city was rebuilt and greatly beautified during the 1520–1566 rule of Sultan Suleiman the Magnificent. Later, the city was modernized by Ottoman rulers, and was brought into the 20th century. During the Ottoman period, Jerusalem was a regional capital of Palestine. From 1917, the British ruled Jerusalem under a League of Nations mandate until 1948. Independence for State of Israel from Mandatory Palestine was proclaimed on May 14, 1948, and Jerusalem was proclaimed capital in 1950, replacing a temporary capital in Tel Aviv. Many Jews migrated to Israel from Diaspora residence around the world, including many survivors of the Holocaust in Europe. Following the 1967 Six-Day War, the entire city was brought under Israeli control. Many countries view the annexation of East Jerusalem by Israel as a violation of international law. Palestinian opposition to Jewish control continues, and the city is divided between Jewish and Palestinian areas by walls. Even many supporters of Israel view the walls as a symbol of intolerance by hardliners in the Israeli government and discrimination against Palestinians.

Major Landmarks. The Old City and its walls are a UNESCO World Heritage Site. There are four quarters to the Old City, the Armenian Quarter, the Christian Quarter, the Muslim Quarter, and the Jewish Quarter. The major sites include the Dome on the Rock, the Wailing Wall, the Al-Aqsa Mosque, and the Church of the Holy Sepulcher. The latter is the most sacred site of Christianity. The highlight of the Armenian Quarter is the Armenian Cathedral and Museum. The most elaborate of the city's gates is the Damascus Gate. At Jaffa Gate is the Tower of David, a museum of city history. Other landmarks are the Biblical Zoo, the Israel Museum where the Dead Sea Scrolls are kept, and Yad Vashem, Israel's Holocaust Museum. The Knesset, the unicameral legislature of Israel, has its seat in the Givat Ram district of Jerusalem.

The Knesset, the Parliament Building of Israel. (Alon Othnay/Shutterstock.com)

Culture and Society. Approximately 64 percent of the population of Jerusalem is Jewish, 32 percent is Muslim, and 2 percent is Christian. The percentage of Jews has been decreasing largely because of in-migration by Muslims and a higher birth rate among Muslims, and because many Jews have been leaving the city in favor of new settlements in the West Bank and elsewhere. For Palestinians, Jerusalem is a source of employment and better health care, education, and other services in comparison to other areas where Palestinians live. As elsewhere in Israel, Jerusalem is a closely guarded city as terrorist attacks from the country's opponents are possible at any time. Palestinians would like to make East Jerusalem the capital of a Palestinian state, but the Israeli government is strongly opposed.

Further Reading

Benvenisti, Meron. *City of Stone: The Hidden History of Jerusalem.* Berkeley: University of California Press, 1996.
Delisle, Guy. *Jerusalem: Chronicles from the Holy City.* Montreal: Drawn and Quarterly, 2012.
Jacobson, Abigail. *From Empire to Empire: Jerusalem between Ottoman and British Rule.* Syracuse, NY: Syracuse University Press, 2011.
Montefiore, Simon Sebag. *Jerusalem: The Biography.* New York: Knopf, 2011.

JUBA

Juba is the capital and largest city of the Republic of South Sudan, a landlocked country in the northeast of Africa that is one of the world's newest independent nations. It was formed on July 9, 2011, with the final break of South Sudan from Sudan. The city is on the White Nile River in the south-center of the country and is also the capital of Central Equatoria, the smallest of South Sudan's 10 states.

The population of Juba is estimated to be 372,410 (2011). It is growing rapidly as the new country gets started and with wealth generated from South Sudan's oil resources.

Historical Overview. The origins of Juba are traced to the nearby town of trading post of Gondokoro, the southern most of the Egyptian garrisons in early 19th-century Sudan. The British explorer Sir Samuel Baker used Gondokoro as a base for his exploration of the African interior in the 1860s and 1870s. Juba was founded in 1922 on the west bank of the White Nile by Greek traders. From 1899 to Sudanese independence in 1956, Juba was in the Anglo-Egyptian Sudan. It was a center of conflict in Sudan's three civil wars afterwards until the eventual break of North and South Sudan into separate countries. Infrastructure in Juba is still in disrepair from the wars. On September 5, 2011, the new government of South Sudan announced a plan to move the capital to a new planned city that is to be built at Ramciel (also spelled Ramshiel) in Lakes State to the north of Juba in the geographical center of the country.

Major Landmarks. As a new capital that is just emerging from war, Juba does not offer much in the way of attractions for tourism, although there are quite a few hotels as South Sudan attracts many foreign aid workers, foreign diplomats, and international business investors. The main sights of the city are the White Nile River, the Nile Bridge, and Jebel Kujur Hill outside the city. The mausoleum of John Garang de Mabior, a South Sudan rebel leader who died in 2005 in a helicopter crash is in Juba.

Culture and Society. South Sudan is a very poor country. Its approximately 8 million residents include as many as 4 million internally displaced people. The health conditions of the population are among the worst in the world, with maternal mortality being the world's highest. The population is comprised of a great many ethnic groups and indigenous languages. English is the official national language, but only a small percentage of the population can speak it and Arabic is more prevalent as a common language. The government has declared that it will support freedom for all religions. The population of South Sudan includes a mix of Christians, Muslims, and indigenous African tribal beliefs. A number of South Sudanese have excelled in athletics, including basketball. Some stars have played successfully in the United States in the National Basketball Association such as the 7'7" (2.31 m) tall Manute Bol.

Further Reading

Achier, Deng Akol Ayay. *Beautiful South Sudan: The Heart of Africa*. Kuajok, South Sudan: Achier Deng Akol Ayay (self-published), 2010.

Benjamin, Barbara Marial. "Relocation, Relocation, Relocation," *Geographical Magazine* 83, no. 12 (December 2011): 26–30.

KABUL

Kabul is the capital and largest city in Afghanistan, officially the Islamic Republic of Afghanistan, a landlocked country in central Asia that has long been wracked by war and terrorist strife. American troops are presently engaged in the conflict in what has become the longest running war for the United States (since 2001), although a phased withdrawal has begun under the administration of President Barack Obama. Kabul is a major base for U.S. forces in the country, as well as for aid workers from abroad. The city is located in the eastern part of the country along the Kabul River and has an elevation of about 5,000 ft (1,800 m). It has a population of more than 3 million, perhaps as many as 4 million.

Historical Overview. Kabul's history goes back some 3,500 years. It occupies a strategic site for control of mountain passes between Central Asia, the Middle East and the European world, and South Asia, and owes its existence and long history of invasions and conflict to this geography. The Islamic conquest began in 642, and in 870 the city was conquered by the Saffarid ruler Ya'qub bins Laith as-Saffa from Zaranj, now a small town in southwestern Afghanistan, and became Moslem. In the 13th century it was all but destroyed by the Mongol invader Genghis Khan. From 1504 to 1526, Kabul was the capital of the Mughal dynasty under Bābur. It remained under Mughal rule until it was conquered by Nāder Shah, the founder of the Persian Afsharid dynasty, in 1738. It became the capital of the emerging state of Afghanistan in 1776. In the 19th century, the city was invaded many times by British-Indian forces during the Anglo-Afghan Wars. These conflicts were rooted in the so-called Great Game in which Great Britain vied with Imperial Russia for control of Central Asia. In 1919, the British Royal Air Force inflicted damage on the city during the Third Anglo-Afghan War. A Marxist government took control of Afghanistan in 1978 in the so-called Saur Revolution. On December 24, 1979, the Soviet Union invaded Afghanistan in an effort to prop up the Communist government against opposition from Mujahidin rebels and used Kabul as its base. For nearly 10 years, the Soviets were caught up in a costly war in Afghanistan until withdrawal in 1989. Fierce fighting continued between rival Afghan groups afterward, and much of Kabul was reduced to ruins. In 1996, the Taliban gained control of the city until a U.S.-led coalition took over in 2001 and helped establish an interim government. The U.S. presence continues in Afghanistan. In 2004, Hamid Karzai won an election to become president and has been at the helm of Afghanistan's government with U.S. support ever since. Kabul is a frequent target of bombings and other terrorist attacks.

Major Landmarks. Some parts of Kabul are still marked by narrow, crooked streets, and bazaars, although most of the historic core have been torn down after war damage and rebuilt along modern lines. The city has many beautiful mosques such as the Abdul Rahman Mosque and the Id Gah Mosque, as well as mausoleums of past rulers and several museums of which the National Museum of Afghanistan built in 1922 is the most prominent. The gardens surrounding the tomb of the Mughal emperor Bābur are especially beautiful. Paghman Gardens is a popular park near the city. Kabul City Center, opened in 2005, is Afghanistan's first modern indoor shopping mall.

Culture and Society. Afghanistan is a country torn by war and Kabul is a city of under constant threat. Many of its residents are migrant refugees from war-torn parts of the country controlled by the Taliban. The ethnic mix reflects the diversity of Afghanistan, although the main population group is Pashtun. Almost all Afghans are Muslims. There are many foreign aid and development workers, consultants and advisors, and other specialists who live in heavily guarded compounds and venture out only with armed escort. Nevertheless, there is voiced optimism that one day the fighting will cease and a good life will return. There are ambitious city planning projects for Kabul that would restore damaged districts, make the city greener, and develop modern, large new districts.

Further Reading

Esser, Daniel. "The City as Arena, Hub, and Prey—Patterns of Violence in Kabul and Karachi," *Environment & Urbanization* 16, no. 2 (2004): 31–38.

Rodriguez, Deborah. *Kabul Beauty School: An American Woman Goes behind the Veil.* New York: Random House, 2007.

Seierstad, Āsne. *The Bookseller of Kabul.* New York: Little, Brown and Company, 2003.

KAMPALA

Kampala is the capital and largest city of Uganda, a landlocked country in East Africa on the north shore of Lake Victoria. The city is located in southern Uganda about 20 miles (32 km) north of Victoria, and about 29 miles (47 km) north of Entebbe, the lakeside city where Kampala's international airport is located. The population of Kampala is about 1.7 million.

Historical Overview. Uganda takes its name from the Buganda Kingdom, one of the traditional kingdoms within Uganda since the 14th century. It is the kingdom of the Ganda people and includes Kampala within its territory. The city became the Bugandan capital in the 19th century, and still includes some landmarks from that period. British explorers searching for the source of the Nile River arrived in the 1860s, followed by both Protestant and Catholic missionaries and, in 1888, agents of the British East Africa Company. In 1894, Uganda was made a British protectorate. An epidemic of sleeping sickness took many lives along the Lake Victoria shoreline in the early 20th century. Uganda gained its independence from Great Britain in 1962. The Buganda Kingdom was abolished after independence, but was then restored in 1993. From 1971 to 1979, Uganda was ruled by the dictator Idi Amin who lavished the country's wealth on himself and carried out killings of as many as 300,000 Ugandan citizens. His regime was ended by military intervention by neighboring Tanzania in cooperation with exiled Ugandans. The country is still not democratic, and subsequent elections for national leadership are said to have been fraudulent. In 1976, Entebbe Airport was the scene of a generally successful rescue by Israeli Special Forces personnel of Israeli hostages being held by terrorists who had hijacked a French airliner that had departed from the airport in Tel Aviv. Kampala is said to have been founded on seven hills, as is the history of other prominent capitals in the world, but its expansion in the decades after independence now covers more hills.

Major Landmarks. Landmarks of the Buganda Kingdom include the Kasubi Tombs, Lubiri Palace, and Bulange, the Buganda parliament building. Kampala's modern Central Business District is on Nakasero Hill where the early-20th-century Grand Imperial Hotel is a notable landmark. Other landmarks include the Nakasero Market, the Ugandan Museum, the Ugandan National Theater, the Parliament of Uganda, and the national independence monument.

Culture and Society. Kampala's main ethnic group consists of the Ganda people from the local Kingdom of Buganda. They comprise an estimated 60 percent of the total population of the metropolitan area. The Banyankole population of Kampala has grown since Yoweri Museveni became president in 1986. The city has many other ethnic groups as well. Christianity is the main religion of Uganda, with Roman Catholics being the most numerous denomination, followed by the Anglican Church of Uganda. Eastern Uganda has many Muslims, some of whom have settled in the capital city. English and Swahili are the official languages, but many other languages are spoken as well. Uganda is generally a poor nation, in part because of many years of corrupt government and violence. There is more poverty in the countryside; hence, many Ugandans have migrated to Kampala and other cities in search of better opportunity. Uganda had one of the world's highest rates of HIV/AIDS infection, but this has fallen markedly thanks to aggressive publicity about AIDS prevention. In recent years, Kampala has earned global notoriety for intolerance about homosexuality and acts of violence against members of sexual minorities.

Further Reading

Wallman, Sandra. *Kampala Women Getting By: Wellbeing in the Time of AIDS.* Athens: Ohio University Press, 1996.

KATHMANDU

Kathmandu is the capital and largest city of the Federal Democratic Republic of Nepal, commonly referred to as simply Nepal, a landlocked country in South Asia that is bordered by China to the north and India to the south. The city is located near the center of the country in the northwestern part of the Kathmandu Valley at an average elevation of about 4,430 ft (1,350 m). The Bagmati and Bishnumati rivers converge near the center of the city. The high Himalayan Mountains of Nepal are to the north of the city, while to the south, Nepal is a flat subtropical jungle area called the Terai that is increasingly being converted to farmlands because of over-crowding elsewhere in the country. The population of Kathmandu is a little more than 1 million. In addition to being the national capital, Kathmandu is Nepal's most important industrial and commercial center, and the starting point for most foreign tourism to the country including Himalayan treks.

Historical Overview. According to myth, in ancient times the Kathmandu Valley is said to have been a lake. It was then drained for settlements, one of which became Kathmandu. Historical record says that the city was part of the early Licchavi Kingdom and was founded in 723 by Raja Ginakamadeva. Its original name was Manju-Patan. The name Kathmandu comes from *kath* meaning "wood" and *mandir* meaning temple or building, and refers to a temple that was made in 1596 by Raja Lachmina Singh from the wood from just one tree. A building that is said to be the original temple still stands in the center of the city. The Malla dynasty followed the Licchavis and ruled the Kathmandu Valley from the 12th to 18th century. Under Malla rule the city was known as Kantipur, a name that is still in occasional use. The city prospered from the agricultural potential of the Kathmandu Valley and from trade from various parts of Nepal, Tibet, and India. The reign of King Jayasthiti Malla (1380–1395) was especially influential in the shaping of Nepalese society and the shaping of Kathmandu's urban development. The Mallas were succeeded by the Shah dynasty in 1768. They ruled from Kathmandu until 2008 when King Gyanendra was deposed by popular revolt and Nepal became a republic.

Major Landmarks. The hub of Kathmandu is the historic Durbar Square and its many palaces and temples including the historic wooden structure that gives the city its name. The square is one of the several UNESCO world heritage sites in Kathmandu. What was once the royal palace of the Shah dynasty was converted in 2009 into the Narayanhiti Palace Museum. Hanuman Dhoka is a complex of structures associated with Malla royalty. Swayambhu is a large Buddhist stupa, one of the most scared religious sites in the country. It is also known as the Monkey Temple, and is on a high hill overlooking the city that affords a fine view of Kathmandu Valley. The Boudha stupa is also a remarkable religious landmark. Pashupatinath is an important Hindu temple. Freak Street is a street of restaurants, inexpensive hotels, cafes, and other businesses that are popular with backpacking Western tourists in search of the special enlightenment that travel to Nepal can bring. New Road is a busy commercial street in the city.

Culture and Society. Kathmandu is a city of a great many temples and other religious shrines, and of many religious festivals. The main religions are Hinduism and Buddhism. There are also minorities of Kirats, Jains, Sikhs, Muslims, Christians, and others. There are many Hindu temples along the Bagmati River, which is considered to be a holy river. Kathmandu is the only place where Buddhism is practiced in Sanskrit. The main ethnic group of Kathmandu comprises the Newar people, the native inhabitants of the Kathmandu Valley. In addition, the city has ethnic minorities from various other districts of Nepal and a mix of caste groups. Although Kathmandu is by far the most developed city in Nepal, with the country's best schools, hospitals, and other institutions, Nepal is a generally a poor country and the city reflects this aspect as well. Many of its residents are impoverished migrants from the countryside who earn meager livelihoods and live in poor conditions.

Further Reading

Rademacher, Anne. *Reigning the River: Urban Ecologies and Political Transformation in Kathmandu*. Durham, NC: Duke University Press, 2011.

Thapa, Rajesh Bahadur, Yuji Murayama and Shailja Ale. "Kathmandu," *Cities: The International Journal of Urban Policy and Planning* 25 (2008): 45–57.

Zurick, David. "Kathmandu" (Kathmandu Valley, Nepal; 1975–2010), *Journal of Cultural Geography* 27, no. 3 (October 2010): 367–78.

KHARTOUM

Khartoum is the capital city of the Republic of Sudan, also called Sudan or North Sudan, a country in the northeast of Africa that borders Egypt to the north and the Red Sea. The city is near the center of the country at the confluence of the White Nile River and the Blue Nile River, from which the main course of the Nile River flows north into Egypt and the Mediterranean Sea. Because of the rivers, the city is divided into three parts and has the nickname "Triangular Capital." The population of Khartoum proper is about 640,000 (2008), while that of the metropolitan area is as large as 5 million because it includes the city's two giant neighbors to the north, Omdurman and Bahri. Omdurman is Sudan's largest city with about 2.5 million inhabitants, while Bahri, also known as Khartoum Bahri or Khartoum North, has over 1 million.

Historical Overview. Khartoum was founded in 1821 by Ibrahim Pasha, the son of the powerful ruler and founder of modern Egypt, Mohammed Ali Pasha who had expanded his realm southward into Sudan. It was established 15 miles (2 km) north of the ruins of the ancient Nile River city of Soba, whose remains were pillaged for construction materials to build Khartoum. The city was at first a military outpost, but it soon evolved into a center of trade and the administrative center for Sudan. Trade in slaves was a principal business in early Khartoum. In 1884–1995, the city was overtaken by troops under the command of Muhammad Ahmad bin Abd Allah, a messianic Islamic leader called a Mahdi, and all of its inhabitants and Anglo-Egyptian defenders were massacred. That incident is known as the Battle of Khartoum or the Siege of Khartoum. British forces under the command of Herbert Kitchener then defeated the Mahdists in a decisive battle at Omdurman north of Khartoum in 1898 and regained the city. The following year, Khartoum became the capital of Anglo-Egyptian Sudan. Under British rule, the rebuilt city was laid out in a grid plan with key diagonal streets that some have observed resembled the pattern of the British Union Jack. Independence for Sudan was achieved in 1956.

In 1973, Khartoum was the site of a terrorist attack by the radical Palestinian paramilitary organization in the Saudi Arabian embassy in which the U.S. ambassador, the U.S. deputy ambassador, and a diplomat from Belgium were killed. From 1991 to 1996 Osama bin Laden lived in the city. In the late 20th century, Khartoum and the towns around it became magnets for refugees from political conflict and environmental disasters in Sudan and neighboring countries, including refugees from a series of civil wars in Sudan and the conflicts in Sudan's Darfur district. On August 20, 1998, a cruise missile attack by the United States destroyed a pharmaceutical factory in Bahri in retaliation for deadly bombings of U.S. embassies in

Tanzania and Kenya that the American government blamed on Al Qaeda. The U.S. position was that the factory was linked to Al Qaeda and manufactured a deadly nerve agent, although those allegations are disputed. In 2011, South Sudan seceded from the rest of the country and became an independent nation.

Major Landmarks. The Corinthia Hotel Khartoum is an architecturally distinctive 5-star hotel in the heart of Khartoum where the White Nile and the Blue Nile come together. It has a curved oval façade that resembles a sailboat, and is sometimes called "Gaddafi's Egg" because the building was funded by the government of Libya. The University of Khartoum was founded in 1902 as Gordon Memorial College and is the oldest university in Sudan. The Souq al Arabi is a large open market in the center of Khartoum. The Great Mosque of Khartoun (Mesjid al-Kabir) is located nearby. Major museums are the Sudan Presidential Palace Museum, the Sudan National Museum, and the Sudan Ethnographic Museum. The ruins of the Al-Shifa pharmaceutical factory in the north of Khartoum have become a tourist attraction.

Culture and Society. The metropolitan area of Khartoum, Omdurman, and Bahri has been growing rapidly because of the many refugees produced by decades of conflicts in Sudan and neighboring countries. The population is highly diverse, as Sudan itself is a nation of 597 tribes and more than 400 languages and dialects. Arabic is the principal common language and is the official language of the country along with English. About 97 percent of the population adheres to Sunni Islam, with Coptic Christians comprising much of the remainder. Because of political instability, civil strife, authoritarian government, and environmental issues such as drought, Sudan is a very poor country. Its human rights record is very poor.

Further Reading

Collins, Robert O. *A History of Modern Sudan.* Cambridge, UK: Cambridge University Press, 2008.

Haywood, Ian. "Khartoum," *Cities: The International Journal of Urban Policy and Planning* 2, no. 3 (1985): 186–97.

Holt, P. M. and M. W. Daly. *A History of the Sudan: From the Coming of Islam to the Present Day.* London: Longman, 2011.

KIEV. *See* Kyiv.

KIGALI

Kigali is the capital and largest city of the Republic of Rwanda, a small country in the interior of Africa just south of the equator. The city is in the very center of Rwanda in an area of hills and valleys. The population of Kigali is just under 1 million.

Historical Overview. Kigali was founded in 1907 when Rwanda was part of German East Africa, a colony comprised of today's Burundi, Rwanda, and the mainland part of Tanzania (then called Tanganyika). It was a commercial and mining center. The German colonial capital was Butare in the south of Rwanda, and the traditional capital where the Royal Palace of Rwanda was located was Nyanza, also in the south. In 1919, after Germany's loss in World War I, the League of Nations assigned the territory of Rwanda and Burundi (Rwanda-Urundi) to Belgium as a mandate territory. The Belgians renamed Butare Astrida in honor of the Belgian Queen Astrid. After Rwanda separated from Burundi and achieved independence in 1962, Kigali was chosen to be the national capital because of its central location.

The struggle for independence in Rwanda was accompanied by the worst conflict and mass killing between the nation's two principal population groups, the Hutu and the Tutsi. The Belgians exacerbated ethnic tensions by specifying Hutu Tutsi as separate ethnic identities on personal identification papers, and by favoring the Hutu as Rwanda prepared for independence. In 1959, the Rwandan Revolution claimed lives on both sides of the conflict, most notably an estimated 20,000–100,000 Tutsis, the main victims of the crisis, and sent as many as 100,000 of the survivors into exile in neighboring countries. That period is also called the Rwandan Wind of Destruction. In 1990, Tutsi refugees in nearby Uganda organized as the Rwandan Patriotic Front (RPF) and invaded the country, starting the Rwandan Civil War. By 1994, the conflict had degenerated into the Rwandan Genocide, in which some 500,000 to 1,000,000 Tutsis were systematically killed. Along with them, the victims included Hutus who opposed the wanton violence and many Twa, a third population group in Rwanda. Much of the conflict was centered on Kigali itself. On July 4, 1995, Tutsi fighters (RPF) won control of Kigali, and by July 18 they controlled the entire country. As a result, some 2 million Hutus fled Rwanda in fear of reprisals by Tutsis for the genocide. A great many of the refugees found themselves in crowded camps in neighboring Zaire (now the Democratic Republic of the Congo) where they faced outbreaks of disease and other hardships. By the end of the 1990s, the conflicts had settled down, and by the 2000s, Rwanda had begun to function more normally with a growing economy and the beginnings of prosperity. After a period of the worst misery, it is now said that Rwanda has had a rebirth.

Major Landmarks. Given the recent tragic history of Rwanda, it is appropriate that the first landmark of the country to be mentioned is the Kigali Genocide Memorial Center. Opened in 2004 on the city of a mass grave of genocide victims, the center is engaged in documentation of the tragedy and public education. There are other museums and memorials in Kigali that are dedicated to the Rwandan Genocide, as well as memorials in cities other than Kigali. A landmark of a different sort is Kigali City Tower, a sleek urban development complex centered on a shopping mall with some 60 retail stores and glass-skinned office tower in the city center near Place de la Constitution, a prominent city square.

Culture and Society. Rwanda has one major ethnic group, the Banyarwanda, within which there are three separate groups, the Hutu, Tutsi, and Twa. The Hutu group comprises the large majority of the national population (about 85%), while the Tutsi group comprises most of the remainder. The Twas form less than 1 percent of the total. The three groups share a common culture and language; therefore, they are not ethnic groups per se but social groups within Rwanda. The language they speak is Kinyarwanda, one of Rwanda's three official languages in addition to English and French. French was introduced to Rwanda by Belgian colonialists; German (from the time of Germany's colonization of the country) has essentially disappeared; while the use of English increased in the late 1990s with the arrival of Tutsi refugees in Rwanda who had spent time in English-speaking Uganda. The majority of Rwandans are Roman Catholics, although their proportion in the population has been declining because of conversions to evangelical Protestantism and to Islam. Rwanda has high birth rates and a demographic structure that is very young.

Further Reading

Kinzer, Stephen. *A Thousand Hills: Rwanda's Rebirth and the Man Who Dreamed It*. Hoboken, NJ: John Wiley & Sons, 2008.

Straus, Scott. *The Order of Genocide: Race, Power, and War in Rwanda*. Ithaca, NY: Cornell University Press, 2006.

KINGSTON

Kingston is the capital and largest city in Jamaica, an island nation in the Caribbean Sea. The city is located on the southeastern coast of the island. Its harbor is protected from the sea by a long sand spit called the Palisadoes on which the city's main airport has been built and at the end of which is the historic down of Port Royal. The urban area consists of an amalgamation Kingston and St. Andrew parishes (local government units) that took place in 1923 and is referred to formally as Greater Kingston, the Corporate Area, or KSA (Kingston—St. Andrew). The population totals about 580,000. Another 250,000 or so residents live near KSA and can also be counted as part of the metropolitan population.

Historical Overview. The origins of Kingston are traced to nearby Port Royal which had been founded in the early 16th century by Spanish settlers. The Spanish had come to the island to look for gold but would end up raising sugarcane using native Arawak Indians as slave laborers. England took control of the area in 1655, and Port Royal became a base for English and Dutch privateers who were hired by England's business interests to attack Spanish ships. In 1692, Port Royal was devastated by a major earthquake, and Kingston was founded as a refuge for survivors. It started to grow quickly after another earthquake disaster in Port Royal in 1703, and became Jamaica's main port and largest city soon thereafter. Settlers arrived from England and prospered in trade and as plantation owners. They imported slaves from Africa as their labor force because the local native population had been decimated by disease. The majority of the population of Kingston and of Jamaica as a whole is descendant from former African slaves. The administrative capital of the English colony, however, was in nearby Spanish Town and not in Kingston, but it was finally moved to the larger city in 1872. A major earthquake devastated the city in 1907 and took nearly 1,000 lives. When Jamaica gained independence in 1962, Kingston became a national capital instead of a colonial capital. In 1966, Kingston hosted the Commonwealth Games.

Major Landmarks. Kingston's modern downtown commercial center is New Kingston in the Uptown section of the city. The Bob Marley Museum in the Trenchtown neighborhood of St. Andrew parish is a popular visitor attraction. It was the home and recording studio of the Jamaican superstar of reggae music, and displays his personal effects and various other memorabilia. The National Gallery of Jamaica displays Jamaican art from historic times to the modern period. The nearby town of Port Royal has been partly restored and beckons visitors for a glimpse at pirate history. Outside of Kingston are Jamaica's Blue Mountains.

Culture and Society. In addition to the African majority population, Kingston has minority East Indian and Chinese ethnic groups, as well as a small Lebanese Christian community and the descendants of European settlers. Most of the population is Protestant Christian. There are also followers of the Rastafari Movement, or Rasta, a new religious movement that took root in Jamaica in the 1930s and worships Haile Selassie I, the former Emperor who ruled Ethiopia in the mid-20th century, as God and the second coming of Jesus Christ. Birth rates are high in Jamaica and most of the city's population is young. Many residents of the city are migrants from the countryside. The city has many poor slum neighborhoods, including Trenchtown, which is closely associated with the history of reggae music. Illegal drugs and drug violence are major problems in Kingston.

Further Reading

Brown-Glaude, Winnifred R. *Higglers in Kingston: Women's Informal Work in Jamaica.* Nashville: Vanderbilt University Press, 2011.
Clarke, Colin G. *Decolonizing the Colonial City: Urbanization and Stratification in Kingston, Jamaica.* Oxford: Oxford University Press, 2006.

KINGSTOWN

Kingstown is the capital and largest city of St. Vincent and the Grenadines, an island country in the Lesser Antilles, an island chain between the Caribbean Sea and the Atlantic Ocean. The country consists of one main island, St. Vincent, that comprises most of the national population, and the northern two-thirds of the Grenadines, a lightly settled chain of small islands to the south. Kingstown is in the south of St. Vincent on Kingstown Bay, and is the island's chief port and business center as well as center of administration. The port ships bananas, coconuts, arrowroot, and other agricultural products to foreign markets, and it is also a base for tourist cruise ships. The city's nickname is "City of Arches" referring to the arches that support the covered sidewalks in the downtown. The population of Kingstown is 25,218 (2005).

Historical Overview. The indigenous people of St. Vincent and the Grenadines were Caribs. French settlers from Martinique gained control of St. Vincent in 1719 and established plantations to grow sugar, coffee, tobacco, and other crops using slaves imported from Africa as labor. The island was ceded to Great Britain in 1763, and was then retaken by France in 1779 until the Treaty of Versailles in 1783 which returned St. Vincent to British control. From 1783 until it was finally put down by military force in 1796, there was rebellion against British rule by Black Caribs (a mix of indigenous people and former African slaves) led by Joseph Chatoyer. Chatoyer was killed on March 14, 1795, during an attack on Kingstown. After the conflict, some 5,000 Black Caribs (also known as Garifuna) were deported from St. Vincent to Roatán, a small island off the coast of Honduras. Slavery was abolished in 1834, after which the plantation labor forces was supplemented by indentured from Portuguese Madeira and from

British South Asia. St. Vincent and the Grenadines gained independence from Great Britain on October 27, 1979.

Major Landmarks. Fort Charlotte was built in 1806 to defend the city from attack by sea and by land by Caribs, and stands on a promontory to the west of the city center. The Botanic Gardens on the city's north side date back to 1765 when they were established as a plant breeding center to provide medicines for British troops and settlers. The two main churches are St. Mary's Roman Catholic Cathedral of the Assumption, a rather eccentric building architecturally, and St. George's Anglican Cathedral.

Culture and Society. About two-thirds of the population of St. Vincent and the Grenadines is black and about one-fifth is of mixed race. East Indians account for about 6 percent, whites (mostly British and Portuguese) 5 percent, and Carib Amerindians about 2 percent. The official language is English, but most Vincentians, as citizens of the country are known, speak a creole language known as Vincentian Creole. The most popular sports in St. Vincent and the Grenadines are cricket, football (soccer), and rugby. Parts of the popular Hollywood film *Pirates of the Caribbean* were filmed in Kingstown.

Further Reading

Child, Vivian., *City of Arches: Memories of an Island Capital, Kingstown, St. Vincent & the Grenadines*. North York, Ontario: Cybercom, 2004.
Sutty, Lesley. *St. Vincent and the Grenadines: An Introduction and Guide*. Oxford: Macmillan Education, 1997.

KINSHASA

Kinshasa is the capital and largest city of the Democratic Republic of the Congo, a large country in central Africa with a narrow strip of territory that reaches the Atlantic Ocean along the lower reaches of the Congo River. The country is often called DR Congo, Congo-Kinshasa, and DRC to distinguish it from its neighbor, the Republic of the Congo (Congo-Brazzaville). Its previous names have been Congo Free State, Belgian Congo, Congo-Léopoldville, Congo-Kinshasa, and Zaire. The city was originally named Léopoldville in honor of Belgian King Leopold II.

Kinshasa is located in the west of the country on the south bank of the Congo River directly opposite Brazzaville, the capital of the Republic of the Congo. The wide stretch of the river between the two capitals is called Malebo Pool. The city is about 190 miles (300 km) above the series of rapids of the lower Congo that are collectively referred to as Livingstone Falls. The population of the Kinshasa metropolitan area is almost 10 million, by some definitions the largest urban agglomeration in Africa, and certainly one of the top three along with Cairo in Egypt and Lagos, Nigeria.

Historical Overview. Although the shores of the Congo River where Kinshasa is situated have been inhabited for hundreds of years, the city itself is fairly recent and was founded in 1881 by Welsh explorer Henry Morton Stanley. He named it Léopoldville in honor of the Belgian king who controlled the vast area that is now the Democratic Republic of the Congo as a personal possession. It became a major river port and gateway into the interior of Africa from the start, as the middle section of the Congo is navigable even for today's large river barges from Kinshasa as far inland as Stanley Falls and the city of Kisangani (formerly Stanleyville), a distance of some 1,000 miles (1,600 km). For much of the 20th century, river steamers plied this stretch of the Congo and many of its tributaries. Kinshasa's ocean port is Matadi. At first, goods being transported between Kinshasa and Matadi had to be transported around Livingstone Falls on foot by porters, but in 1898 a railway was built to replace the portage and Kinshasa's economic advantage was greatly enhanced. In 1908, the Congo Free State, as Leopold II's personal territory was known, was transferred to Belgium and became the Belgian Congo. Leopold made an enormous personal fortune from rubber and other resources in the Congo, but his rule (1865–1910) was exceptionally cruel and bloody, and may have resulted in as many as 10 million Congolese deaths (about one-half the total population of the time) as a consequence of forced labor and abject poverty. In 1920, Leopoldville, as the city was still named at the time, was made the capital of the Belgian Congo, replacing the river-mouth city of Boma.

Independence for the Congo was achieved in 1960, but was followed by internal conflict that resulted in the separation of the former colony into two countries, and further conflict within what became the Democratic Republic of the Congo over control of government and the military, valuable natural resources, and foreign intervention. From 1965 to 1997, DR Congo was ruled by strongman Mobutu Sese Seko with support from the United States because of his strong anti-Communist stance, and from 1971 until the end of his rule the country was named Zaire as part of Mobutu's campaign to Africanize place names. It is he who changed the name of Leopoldville to Kinshasa. His rule was a kleptocracy that plunged a country rich in natural resources into even greater poverty. As a result, the city has a very high crime rate and many street children. In 1974, Kinshasa hosted the world heavyweight championship boxing match between Muhammad Ali and George Forman that was promoted as "The Rumble in the Jungle."

Major Landmarks. For a city so big, Kinshasa has comparatively few landmarks to show off. Nevertheless, there is a monument to Congolese independence leader Patrice Lumumba, the University of Kinshasa, the Kinshasa Museum, and the Kinshasa Fine Arts Academy. The Stadium of the Martyrs is the national stadium. The main commercial street of the city is Boulevard of June 30th. Lola ya Bonobo, which is the world's only sanctuary for orphaned bonobos, was opened in 1994 in a forested area outside Kinshasa.

Culture and Society. The population of Kinshasa includes a great many Congolese ethnic groups and languages, the latter including Lingala, Kikongo, Swahili, and Tshiluba. French is the official language of the country, which in effect makes Kinshasa the world's largest Francophone city, although only some of the city's residents actually speak the language and Lingala is most widely spoken. About 96 percent of the Congolese population is Christian. Despite crime and poverty, Kinshasa is becoming a thriving commercial and industrial center. There is currently considerable construction underway in the city that is sponsored by Chinese companies and investors. Kinshasa is the principal cultural and intellectual center of central Africa and has an especially lively music scene and galleries that sell outstanding locally made art.

Further Reading

Freund, Bill. "City and Nation in an African Context: National Identity in Kinshasa," *Journal of Urban History* 38, no. 5 (2012): 901–13.
Hochschild, Adam. *King Leopold's Ghost: A Story of Greed, Terror, and Heroism in Central Africa*. New York: Mariner, 1999.
Plissart, Marie-Françoise and Filip De Boeck. *Kinshasa: Tales of the Invisible City*. Antwerp, Belgium: Ludion, 2005.
Wright, Herbert. *Instant Cities,* 40–43. London: Black Dog Publishers, 2008.

KUALA LUMPUR

Kuala Lumpur is the capital and largest city of Malaysia, a country in Southeast Asia that occupies the southern tip of the Malay Peninsula and part of the north of the island of Borneo. The city is on the Malay Peninsula, where most Malaysians live. It is in the west-center of that part of the country, where the Klang and Gombak rivers come together, in a broad valley called the Klang Valley. The meaning of "Kuala Lumpur" is "muddy confluence" in the Malay language. The city has a population of about 1.6 million people and is the focus of a huge and fast-growing metropolitan area with a population of about 7.3 million. Since 1999, national government functions have been shared between Kuala Lumpur and the planned new city of Putrajaya located about 16 miles (25 km) to the south. Putrajaya has the executive branch of Malaysia's government and most of the judiciary, but Kuala Lumpur retains some of the judiciary, houses the Parliament, and is the seat of the King of Malaysia. Putrajaya is promoted as a "Garden City" and an "Intelligent City," and is part of a much larger development of new cities and high-technology outside Kuala Lumpur called the Multimedia Super Corridor. Kuala Lumpur is often referred to as simply "KL."

Historical Overview. Much of what is Malaysia today was part of the British Empire from the establishment of the Straits Settlements in 1826. The origin of Kuala Lumpur is traced to Chinese tin miners who settled in the area in the 1850s and opened a trading post at the muddy confluence of the Klang and Gombak Rivers. There was considerable conflict between Chinese miners for control of the tin business until the 1870s when Yap Ah Loy, the Chinese leader of Kuala Lumpur restored order and began to develop the rowdy settlement into a proper city. In 1880, Kuala Lumpur replaced Klang as the capital of Selangor State. Frank Swettenham, the British Resident of Selangor from 1882, required that the city be built in brick and tile rather than wood and thatch as a precaution against fire and malaria. In 1896, Kuala Lumpur became the capital of the Federated Malay States. From 1942 to 1945 it was occupied by the Japanese in World War II. After the war a Communist insurgency was put down in Malaysia. The country achieved independence from Great Britain in 1957, with Kuala Lumpur remaining as capital. On May 13, 1969, there was rioting in the city between the Malay and Chinese ethnic groups, resulting in controversial agreements called the New Economic Policy about the distribution of political power and economic opportunity across ethnic lines. The economy of Malaysia grew rapidly during the administration of Prime Minister Tun Dr. Mahathir bin Mohamad (1981–2003), and Kuala Lumpur was greatly developed with new tall buildings and monuments, and many new residential and commercial areas. Putrajaya was also developed at this time.

Major Landmarks. Jamek Mosque is at the confluence of the Klang and Gombok Rivers. Behind it are the modern skyscrapers of downtown Kuala Lumpur. The National Mosque (Masjid Negara) was built in 1965 and reflects the bold ambitions of the country after independence. The most iconic buildings in the city are the Petronas Towers, two twin office towers that were the tallest buildings in the world from 1998 to 2004. Both buildings have 88 floors above ground and a height of 1,242 ft (379 m). Other landmarks are Kuala Lumpur Tower, the TM Tower, the historic Kuala Lumpur central train station, the former offices of the British Colonial Secretariat, the Royal Selangor Club, and Merdeka Square (Independence Square). The city also has a spectacular new airport, many beautiful ultramodern shopping malls, and a landmark Chinatown neighborhood centered on busy Petaling Street.

Culture and Society. Kuala Lumpur has a fast-growing middle class and a rising consumer culture. Conflicts simmer between the more prosperous Chinese population of the city and Malays, the indigenous ethnic group. Malays constitute 44.2 percent of the population, Chinese 43.2 percent, and Indians 10.3 percent. There are also foreign low-skilled workers from Indonesia, Myanmar, Nepal, Bangladesh, and the Philippines in the city who immigrated because of jobs. Islam is the main religion of the country among the Malay population. Some believers want Malaysia to become

Putra Mosque in Putra Jaya, Malaysia's new government center on the outskirts of Kuala Lumpur. (Szefei/Dreamstime.com)

an Islamic state and adhere strictly to Islamic law. Most Chinese are Buddhists and most Indians practice Hinduism. The Batu Caves outside Kuala Lumpur are an important religious shrine for Hindus, as well as an interesting landmark for tourists. The principal Language is Bahasa Malaysia (Malay), but other languages such as Chinese and various Indian languages are also spoken. English is widely understood.

Further Reading

Bunnell, T., P.A. Barter, and S. Morshidi. "Kuala Lumpur Metropolitan Area: A Globalizing City-Region," *Cities: The International Journal of Urban Policy and Planning* 19, no. 5 (2002): 357–70.

Goh, Benh-Lan and David Liauw. "Post-Colonial Projects of a National Culture: Kuala Lumpur and Putra Jaya." *City: Analysis of Urban Trends, Culture, Theory, Policy, Action* 13, no. 1 (2009): 71–79.

Sardar, Ziauddin. *The Consumption of Kuala Lumpur*. London: Reaktion Books, 2000.

KUWAIT CITY

Kuwait City is the capital and largest city of the State of Kuwait, commonly called Kuwait, a sovereign Arab state in western Asia in the northeastern part of the Arabian Peninsula and at the northwestern head of the Persian Gulf. The city and the country are rich with oil, Kuwait having the fifth largest petroleum reserves in the world. The country ranks fifth globally in income per capita. Kuwait City is located in the heart of the country on the southern shore of Kuwait Bay, a natural deepwater harbor in the Persian Gulf. In addition to being the national capital, the city is a major port, financial center, and regional hub for international tourism and shopping. The population of Kuwait City is about 2.4 million, while that of the State of Kuwait as a whole is about 3.6 million.

Historical Overview. The history of Kuwait City is traced to the early 18th century when the members of the Al-Sabāh clan migrated from what is today Saudi Arabia and settled there. In 1756, Sabah I bin Jaber became the first Emir of Kuwait. The House of Sabah continues to be the ruling family of Kuwait. The settlement prospered as a port and trading center. Its pearl industry was the largest in the world until Japan introduced pearl farming. The first walls of Kuwait City were built in 1760. In the 19th century, Kuwait resisted annexation by the Ottoman Empire, entering into a treaty in 1899 with the United Kingdom that provided military protection. Oil was discovered in 1936, transforming the economy and attracting foreign investment and foreign workers. By 1952, Kuwait became the largest exporter of oil in the Persian Gulf region. Its living standards were greatly elevated as a result.

Kuwait became an independent state on June 19, 1961. The country was a major financial supporter of Iraq during the 1980–1988 Iran–Iraq War. Afterwards Kuwait and Iraq entered into disputes about the Iraq's financial debt to Kuwait, about the rate at which Kuwait was producing oil, and about slant drilling by Kuwait to withdraw oil from under Iraqi territory in the Rumaila Oil Field. As a result, Iraq, led by Saddam Hussein invaded and annexed Kuwait on August 2, 1990. This led the United States, led by President George H.W. Bush, and a coalition of other nations to launch Operation Desert Storm to liberate Kuwait. The war lasted until February 28, 1991, and resulted in a convincing military victory for the American side. Kuwait paid $17 billion to the coalition forces for their efforts. Kuwait City was heavily damaged during the Iraqi occupation and the fighting that followed, and was subsequently rebuilt.

Major Landmarks. The most iconic landmark of Kuwait City is Kuwait Towers, a complex of three distinctive towers. The tallest is 583 ft high (178 m) and is a major tourist attraction with two spheres, one higher than the other, that offer great views of the city and various restaurants. A second tower with a single sphere is 270 ft high (82 m) and is for water storage. The third tower is a spike that provides lighting for the other two. Another icon of the city is Liberation Tower, one of the tallest telecommunications towers in the world. Other landmarks in Kuwait City are the Grand Mosque, the War Museum, the National Museum, and Seif Palace. The city's fish market is also a major landmark. In addition, Kuwait City has numerous popular shopping malls such as Souk Sharq, Marina Mall, and The Avenues, as well as traditional markets such as Souk Al Mubarakiya. There are also family-fun amusement parks such as Kuwait Aqua Park.

Culture and Society. Kuwait is an Islamic society, with about 85 percent of the population Muslim. About 60 percent of the Muslims are Sunni and 40 percent Shi'as. Arabic is the official language, with the Gulf dialect of Arabic being the most common colloquial dialect. English is widely spoken too, especially in Kuwait City where it is a major language for business. The population of Kuwait is international, and Kuwaiti citizens are a minority in their own country. This is because the country has many foreign workers, including those who have come to the country to do the jobs of hard labor and in service economy, as well as many other foreigners who are employed as engineers and executives in the oil industry, or in finance, trade, and other businesses. In 2008, 68.4 percent of the population of Kuwait (the country) consisted of expatriates. About 57 percent of the population is Arab and 39 percent area from South or East Asia. Indians are the most numerous foreign group, followed by Egyptians, Syrians, and Iranians. Much of the Kuwaiti national population lives very comfortably, as the country is quite wealthy, while many of the foreign workers from poor countries live in crowded housing and enjoy few of the country's benefits. In recent years, many Kuwaiti women have spoken out in favor of equal rights for women in a society that is male-dominant and bound by old traditions.

Further Reading

Al-Mughni, Haya. *Women in Kuwait: The Politics of Gender*. London: Saqi Books, 2000.

Casey, Michael S. *The History of Kuwait*. Westport, CT: Greenwood Press, 2007.

Haywood, Ian. "Kuwait," *Cities: The International Journal of Urban Policy and Planning* 6, no. 1 (1989): 2–16.

KYIV

Kyiv is the capital and largest city in Ukraine, the largest country wholly within the confines of Europe. The country is in the east-center of the continent, and the city is located in the north-center of Ukraine on both banks of the Dnipro River that divides Kyiv into left bank and right bank districts. The historic core of Kyiv is high on bluffs overlooking the wide river on the right bank, while the left bank is mostly a residential zone that was first urbanized in the late 20th century. In addition to being the capital, Kyiv is the main business and financial center of Ukraine, an important center of the arts, education, and media, and a growing center for foreign tourism attracted by the city's charms and rich history. The city is making a transition from urban patterns of the socialist-Soviet period to those of a capitalism-based postsocialist society. Because the Ukrainian land is rich with potential and the city is rich with history and blessed with beautiful parks and riverbanks in the very center, Kyiv could become one of the great capital cities of Europe. At present, however, it suffers from too much deteriorating housing, infrastructure and unplanned urban development that damages environment and historical urban fabric. The population of Kyiv is approximately 2.8 million, while that of the metropolitan area approaches 3.7 million (2010). The old spelling for the city "Kiev" comes from the Russian language and is now out of date.

Historical Overview. According to a legend described in a document that was written in Kyiv in about 1113 called the *Primary Chronicle*, Kyiv was founded by the leader of a Slavic tribe who was named Kyi, after whom the city is named, and his younger brothers Schek and Khoriv, and their sister Lybid. It started to become a city in the ninth century, and from the late ninth century to the 13th century was capital of a large and prosperous princely state Kyivan Rus' that covered most today's Ukraine, Belarus, the historic heart of Russia, and the far north. The city was at the center of trade routes between the Baltic and Black Seas, and during its zenith in the 11th century, was among the largest cities in Europe with 50,000 inhabitants. In 988, the country was Christianized with a mass baptism in a tributary of the Dnipro in Kyiv by Prince Volodymyr the Great, one of several enlightened rulers of Rus'. In 1240, the city was destroyed by a Mongol invasion. The empire of Rus' disintegrated, but is considered the antecedent of the Ukrainian state and also of Russia to the north. At various times afterward, Kyiv was ruled by the Grand Duchy of Lithuania and the Polish-Lithuanian Commonwealth, and in the mid-17th century became part of the Russian Empire. By that time, Kyiv had emerged as a major center of the Orthodox faith and education, and hub of trade in the heart of the fertile farmland of Ukraine. The famous Kyiv Mohyla Academy, an early center of learning, was founded in the city in 1632. The city industrialized in the 19th century. Sugar refining and grain milling were especially important and was the basis for wealth that funded sumptuous mansions for industrialist-magnates and their families and great public institutions for the city. From 1917 to 1921, Kyiv was the capital of the independent Ukrainian National Republic.

Kyiv was capital of the Ukrainian Soviet Socialist Republic from 1934 until the demise of the Soviet Union in 1991 (prior to then, the capital of Soviet Ukraine was Kharkiv). Industrialization continued during Soviet rule, as did Kyiv's role as a center of education and culture. However, religion was all but destroyed and a great many of Kyiv's beautiful historical churches were looted and then dynamited, or were converted to secular uses. Most of the city's synagogues were also closed. In 1932–1933, Kyiv and much of the rest of Ukraine, as well as agricultural areas in adjacent parts of Russia, experienced a manmade famine that took as many as 10 million lives. Soviet authorities engineered this crime during the collectivization of agriculture in order to eradicate private land ownership and very deliberately as documented with historic evidence, to suppress Ukrainian national identity and the threats that it might pose to Russian-Soviet rule. During World War II, Kyiv was occupied by the Nazis and sustained enormous damage and losses of life. The city was later awarded the title "Hero City" by Soviet authorities. The catastrophic Chernobyl nuclear disaster took place north of Kyiv on August 26, 1986, and contaminated the city and nearby areas with radiation. On August 24, 1991, Ukraine declared its independence from the Soviet Union, and Kyiv became the capital of an independent nation.

Major Landmarks. The center of Kyiv is the Independence Square that is marked by a towering Independence Monument where a large statue of Vladimir Lenin once stood, and the main street is an iconic street named Khreschatik. There are two UNESCO World Heritage Sites in the city, the impressive Pecherska Lavra religious complex (Monastery of the Caves) overlooking the Dnipro River, and the beautiful St. Sophia Cathedral complex whose history dates back to the old walled city of Prince Yaroslav the Wise who ruled Kyivan Rus from 1019 to 1054. There are also historic religious sites in the Podil neighborhood at river level. The district was also known for commerce and trade, and is the city of the modern-day National University of Kyiv Mohyla Academy. Other major landmarks in Kyiv include the Kyiv Opera, the statue of Prince Volodymyr the Great overlooking the Dnipro, the Museum of the Great Patriotic War, the statue of poet Taras Shevchenko opposite the main building of the national university named after him, and the Holodomor (artificial famine of 1932–1933) Monument. Babyn Yar is an important monument to the Jews and Ukrainians who were murdered in Kyiv by the Nazis. Andriyivskyi Uzviz is a funky street of galleries, cafés, museums, and vendors of art, handicrafts, and trinkets for tourists that winds its way from the historic upper city to historic Podil below. The center of Kyiv has many wonderful parks and many miles of beaches along the river.

Kyiv: Punctuations on the Bluffs

The high bluffs on the right bank of Kyiv overlooking the Dnipro River have long been a place for those in power to punctuate their authority through architecture. In the 10th century, after Prince Volodymyr the Great had his subjects baptized into the Orthodox faith, he capped the bluffs with golden-domed churches that echoed the architecture of Byzantium and its spectacular Hagia Sophia. Centuries later, other rulers and prominent citizens rebuilt churches that the Mongols had destroyed and erected beautiful new churches in the Baroque style to show their wealth and be right with God. In the 1930s, during the Communist era, many historic churches were destroyed and symbols of Communist Party authority were put in their place. Later, the Soviet government erected an enormous monument to the Great Patriot War against the Germans and another, much hated, monument that symbolized the bonds of Ukraine to Russia. After independence in 1991, new monuments were erected. Those same bluffs are now also capped with expensive new high-rise apartment buildings for the post-Communist rich. Many Kyivans call these buildings monsters because they have illegally taken parkland from the city.

Pecherska Lavra, the historic Monastery of the Caves, on the bluffs of Kyiv above the Dnipro River. It is a UNESCO World Heritage Site. (Photo courtesy of Roman Cybriwsky)

Culture and Society. The official language of independent Ukraine is Ukrainian, although Russian is still widely spoken in Kyiv as a vestige of centuries of colonialism. Many Kyivans also speak *surzhik*, a blend that mixes Russian and Ukrainian. The Ukrainian presence in the city has been increasing since 1991. Also since independence, many churches, synagogues, and mosques have been constructed in Kyiv. Kyiv once had a very large and influential Jewish minority, but emigration beginning in the late 19th century and then the murderous Holocaust of World War II greatly reduced the Jewish population. Today's Kyiv shows widened gaps between newly rich residents and the majority of the population who struggle to make a living in a competitive and often corrupt economy. There are many sex tourists who come to Kyiv from Turkey, the Middle East, and other countries to take advantage of poverty in Ukraine and the reputed beauty of its young women. The activist women's organization FEMEN has been a leader in combating this plague. The organization Save Old Kyiv is among those that have battled against illegal urban development in the city and the destruction of historic architecture by builders.

Further Reading

Hamm, Michael F. *Kiev: A Portrait, 1800–1917.* Princeton, NJ: Princeton University Press, 1993.

Meir, Natan M. *Kiev: Jewish Metropolis: A History, 1859–1914.* Bloomington: Indiana University Press, 2010.

LA PAZ

La Paz is the second-largest city after Santa Cruz de la Sierra and is the de facto capital of Bolivia (official name: the Plurinational State of Bolivia), a landlocked country in central South America. The official capital of the country is Sucre, located about 250 miles (402 km) southeast of La Paz, but La Paz has more offices of national government and is, therefore, considered to be the country's "other" capital. La Paz is built in a bowl amidst steep hills in the western part of Bolivia in La Paz Department and has an elevation that ranges from roughly 9,800 to 13,500 ft (3,000 to 4,100 m), making the city the world's highest capital city. The population of La Paz was officially last counted in 2001 at 877,363. It is undoubtedly larger now, possibly over 1 million in population. The La Paz metropolitan area, which includes the cities of El Alto and Viacha in addition to La Paz, has a population of at least 2.4 million.

Historical Overview. La Paz was founded by Spanish conquistador Captain Alonso de Mendoza in 1548 on the site of a native Inca settlement that was called Laja. It was given the name Nuestra Señora de La Paz, meaning "Our Lady of Peace." It was laid out in 1559 by Juan Gutierrez Paniagua in the standard Spanish colonial style, notably a grid plan for streets, a central public square surrounded by the main church and government and military buildings, and other public squares in the grid. In 1781, a group of indigenous Aymara people under the leadership of Tupac Katari rebelled against Spanish rule and laid siege to La Paz for six months, destroying churches and government buildings in their fight for independence. Another native rebellion took place in 1811. Independence came in 1825. At that time, the city was renamed La Paz de Ayacucho in commemoration of a decisive battle that resulted in Bolivia's independence. The city was made the de facto capital of Bolivia in 1898, shifting many of the functions of government from the historic capital of Sucre. In 1998, La Paz was devastated by a major earthquake that took hundreds of lives and destroyed thousands of buildings.

Major Landmarks. The heart of La Paz is Plaza Murillo. Named for Pedro Murillo, one of the heroes of Bolivia's fight for independence, the square was originally named Plaza Mayor and La Plaza de los Españoles, and then Plaza de Armas. It is the central square of the historic Spanish city and is flanked by prominent buildings that include the Presidential Palace, the National Congress of Bolivia, and the Cathedral of La Paz. The former home of Pedro Domingo Murillo is now one of the many museums in the city. Other museums are Museum San Francisco, the National Museum of Archeology, the Ethnography and Folklore Museum, the Museum of Precious Metals, the Museum of Musical Instruments, and the Coca Museum. The main tourist strip is the Sagarnaga Street. Also popular with tourists is the nearby Witches' Market, which is more-or-less what it sounds like. Valle de Luna is a park of spectacular moon-like weathered rock formations outside the city.

Culture and Society. The population of Bolivia is diverse ethnically and includes approximately three dozen native Indian groups that constitute about 55 percent of the nation's people. The most numerous among the natives are the Quechuas and Aymaras. Mestizos (mixed and Amerindian and white) make up about 25 percent of the population and whites about 20 percent. The white population is made up mainly of *criollos*, the descendants of early Spanish colonists. They continue to make up much of the aristocracy of the country. Bolivia as a whole is quite poor, and many of the Amerindians in La Paz and other cities live in squatter communities. The official languages of Bolivia are Spanish, Quechua, and Aymara, plus 34 other native languages. About three-quarters of Bolivians are Roman Catholics. Protestantism and traditional Inca religious practices are on the rise. The culture of Bolivia is distinguished by the folklore of its native people, including textiles and other crafts, and a rich musical tradition.

See also: Sucre.

Further Reading

Arbona, Juan M. and Benjamin Kohl. "La Paz—El Alto," *Cities: The International Journal of Urban Policy and Planning* 21, no. 3 (2004): 255–65.

Kohl, Benjamin and Linda Farthing. *Impasse in Bolivia: Neoliberal Hegemony and Popular Resistance.* London: Zed Books, 2006.

Lazar, Sian. *El Alto: Rebel City: Self and Citizenship in Andean Bolivia.* Durham: Duke University Press, 2008.

LIBREVILLE

Libreville is the capital and largest city in Gabon, a small country in west-central Africa that faces the Gulf of Guinea of the Atlantic Ocean. The equator runs through the center of the country, and Libreville itself is just north of the equator at 0°23′ N latitude. The city is in northwest Gabon at the mouth of the Komo River and is Gabon's principal port as well as center of government. Export of oil and timber are important elements of the city's economy. There is uncertainty about the oil economy in the future because of dwindling resources. The city also has sawmills and a shipbuilding industry. The population of Libreville was 578,156 in 2005, more than one-third that of Gabon as a whole.

Historical Overview. Libreville was founded in 1848 by slaves who were freed from a Brazilian-Portuguese slave ship that had been captured by the French navy. They named the city Libreville, which is French for "Freetown," in imitation of Freetown in Sierra Leone that had been founded years earlier by freed slaves from the United States. Gabon was colonized by the French and was part of the federation called French Equatorial Africa. Libreville was the colony's main port. In 1940, the city was the site of the Battle of Gabon in which Free French forces led by General Charles de Gaulle defeated the forces of Vichy France, an ally of Nazi Germany, and gained control of France's colonies in Africa. Independence from France was achieved on August 17, 1960.

Léon M'Ba was elected the first president of Gabon in 1961. He ruled auto-cratically with support from France that maintained large business interests in the country, dissolved the National Assembly, banned rival political parties, and sup-pressed freedoms of expression. When he died in 1967, Omar Bongo Ondimba succeeded him as president for multiple terms until his death in 2009. His 42 years in office made Bongo the long-serving ruler of any independent African nation. He was also unpopular and continually had to maneuver to win elections and stay in power. Elites in Gabon prospered because of the nation's oil wealth, but it is argu-able whether much progress was made during Bongo's years to improve the lives of most Gabonese. Bongo's son Ali Bongo was elected to the presidency in 2009.

Major Landmarks. Notable landmarks in Libreville include the National Museum of Arts and Traditions, St. Marie's Cathedral, and Omar Bongo University, as well as Gabon's government buildings. There is a row of statues along the city's seafront. The main boulevard along the waterfront, *Boulevard de bord de mer*, is increasingly lined with expensive stores and restaurants, while behind it the central city's Louis quarter is popular for nightclubs and restaurants. The main markets of the city are Marché Artisanal and Marché Mont-Bouet. Arboretun de Sibang offers a unique excursion into an equatorial jungle. Akanda National Park is located just outside the city. Pointe Denis is a beach resort island just a short boat ride away from central Libreville.

Culture and Society. Gabonese people are mostly of Bantu origin and from at least 40 different ethnic and linguistic groups. The Fang and the Nzebi are the two most populous groups. French is the common language. Some 80 percent of Gabonese can speak French and 30 percent of the population of Libreville are native speakers. Most people are Roman Catholics or identify with a Protestant religious group. Gabon has a rich tradition of folklore and mythology that has been passed down from generation to generation by story tellers. A special art form for which Gabon is known is beautifully decorated masks made from rare woods that are worn at weddings, births, and funerals.

Further Reading

Gardinier, David E. *Historical Dictionary of Gabon*. Lanham, MD: Scarecrow Press, 2006.
Rich, Jeremy. *A Workman is Worthy of His Meat: Food and Colonialism in the Gabon Estuary*. Lincoln: University of Nebraska Press, 2009.
Yates, Douglas Andrew. *The Rentier State in Africa: Oil Dependency and Neocolonialism in the Republic of Gabon*. Trenton, NJ: Africa World Press, 1996.

LILONGWE

Lilongwe is the capital and largest city of Malawi, a small landlocked country in southeastern Africa that is elongated north-south along the shores of Lake Malawi that forms much of the country's eastern border. The capital city is on the Lilongwe River near the center of the country, close to its borders with Zambia and Mozam-bique, on the country's main north-south highway, M1. Its population is approxi-mately 781,000 (2012). Although Lilongwe is the political capital, the financial and commercial hub of Malawi is Blantyre, a city of about 732,000 inhabitants in the southern part of the country. The two cities are closely linked by frequent travel, as well as via telecommunications and electronically. Some sources list Blantyre as Malawi's largest city, the discrepancy coming largely because of lack of accurate population data.

Historical Overview. Lilongwe was a small village until the early 20th century when it became an administrative center in Great Britain's colony, the Nyasaland Protectorate. The country's name was changed to Malawi when independence was achieved in 1964. Zomba, in the south of the country, was the first capital. It had earlier been capital of British Central Africa and then Nyasaland. Lilongwe became Malawi's capital in 1975. Hastings Banda was a strong-armed president from 1964 until he was deposed in 1994. The city grew rapidly after it became capital, and is now the headquarters for many NGOs, United Nations–aided organizations, and the foreign aid programs of the United Kingdom and the United States that are active in Malawi. The country is one of the world's least developed nations and receives considerable assistance from abroad.

Major Landmarks. Lilongwe is not noted for landmarks or tourist attractions, although it does have the Lilongwe Wildlife Centre between Old Town and New Town, and the Kumbali Cultural Centre with performances of Malawian dancing and drum music. There are government buildings, other offices, and hotels in New Town, while the Old Town Mall is a comparatively upscale shopping area in Old Town.

Culture and Society. Malawi is a very poor country, and most of Lilongwe's Malawian citizens are poor as well. Foreign aid workers, diplomats, and other expatriate residents tend to be at the upper end of the city's income spectrum. The Malawian population is of many tribal backgrounds and languages, with various artistic expressions through dance and music. The official languages are English and Chichewa. About 80 percent of the population is Christian, with Roman Catholics, the Church of Central Africa, and Presbyterians being most numerous, and about 13 percent is Moslem. Infant mortality rates are high in Malawi and life expectancy low. There is a high rate of HIV/AIDS infections among the adult population.

Further Reading

Harrigan, Jane. *From Dictatorship to Democracy: Economic Policy in Malawi, 1964–2000*. Aldershot, UK: Ashgate, 2001.

Kalinga, Owen J. M. *Historical Dictionary of Malawi*. Lanham, MD: Scarecrow Press, 2001.

Power, Joey. *Political Culture and Nationalism in Malawi: Building Kwacha*. Rochester, NY: University of Rochester Press, 2010.

LIMA

Lima is the capital and largest city in Peru, a country on the Pacific Coast of South America. It is located in the valley of the Rimac River about midway along the Peruvian coast, a short distance inland from the sea. The neighboring city of Callao serves as Lima's port. The climate of the region is dry, and Lima is the second-largest city in the world that is located in a desert, the largest being Cairo in Egypt. The population of Lima is approximately 9 million, nearly one-third of Peru as a whole. The city is fourth in population in South America after Sao Paulo (Brazil), Buenos

Aires (Argentina), and Rio de Janeiro (Brazil). The city is an active earthquake zone and has periodically suffered severe damage and loss of life, including in the years 1746, 1940, and October 2011.

Historical Overview. The site of Lima was initially settled by native Inca people, but the city itself was founded after the 1532 bloody conquest of the Inca Empire by Spanish conquistador Francisco Pizarro. In 1535, he chose the site to be his capital, which he named Ciudad de los Reyes (City of the Kings) in honor of the Roman Catholic feast of the Epiphany. The name Lima came into use a short time later. It is from native origin, but the precise meaning is contested. In 1543, the city was designated as the capital of the Viceroyalty of Peru. And because of Spanish plunder of the continent's wealth it grew as a trade center for Spanish settlement in South America. Because of pirate attacks, defensive walls were constructed between 1684 and 1687 to enclose the city. The form of the city was typical of Spanish urbanism in the New World: a grid pattern street plan and a square plaza in the urban center surrounded by a church or cathedral and various government buildings.

Peru declared independence from Spain in 1821, after which Lima became a national capital. The city prospered after 1850 with export of guano resources. There was war with neighboring Chile between 1879 and 1883, during which time Lima was occupied and ransacked by Chilean troops. In the 20th century, Lima developed even more into a center of trade and finance, as well as a center of industrial production. Many migrants came to the city from the Andean Mountains and developed squatter communities called *pueblos jóvenes* at the city's margins. A prominent example is Villa El Salvador that grew up in the 1970s in a desert just south of the city.

Major Landmarks. The historic core of Lima is a UNESCO World Heritage Site and centers on Plaza Mayor, the Lima Cathedral, and Peru's historic government buildings. The coffin of Pizarro is on display in the Cathedral. The Church of San Francisco is a beautiful historic church in the center of the city with catacombs underneath. Important museums include the National Museum of the Archaeology, Anthropology and History of Peru, the Museum of Natural History, the Museum of the Nation, and the Museum of Art of Lima. There is also the Larco Museum, which is a privately owned museum of pre-Columbian art, and the Museum of Gold. Larcomar is a popular shopping mall built atop a high bluff overlooking the city's beaches and the Pacific Ocean. In 2011, a 121-ft-high (37 m) statue of Jesus Christ called "the Christ of the Pacific" was unveiled in Lima. It was inspired by the famous Christ the Redeemer Statue in Rio de Janeiro, and has caused considerable controversy because of expenditure of public funds to construct a religious symbol.

Culture and Society. The city of Lima is sharply divided between haves and have-nots, with upscale districts near the beaches such as Miraflores and San Isidro contrasting with the poverty of migrant squatter settlements at the city's margins. The privileged neighborhoods are comprised mostly of European Peruvians and *mestizo* population (mixed Amerindian and European), while the poorest areas are mostly Amerindian (Aymara and Quechua). The rich people's neighborhoods are all heavily fortified with security guards, high walls around private homes, and surveillance cameras. The city also has minority groups of Afro-Peruvians, Japanese Peruvians, and Chinese Peruvians. All three groups descend from earlier labor migrations from abroad. Lima has the largest Chinese-origin population in Latin America and a booming Chinatown district called Calle Capon. The main faith of the city is Roman Catholicism. There are many churches and religious festivals in the city.

Further Reading

Gandolfo, Daniella. *The City at Its Limits: Taboo, Transgression, and Urban Renewal in Lima.* Chicago: University of Chicago Press, 2009.
Higgings, James. *Lima: A Cultural History.* Oxford: Oxford University Press, 2005.
Leonard, John B. "Lima," *Cities: The International Journal of Urban Policy and Planning* 17, no. 6 (2000): 433–45.

LISBON

Lisbon is the capital and largest city of the Portuguese Republic, better known as Portugal, a country on the Iberian Peninsula at the western-most end of the European continent. The city is located on the Atlantic Ocean coast of the country at the mouth of the Tagus River. Like a handful of other famous cities in the world, Lisbon is said to be built on seven hills. The population of Lisbon is about 548,000 (2011), while that of the metropolitan area is approximately 3 million. In addition to being the administrative center of Portugal, Lisbon is a significant center of trade and commerce, arts, education, and international tourism.

Historical Overview. Archaeological evidence indicates that Lisbon's sheltered natural harbor at the mouth of the Tagus River may have been base for Phoenician trading ships as far back as 1200 BC. The Romans conquered the Iberian Peninsula in the second-century BC, and Olisipo, as Lisbon was known to the Romans, was annexed to the Roman Empire as part of the province of Lusitania. The Romans built many temples in the city, a necropolis, baths, and a theater. Christianity appeared in the third century; there were persecutions by the Romans of converts. The first Christian bishop was Potamius in about 356 AD. In 711, Lisbon was captured by Muslim invaders from North Africa and was an Islamic city for more than three centuries afterward. Roman temples were converted to mosques and many mosques were built, but there was religious freedom for Christians and Jews, albeit with a lower social status, and the city prospered. Crusaders took Lisbon back in 1108 but held it for only three years. A second reconquest led by Alfonso I in 1147 drove the Muslims from Iberia, and Lisbon once again became a Christian city. Remaining Muslims either converted to Christianity or left, and mosques were either destroyed or remade into churches.

During the Age of Discovery from the 15th to the 17th centuries, Portugal was a leading seafaring nation, with explorations along the coasts of Africa, Asia, and South America. Lisbon entered a golden age from the wealth that resulted, as resources and other valuables from abroad greatly enriched Portuguese traders and royalty. Lisbon's wealth was derived as well from the slave trade, both to Portugal itself as well across the Atlantic from Africa to Portuguese Brazil, and from the tropical plantations where captive slaves labored. Gold and diamonds from Brazil funded Lisbon's urban development, including the construction of beautiful churches. During the Portuguese Inquisition of the 16th and 17th centuries, Jews were forced to convert to Christianity and many of those who were discovered to practice Judaism in secret were killed. On November 1, 1755, an enormous earth- quake rocked Lisbon and was then followed by outbreaks of fire and three killer tsunamis in succession. About 85 percent of the city was destroyed and as many as 30,000–40,000 lives from a population of more than 200,000 were lost. Re- construction of the city after the earthquake under the direction of Sebastião José de Carvalho e Melo (the Marquês de Pombal) resulted in modernization of urban form and construction of large rectangular city squares.

Lisbon expanded and grew with industrial development in the 19th and early 20th centuries. In 1910, the First Portuguese Republic was formed. From 1926 to 1974, Portugal was an authoritarian state called Estado Novo (New State) or the Second Republic. In 1974, the Carnation Revolution marked the end of Portugal's colonial wars in Africa and Asia, and opened the way for democratic government.

Major Landmarks. The Castle of São Jorge is a Moorish citadel for medieval times that stands on the highest hill in central Lisbon and overlooks the city. It is one of Lisbon's principal tourist attractions and symbols. Belém Tower is an iconic defense tower built in the early 16th century on the banks of the Tagus River. Nearby is the 16th-century Jerónimos Monastery. Both the monastery and the river-side fortification are UNESCO World Heritage Sites. The Cathedral of St. Mary Major is a historic Roman Catholic church whose construction began in 1147. Important museums include the Maritime Museum and the National Archaeology Museum. Praça do Comércio (Commerce Square) is commonly known as Terreiro do Paço (Palace Square) because it was the location of the Royal Palace until it was destroyed by the earthquake of 1977. A statue of King José I on horseback dominates the square. Cristo-Rei (Christ the King) is a 260-ft-high (79 m) monument capped by a statue of Jesus Christ with outstretched arms atop a cliff overlooking parts of Lisbon and the Tagus River. It was built in 1959 and was inspired by a monument with the same theme that overlooks Rio de Janeiro in Brazil. The 25th of April Bridge built in 1966 and the Vasco da Gama Bridge are two landmark spans across the wide Tagus estuary that link distant parts of the Lisbon metropolis. The Vasco da Gama Tower is a 476-ft-high (145 m) observation tower adjacent to a new skyscraper beside the Tagus River that was constructed in 2010.

Culture and Society. The main minority groups in Lisbon include Brazilians, Cape Verdeans, Goans, and Angolans from former Portuguese colonies, and Ukrainians from eastern Europe who have arrived recently to work in construction and service industries. Most Portuguese are Roman Catholics. The official language of the country is Portuguese. Football (soccer) is far and away the most popular sport. Lisbon has an active cultural scene with many museums, concerts, and theater performances, film festivals, and a large gay pride festival, among other events, as well as inviting cafés, restaurants, and nightclubs.

Further Reading

Alden, J. and A. Pires. "Lisbon: Strategic Planning for a Capital City," *Cities: The International Journal of Urban Policy and Planning* 13, no. 1 (1996): 25–36.

Christiana, Kerry. *Lisbon Encounter.* Victoria, Australia: Lonely Planet Publications, 2009.

Oliveira, Vítor and Paulo Pinho. "Lisbon," *Cities: The International Journal of Urban Policy and Planning* 27 (2010): 405–19.

Shrady, Nicholas. *The Last Day: Wrath, Ruin, and Reason in the Great Lisbon Earthquake of 1755*. New York: Viking, 2008.

LJUBLJANA

Ljubljana is the capital and largest city of the Republic of Slovenia, commonly called Slovenia, a country in south-central Europe at the northeast coast of the Adriatic Sea. The city is located in the center of the country in the Ljubljana Basin, and has historically been at the crossroads of trade. In addition to being national capital, Ljubljana is the main commercial, transportation, cultural, and education center of Slovenia. The population is about 272,000.

Historical Overview. Archaeological evidence indicates that Ljubljana's site has been settled since about 2000 BC. Remains of ancient dwellings built on piles in the nearby Municipality of Ig have been designated as a UNESCO World Heritage Site. In about 50 BC, the Roman built a fortress on the site. From that start, a permanent settlement called Aemona (Emona) developed. The Huns destroyed the town in 452, as did other marauders later. The ancestors of today's Slovenes settled the area in the ninth century and came to be ruled by the Frankish Kingdom. Ljubljana itself was founded in about 1220, maybe earlier. In 1335, the town came under Hapsburg rule which lasted until the end of World War I in 1918 when Austria-Hungary was dissolved and Ljubljana became part of the Kingdom Serbs, Croats, and Slovenes, and predecessor to Yugoslavia. The only interlude of Hapsburg rule was 1809–1813 when Ljubljana was part of the Napoleonic First French Empire and capital of the Illyrian Provinces. At that time, the city was known as Laibach. The city prospered when the first rail link with Vienna was inaugurated in 1849, and then again when the line was extended to Trieste, part on the Adriatic Sea, in 1857. An earthquake damaged much of the city in 1895, leading to reconstruction in the style of Vienna's architecture. Both World War I and World War II damaged the city. In 1918, after World War I, Ljubljana became part of the independent Kingdom of Slovenes, Croats, and Serbs that emerged with the demise of Austria-Hungary, a precursor of Yugoslavia. In World War II, the city was occupied first by Italian fascists and then by the Nazis who encircled the city with a barbed-wire fence, as Ljubljana was the center of strong resistance movements. After World War II, Ljubljana became capital of the Socialist Republic of Slovenia within Communist Yugoslavia. In 1991, Slovenia achieved independence, and in 2004 the country joined the European Union.

Major Landmarks. Ljubljana Castle built in the 12th century stands on a hill overlooking the river. The city's historic Town Hall stands on Town Square in the center of which is a replica of the 18th-century Robba Fountain, also known as the Fountain of the Three Rivers of Carniola. The original fountain was moved for display inside the National Gallery of Slovenia in 2006. Near Town Hall and the Ljubljana Central Market is St. Nicholas Cathedral. Other famous churches in Ljubljana are the Franciscan Church of the Annunciation, St. Peter's Church, and the Serbian Orthodox Church of Saints Cyril and Methodius. Nebotičnik, a word meaning "skyscraper," is a 13-storey high-rise office and shops building that was built in the early 1930s and is one of the city's most distinctive and best-known structures. Čop Street, named after the 19th-century writer Matija Čop, is a downtown pedestrian street with shops, restaurants, and crowds of residents and tourists alike.

Culture and Society. Ljubljana is a city with a thriving cultural life, including museums, theater, concert halls, fine universities, and popular festivals such as the Ljubljana Summer Festival. Its region grows grapes, and Ljubljana is sometimes referred to as "a city of vine and wine." There is a youthful feel to Ljubljana because of the city's many universities and large numbers of students. According to the census of 2002, 39.2 percent of the residents of Ljubljana were Roman Catholics, 30.4 percent said that they were believers who did not affiliate with any organized religion or did not reply, and 19.2 percent said they were atheist. About 5.5 percent were Eastern Orthodox and 5 percent were Moslem. About 80 percent of the population is ethnically Slovenes (or Slovenians), and Slovene is the official national language.

Further Reading

Balažič, Gregor. "Marshaling Tito: A Plan for Socialist Cultural Heritage Tourism in Slovenia," *Focus on Geography (American Geographical Society)* 54, no. 3 (2011): 103–8.

Ferfila, Bogomil. *Slovenia's Transition: From Medieval Roots to the European Union.* Boulder, CO: Lexington Books, 2010

Gašperič, Primož. "The Expansion of Ljubljana onto the Ljubljansko Barke Moor," *Acta Geografica Slovenika* 44, no. 2 (2004): 7–33.

Hrast, Maša Filipovič and Vesna Dolničar. "Sense of Community and the Importance of Values: Comparison of Two Neighborhoods in Slovenia," *Journal of Urban Affairs* 34, no. 3 (2012): 317–36.

LOBAMBA. *See* Mbabane.

LOMÉ

Lomé is the capital and largest city of Togo, a small sliver-shaped country with a narrow coastline along the Bight of Benin on the Gulf of Guinea in West Africa. The city is located in the south of the country on the coast near its western border with Ghana, and is Togo's main port. It exports coffee, cocoa, copra, and palm kernels. Lomé also has an oil refinery. Geographically, the city had squeezed between the oceanfront and a lagoon to the north, but more recent expansion resulted in the filling in of parts of the lagoon and growth of shantytowns to the north. The population of Lomé is 737,751, more than one-tenth that of Togo as a whole.

Historical Overview. The coast along which Lomé is located was called the Slave Coast because of the slave trade that took place there from the 16th into the 19th centuries. In 1884, Togo became a protectorate of Germany, and in 1905 it became a German colony called Togoland. After World War I, Germany lost this colony and it became a League of Nations Mandate administered by Great Britain and France. It was eventually divided between Britain and France, with British Togoland becoming a part of the former British colony that became Ghana in 1957 and French Togoland becoming part of the French Union in 1959. Togo became independent from France in 1960. In the ensuing decades the country saw considerable political instability and violence, political assassinations, coups, and dishonest elections. Togo has a very poor human rights record. In the meantime, in 1975 Lomé has played an historic role as the venue of the Lomé Convention in which an important trade and aid agreement was signed between the European Union and 71 developing countries of Africa, the Caribbean and the Pacific region. There have since been three more Lomé Conventions to update and renegotiate the original agreements. On July 7, 1999, the Lomé Peace Accord was signed in the city in an attempt to put an end to a civil war in nearby Sierra Leone.

Major Landmarks. Independence Monument is an important Lomé landmark. Nearby is the Rally of the Togolese People, an important convention center, inside of which is the Togo National Museum. Lomé Grand Market has three floor levels and sells a wide variety of goods. Akodessewa Market sells many fetish items. Another landmark is the Gothic-style Cathedral of the Sacred Heart.

Culture and Society. The population of Lomé is diverse and includes members of various tribes including Ewe, Mine, and Kabre, among others. Many tribal languages are spoken in the city, although French is the official language. About one-half of the population practices indigenous beliefs, while the rest are nearly equally divided between Muslims and Christians. Most of the population is generally very poor, and includes many refugees from instability and violence in other parts of Togo and in nearby countries.

Football (soccer) is enormously popular in Togo. The country's national team has enjoyed some success in recent years in competitions against other African teams and qualified for the World Cup competition in 2006. However, in 2010 there was a terrorist attack against a bus carrying team members to a match. Three members of the squad were killed.

Further Reading

Decalo, Samuel. *Historical Dictionary of Togo.* Lanham, MD: Scarecrow Press, 1996.
Lyons, Michael and Alkison Brown. "Has Mercantilism Reduced Urban Poverty in SSA? Perceptions of Boom, Bust, and the China–Africa Trade in Lomé and Bamako," *World Development* 38, no. 5 (2010): 771–82.

LONDON

London is the capital and largest city of the United Kingdom, a country in northwestern Europe on the island of Great Britain and the northeastern part of the island of Ireland in the North Atlantic Ocean and North Sea. The country also includes many smaller islands off the shores of the two larger ones. The country's name is officially the United Kingdom of Great Britain and Northern Ireland, but it is called the United Kingdom or the U.K. for short. The United Kingdom is a country in its own right, and consists of four countries: England, of which London is also the capital; Scotland (capital: Edinburgh); Wales (capital: Cardiff); and Northern Ireland (capital: Belfast). The United Kingdom is a unitary parliamentary constitutional monarchy. The Parliament meets in London's Palace of Westminster, while the prime minister of the United Kingdom resides at 10 Downing Street. The monarch of the United Kingdom, currently Queen Elizabeth II since 1952, resides in London's Buckingham Palace and has the constitutional "right to be consulted, the right to encourage, and the right to warn."

London is located on the Thames River in the southeast of the United Kingdom (and the southeast of both England and the island of Great Britain). With a population of 7,825,200 and a metropolitan area population of 13,709,000, it is the largest city in Europe and centers Europe's largest metropolitan area. It is also one of the largest cities and the focus of one of the largest metropolitan areas in the world. In fact, from about 1831 to 1925, London was the largest city in the world. In addition to its role as the capital of the England and the United Kingdom, it was also once the capital of the largest empire in world history, the British Empire at its peak in the late 19th and early 20th centuries. At that time, London was said to rule the world.

Now, in addition to its administrative role in the United Kingdom, London is still a leading global city. It is, along with New York City, one of the top two financial centers in the world, as well as one of the most important corporate centers, and centers of culture, media, the arts, education, entertainment, fashion, professional services, and many other activities. Its main airport, Heathrow, is the world's busiest as measured by numbers of passengers, and the city is one of the world's leading attractions for global tourism. In summer 2012, London has hosted the Olympics Games, the first city in the world to have hosted the event three times.

Historical Overview. London's history goes back to 43 AD, when it was founded as an outpost of the Roman Empire called Londinium. In about 100 it became the capital of the Roman province of Britannia. The original settlement declined with the early fifth-century fall of the Roman Empire, and in the sixth century a new, Anglo-Saxon settlement called Lundenwic was founded nearby in what is today the district if Covent Garden. The original city was reinhabited in the 10th century because its walls provided protection from invasions by Vikings. The city's status as the England's center of power was affirmed in the 11th century during the reigns of King Edward the Confessor and his successor William, Duke of Normandy, also known as William the Conqueror. In the Middle Ages, the City of Westminster was the seat of government, while its neighbor, the City of London, was the principal commercial center and the abode of most of the population. By 1300, the population of the city had risen to 100,000, but the Black Death in the middle of the 14th century wiped out about a third of its population.

By the 17th century, London had become a major mercantile center and the center of an expanding global empire. The population rose rapidly with migration from the countryside and abroad, although the Great Plague of 1665–1666 took about 100,000 lives, a fifth of the total of the time. The year 1666 also saw the Great Fire of London which devastated the historic wooden city and brought in a rebuilding in brick and stone and the construction of new districts at lower density. In the 19th century, London became one of the first great industrial cities in the world, but despite the profits made in its factories, much of the city was characterized by bleak overcrowded slums, high crime, and severe air pollution. During World War II, London was bombed by German planes, resulting in the destruction of many neighborhoods and some 30,000 lost lives. After the war, London hosted the 1948 Olympics, and after a deadly episode of air pollution in December 1952 called the Great Smog, turned attention to cleaning the air and other environmental issues. Later in the 20th century, the social profile of the city changed with explosions of youth culture and settlement by immigrant groups from the former British Empire. Starting in the 1980s, the city saw many large-scale urban redevelopment projects such as the Canary Wharf development and other urban improvements.

Major Landmarks. London has hundreds of landmarks to note, including four UNESCO World Heritage Sites: the Tower of London, Kew Gardens, the area of Westminster Abbey, the Palace of Westminster and St. Margaret's Church, and the district of Greenwich where the Royal Observatory is located. Other landmarks include Buckingham Palace, the Royal Albert Hall, Trafalgar Square, Hyde Park, Regent's Park, St. Paul's Cathedral, Tower Bridge, Wembley Stadium, the British Museum, the Victoria and Albert Museum, and the National Gallery. More recent landmarks are the London Eye (a giant Ferris wheel beside the River Thames), the Tate Modern (a gallery of modern art), a spectacular City Hall building, the distinctive office building at 20 St. Mary Axe that is called the Gherkin among other nicknames, and the O2 Arena where the Millennium Dome once stood.

Culture and Society. London is one of the most ethnically diverse cities in the world and continues to be one of the world's most popular magnets for immigrants from overseas. Listed from higher to lower in their approximate order of population

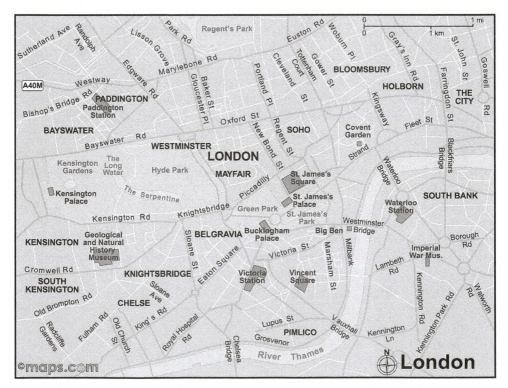

London, England (Maps.com)

Big Ben and the British Houses of Parliament as seen from across the Thames River. (Photo courtesy of Roman Cybriwsky)

totals, the main minority groups are Indians, Irish, Bangladeshis, Jamaicans, Nigerians, Pakistanis, Kenyans, Sri Lankans, Ghanaians, Cypriots, South Africans, Americans, and Australians. The city also has considerable religious diversity, with many Hindus, Moslems, Sikhs, and Jews in addition to the majority Christian population.

London has one of the world's liveliest cultural scenes, with a great many theaters, concert halls, nightclubs, music pubs, art galleries, museums, and other attractions. Likewise, it is a major center of sports, with popular venues and enthusiastic followings for football (soccer), cricket, rugby, and tennis, among other sports. More than 35,000 runners compete annually in the London Marathon.

Further Reading

Ackroyd, Peter. *London: The Biography*. New York: Anchor Books, 2003.
Weinreb, Ben, Christopher Hibbert, Jukia Keay, and John Keay. *The London Encyclopedia* (3rd Edition). London: Macmillan UK, 2010.
Wilson, A. N. *London: A History*. New York: Modern Library/Random House, 2006.

LUANDA

Luanda is the capital and largest city of the Republic of Angola, a country on the Atlantic Ocean coast of southern Africa. The city is on the coast itself, in the country's northwest. Its population is estimated at more than 5 million, the product of extraordinary recent growth and nearly one-quarter that of Angola as a whole. It is one of the most populous cities on the African continent. In addition to being the government center of Angola, the city is the country's main seaport and industrial center, and the hub of the nation's economy. The official name of the city is Luanda is São Paulo de Assunção de Loanda.

Historical Overview. Although there were indigenous African settlements in the area beforehand, the founding of Luanda is recognized officially as January 25, 1576, by Portuguese explorer Paulo Dias de Novais who arrived by ship with 100 settlers and 400 soldiers. A fort was built in 1618, and then two more, one in 1634, and one in 1765–1766. The city's principal role was in the trade of African slaves. This terrible business began on the Angola coast even before Luanda was founded and lasted in Luanda until 1836. Initially, the slaves were taken from their villages by Portuguese soldiers and traders directly, but eventually they were brought by Africans themselves to the Portuguese in the Luanda market. Typically, the slaves were captive members of other tribes. From Luanda the captives were taken across the Atlantic Ocean to the Portuguese colony of Brazil where there was great demand for labor on sugar plantations. After the end of the slave trade, the port of Luanda exported palm and peanut oil from Angola, coffee, cocoa, timber, and other products of the land. Portugal's colonial claim to Angola was ratified at the Berlin Conference in 1885. By that time, Luanda was a prosperous port city with a mostly Portuguese population.

Angolan independence was won in 1975 after the bloody 1961–1975 Angolan War for Independence. Afterwards, the country was in civil war more or less continuously until 2002. The decades of fighting impoverished the country and prevented economic development. However, more than a decade of peace and political stability has now brought rapid economic growth to Angola and considerable wealth to concentrated hands. The chief industries are oil, diamonds, and mineral extraction. Luanda has developed extraordinarily rapidly, as young Angolans leave the countryside to make new lives in a booming capital city. The city itself is being rebuilt after the years of war and is taking on a modern and prosperous look. Construction employs many male workers who have migrated to Luanda. The People's Republic of China and Brazil are now major investors in Angolan industrial development and construction of infrastructure.

Major Landmarks. Luanda faces the sea at the beautiful curved natural harbor of Bahia de Luanda, Luanda Bay. Along the bay there are beaches and the busy Avenue of the 4th of February. The road is increasingly lined with new high-rise buildings. The historic core of Luanda, Baixa de Luanda, is just behind. Fortaleza de São Miguel, an historic fort built that goes back to the earliest years of the city, is on a hilltop nearby and is an important visitor attraction. It was used mostly to hold the slaves captive. The National Museum of Slavery is a related landmark. Other museums are the National Museum of Natural history and the National Museum of Anthropology. The Agostinho Neto Mausoleum is a towering obelisk dedicated to the memory of independent Angola's first president and leader of the fight for freedom from colonial rule.

Culture and Society. Despite rapid economic growth and a fast-developing modern skyline, Luanda is still mostly a poor city. There are many squatter slums at the margins of the city and conflicts when squatters are evicted to make way for new construction. Much of the population of the city is recently arrived from the countryside. More than one-half the population is 15 years of age or younger. The main ethnic groups are Ambundu, Ovumbundu, and Bakongo. There are other African ethnicities as well, plus small minorities of residents of Portuguese origin and from Brazil. Recently, Chinese have settled in Luanda. The common language is Portuguese, but tribal languages are also used. Most Angolans are Christian. There are many denominations, but Roman Catholics are most numerous. In 2011, Leila Lopes, Miss Angola 2011, was crowned Miss Universe, an achievement that is the source of great national pride in Angola.

Further Reading

Jenkins, Paul, Paul Robson, and Allan Cain. "Luanda," *Cities: The International Journal of Urban Policy and Planning* 19, no. 2 (2001): 139–50.

Moorman, Marissa J. *Intonations: A Social History of Music and Nation in Luanda, Angola from 1945 to Recent Times.* Athens: Ohio University Press, 2008.

Pitcher, M. Anne. "Cars are Killing Luanda: Cronyism, Consumerism, and Other Assaults on Angola's Postwar Capital City," in Martin J. Murphy and Garth A. Myers, eds., *Cities in Contemporary Africa*, 173–94. New York: Palgrave Macmillan, 2006.

LUSAKA

Lusaka is the capital and largest city in Zambia, a landlocked, mineral-rich country in southern Africa. The city is in the south-central part of Zambia in the central plateau of the country, and is the national transportation hub as well as seat of government. It is one of the most dynamic cities in southern Africa, with opportunities for increased prosperity because of recent government and economic reforms, rich farmland, and mineral resources in the Copperbelt Province in the north. The headquarters of the Common Market for Eastern and Southern Africa (COMESA) is located in Lusaka. Lusaka's population is about 1.7 million.

Historical Overview. Lusaka was originally a small village named after its chief Lusakaa. In the late 19th century, British settlers arrived via South Africa, attracted by the farming frontier and the prospect of minerals exploitation. From 1911, the colony was called Northern Rhodesia, after Cecil Rhodes, the famous English colonialist, industrialist, and mining baron. The central location and natural transportation corridors of Lusaka made it into a crossroad for British penetration into the region. A railroad was built in 1905, and in 1935 Lusaka was designated the capital of Northern Rhodesia, replacing Livingstone on the Zambezi at the country's southern border. Independence from Great Britain was achieved in 1964 and the country was named Zambia after the Zambezi River, with Lusaka continuing as capital.

Major Landmarks. Lusaka's principal commercial center is along Cairo Road, along which are several multistory bank and office towers and major hotels. The government district of Lusaka lies to the east, with the Zambian National Assembly and other government buildings. At Cathedral Hill is the Anglican Cathedral of the Holy Cross, which has beautiful stained glass windows. Other landmarks include the Lusaka National Museum, the Political Museum, the Zintu Community Museum, and the Freedom Statue which shows a male figure breaking his chains.

Along Great East road are the city's largest shopping malls, Arcades Mall and Manda Hill Mall. The University of Zambia is another landmark in the city. The Munda Wanga Environmental Park, opened in 1956, is a popular wildlife sanctuary, botanical garden, and environmental educational center. In Kabwata, a working-class suburb of Lusaka, is the Kabwata Cultural Center and a popular local market.

Culture and Society. Zambia's population consists of many different African ethnic groups, including Nyanja-Chewa, Benba, Tonga, Tumbuka, Lunda, Luvale, Kaonde, Nkoya, and Lozi. There are also many foreign expatriates, including English and other Europeans, and a minority of Zambian citizens with European heritage. Many of the Europeans are attached to aid organizations in Zambia. The official language of Zambia is English, but African languages are also commonly spoken. In Lusaka the principal African language is Nyanja. About 87 percent of Zambia is Christian, the official national religion according to the constitution, with Anglicans and Roman Catholics being most numerous. The most popular national sports are football (soccer) and rugby.

Further Reading

Gough, Katherine V. "'Moving Around': The Social and Spatial Mobility of Youth in Lusaka," *Geografiska Annaler, Series B: Human Geography* 90, no. 3 (2008): 243–55.

Hansen, Karen Tranberg. "Gender and Housing: The Case of Domestic Service in Lusaka, Zambia," *Africa* 62, no. 2 (1992): 248–65.

Myers, Garth A. "The Unauthorized City: Late Colonial Lusaka and Postcolonial Geography," *Singapore Journal of Tropical Geography* 27, no. 3 (2006): 289–303.

LUXEMBOURG

Luxembourg is the capital and largest city in the Grand Duchy of Luxembourg, also called Luxembourg just like the capital city, a small landlocked country in the heart of western Europe that is bordered by Germany, France, and Belgium. The city, which is also known as Luxembourg City, is located in the southern part of the country at the confluence of the Alzette and Pétrusse Rivers. The population of Luxembourg City is about 94,000 and that of the metropolitan area is about 104,000.

Historical Overview. The history of Luxembourg City can be traced to a Roman era guard tower at a strategic road crossing, and then a castle that was built after 963 by Siegfried I of Ardennes. A church was constructed in 987 by Egbert, Archbishop of Trier, and then a marketplace was developed. The central location of Luxembourg gave it strategic military value, so the town's fortifications were expanded by successive rulers in the Middle Ages, most especially during the reign of John the Blind, Count of Luxembourg, in the 1340s. In the 14th and 15th centuries, three members of the House of Luxembourg ruled as emperors of the Holy Roman Empire. A crisis of succession caused the principality to be sold to the House of Burgundy in 1437. It then passed to the Hapsburg Empire in 1477. In the middle of the 16th century, Luxembourg came under the control of the Spanish Hapsburgs and was part of the Spanish Netherlands. Later, Luxembourg passed from French control to the Austrian Hapsburgs, and then back to France again, then to Prussia, and then to the Netherlands. Following a peace treaty in

1867, many of the city's fortifications were dismantled over the next 16 years. The fortifications were so elaborate that Luxembourg was referred to as "the Gibraltar of the North." Luxembourg's independence was achieved in 1890. The country was occupied by Germany in both World War I and World War II. The Nazis actually annexed Luxembourg to Germany, but a Luxembourg government in exile in Great Britain fought with the allies against Nazi Germany. After the war, the country regained independence and became a founding member of the United Nations. In 1949, Luxembourg joined NATO, and in 1957 it was one of six countries to embark on the beginnings of the European Union. Luxembourg City has since been the seat of various offices of the European Union and venue for many important conferences and EU proceedings.

Major Landmarks. Luxembourg City is a scenic and very historic city on a promontory overlooking the river confluence below. The old city district is a UNESCO World Heritage Site. Specific landmarks include the Gothic Revival Notre-Dame Cathedral from the early 17th century, reconstructed fortifications, and the Grand Ducal Palace. The Casemates Bock is a series of underground fortifications from the 18th century. The Luxembourg American Cemetery and Memorial has the final resting places of more than 5,000 U.S. soldiers from World War II, including that of General George Patton. MUDAM (Musée d'Art Moderne Grand-Duc Jean) is the city's principal art museum.

Culture and Society. Luxembourg has one of the world's highest GDP per capita and ranks as one of the world's best cities in terms of quality of living and personal safety. The city is clean, orderly, and comfortable. The people of Luxembourg are called Luxembourgers. Immigrant minorities in the country include migrants from neighboring countries, as well as from Portugal and the countries that were once parts of Yugoslavia. The official languages are Luxembourgish, French, and German. Luxembourg law does not permit gathering data about religious affiliation. However, it is estimated that about 87 percent of the population is Roman Catholic.

Further Reading

Fetzer, Joel S. *Luxembourg as an Immigrant Success Story: The Grand Duchy in Pan-European Perspective*. Lanham, MD: Lexington Books, 2011.

Newcomer, James. *The Grand Duchy of Luxembourg: The Evolution of Nationhood 983 A.D. to 1983*. Lanham, MD: University Press of America, 1984.

MADRID

Madrid is the capital and largest city of Spain, a country on the Iberian Peninsula at the western tip of Europe. The city is located in the center of the country on the Manzanares River. Its population is approximately 3.3 million, while that of its metropolitan area is about 6.3 million. Madrid is the third-largest city in Europe after London and Berlin, and its metropolitan area is Europe's third-largest as well, after those of London and Paris. Residents of Madrid are referred to as Madrilenians.

Historical Overview. Madrid was founded in 854 as a Moorish garrison strategically between the Muslim Iberian state of Al-Andalus and the European Christian Kingdoms of Léon and Castile, and later became part of the Taifa of Toledo. After the Christian conquest of Toledo in 1085, Madrid was integrated into the Kingdom of Castile. In 1561, King Felipe II moved his court from Toledo to Madrid, making the city Spain's capital. The city evolved into a major center of arts and learning, and a bastion of the Roman Catholic faith, as well as a city of beautiful palaces and other monumental architecture. By the middle of the 17th century, the population of the city had grown to about 100,000. The Bourbon King Carlos III, who ruled Spain from 1759 to 1788, modernized the city by investing heavily in public works and in the construction of many monuments and cultural institutions. Madrid was occupied by French troops in 1808 during the Napoleonic Wars, and Madrilenians put up a brave resistance that led to a war of independence. After the French were pushed out of Spain in 1814 and King Ferdinand IV returned to the throne, he bestowed the title "heroic" on the city of Madrid. The wars of independence that were fought against Spain by colonies in the New World were costly to Madrid and nearly bankrupted Spain. Nevertheless, Madrid continued to grow and developed into a modern European cultural and economic capital. In the Spanish Civil War of 1936–1939, Madrid was a stronghold of the Republican side and was heavily damaged. It was during this conflict that the city became the first European city in history to be bombed from the air. Dictator General Francisco Franco controlled Spain from Madrid from the time of his victory in the Civil War in 1939 until his death in 1975. During that time, he presided over a period of economic boom from 1959 to 1973. The city expanded greatly and there was much construction and redevelopment. Migrants from Spain's rural areas came to the city and settled industrial areas in the south. The north and west of the city, meanwhile, became the city's middle-class area. In 1981, mass protests by Madrid citizens put down an attempted coup. Madrid is one of the great capitals of Europe: it is a busy center of industry, finance, arts and culture, higher education, and tourism, as well as an important transportation hub.

Major Landmarks. The heart of Madrid is Puerta del Sol, a busy plaza that is the hub of city transportation, a popular gathering place, and venue for demonstrations and public celebrations. At its center is an equestrian statue of King Carlos III. The Royal Post Office is another landmark at this square, as is a statue of a bear climbing a *madroño* tree that is a symbol of Madrid. Another landmark square in the center is Plaza Mayor. In its center is a statue of King Philip III. At the Plaza de España are two of the tallest buildings in Madrid, the Torre de Madrid and Edificio España. The Royal Palace is one of the city's most beautiful buildings. Other prominent structures are the Teatro Real (Royal Theater), the National Library, Palacio de Cibeles (Madrid's City Hall), and the Metropolis Building on Alcalá Street. Nearby is Puerta de Alcalá, the grand entrance to beautiful Retiro Park. The Golden Triangle of Art on Paseo del Prado features three outstanding museums, the Prado, the Reina Sofia Museum where Pablo Picasso's *Guernica* about the Spanish Civil War hangs, and the Thyssen-Bornemisza Museum. The city also has many other museums and cultural institutions. A landmark from recent times is the CTBA, the Cuatra Torres Business Area, the Four Towers Business Area, and modern office district that is centered on four prominent skyscrapers: Torre Espacio, Torre de Cristal, Torre Sacyr Vallehermoso, and Torre Caja Madrid.

Culture and Society. Madrid is one of the world's liveliest cities, with a bustling nightlife, many fine restaurants and cafés, and a huge range of theaters, concert facilities, sports arenas, and other attractions. The football (soccer) club Real Madrid is one of the best-known sports teams in the world. Other popular sports are basketball and tennis. The bullfighting ring Las Ventas is considered to be the world's center of bullfighting. The population of Madrid is becoming increasingly diverse with immigration from foreign countries. Approximately 84 percent of Madrilenians are Spanish, and migrants from other countries in Europe, North and South America, Africa, and even Asia form 16 percent of Madrilenians. The most numerous ethnic populations are from Spain's former colonies in South and Middle America. The main faith is Roman Catholicism. The population of Muslim immigrants from North Africa and Balkan Europe is increasing.

Further Reading

dal Cin, Adriana, Javier de Mesones and Jonas Figueroa, "Madrid," *Cities: The International Journal of Urban Policy and Planning* 11, no. 5 (1994): 283–91.

Nash, Elizabeth, *Madrid: A Cultural and Literary Companion.* Brooklyn, NY: Interlink Books, 2001.

Stapell, Hamilton, *Remaking Madrid: Culture, Politics, and Identity after Franco.* New York: Palgrave Macmillan, 2010.

MAJURO

Majuro is the capital and main population center of the Republic of the Marshall Islands, a country of 29 atolls, 24 of which are inhabited, plus five isolated small islands in the Micronesian zone of the Pacific Ocean. It is located on Majuro Atoll, one of the atolls in the Ratak ("Sunrise") chain of the Marshall Islands, and is comprised of three main population centers, the so-called D-U-D communities of Delap, Uliga, and Djarrit as named south to north. About 30 miles (48 km) away at the western end of the atoll is the community of Laura, a growing residential area with a popular beach. The main part of Majuro includes the principal shopping area of the Marshall Islands, the nation's airport, several hotels, and a port. The population of Majuro is about 25,400 (2004), approximately 45 percent that of the Marshall islands as a whole.

Historical Overview. Micronesians settled the Marshall Islands in the second-millennium BC. The first Europeans to arrive were the Spaniards in 1526 commanded by Alonso de Salazar. In 1788, John Charles Marshall and Thomas Gilbert, two captains of the English East India Company came to the islands, after which the islands were named Marshall. Until 1884, the islands were under Spanish sovereignty as part of the Spanish East Indies. They were then sold to Germany and became part of the protectorate of German New Guinea. Japan gained control of the islands in 1914 and administered them until World War II when the United States invaded and occupied the islands in 1944. The islands then became a United States Trust Territory until they gained independence on October 21, 1986. With American involvement, Majuro became the administrative center of the Marshall Islands, taking the place of Jaluit Atoll in the island group's Ralik ("Sunset") chain. From 1946 to 1958, the United States used the islands for nuclear weapons testing, completely destroying one of the islands on one of the atolls and making the Marshall Islands as a whole what the U.S. Atomic Energy Commission called in 1956 "by far the most contaminated place in the world." Subsequently, the U.S. government has paid $759 million to Marshall Islanders as compensation for their exposure to radiation from the weapons testing.

Major Landmarks. Landmarks in Majuro include the Alele Museum and Library, the Marshalls Handicraft Shop, and the U.S. Peace Corps Office in Uliga, the Water Tower in Darrit, and Marshall Islands Visitors Center and copra plant in Delap. Laura Beach is the most popular tourist destination.

Culture and Society. Marshallese are Micronesian people who arrived on the islands centuries ago from Asia. A minority of islanders have at least some recent Japanese ancestry. The official language of the islands is Marshallese, a Malayo-Polynesian language that is spoken in two major dialects, western and eastern, on the islands. English is widely spoken as well. The large majority of Marshall Islanders are Christians, with the United Church of Christ being the largest denomination. Marshall Islanders were once skilled navigators. They used the stars and stick-and-shell charts, but these skills have declined.

Further Reading

Kupferman, David. "The Republic of the Marshall Islands since 1990," *Journal of Pacific History* 46, no. 1 (2011): 75–88.

Woodward, Colin. "You Can't Go Home Again," *Bulletin of the Atomic Scientists* 54, no. 5 (1998): 10–13.

Yamaguchi, Toru, Hajime Kiyanne, and Horoya Yamano. "Archaeological Investigation of the Landscape History of an Oceanic Atoll: Majuro, Marchall Islands," *Pacific Science* 63, no. 4 (2009): 537–65.

MALABO

Malabo is the capital of the Republic of Equatorial Guinea, a small, poor, corrupt, and greatly underdeveloped country on the Atlantic coast of equatorial Africa. The country is made up of two parts, a mainland section called Rio Muni and five small islands in the Bight of Biafra of the Gulf of Guinea off the coast. Malabo is at the northern tip of Bioko Island, the largest of the five islands, and is about 25 miles (40 km) from the coast of Rio Muni's neighbor to the north, Cameroon. Bioko Island's previous name was Fernando Po. The population of Malabo is about 156,000 (2005), the second-largest total for a city in Equatorial Guinea after Bata, a city on the coast of the mainland.

Historical Overview. Indigenous African peoples in the territory of Equatorial Guinea included Bantu tribes, the Fang, and the Bubi. The first Europeans were Portuguese who came in 1472 under the leadership of navigator Fernando Po. He discovered the island of Bioko for Portugal and named it Formosa, meaning "beautiful," but in 1474 the island was made a Portuguese colony and given the name Fernando Po. In 1778, the island and other African territory were ceded to Spain in exchange for territory for Portugal in South America, and from then until 1810 the island was governed from Spanish Buenos Aires, now the capital of Argentina. The British leased the island in 1827 and established Malabo, which they named Port Clarence, as a base from which to fight the slave trade. In 1843, Great Britain moved its base to Sierra Leone on the African mainland and the territory reverted to Spain in the following year. Malabo was renamed Santa Isabel. By 1900, Rio Muni became a Spanish colony. Later the islands and Rio Muni were united into a colony of Spain called Spanish Guinea. Independence came in 1968. Francisco Macías Nguema was elected the first president. In 1969, he chose Santa Isabel to be the country's capital over the colonial capital Bata, and renamed it Malabo, an African name. His rule was one of terror and resulted in about one-third of the population either exiled or killed. Tens of thousands on Bubi on Bioko Island were murdered in an attempt to exterminate the tribe, and were replaced with Macías's own Fang ethnic group brought in from the mainland. In 1979, Macías was deposed in a bloody coup led by Teodoro Obiang. Despite at least 12 coup attempts against him, Obiang has been leader of the country ever since. The killings no longer go on, but there is general unrest and great poverty. Oil production since 1997 has brought great wealth to the inside circle in Equatorial Guinea, but the country as a whole remains undeveloped.

Major Landmarks. Notable buildings are Malabo Cathedral, Malabo Government Building, and Malabo Court House. The campus of the Universidad Nacional de Guinea Ecuatorial has a statue of the current president of the country, Teodoro Obiang. Visitors might find the Malabo Bush Meat Market to be interesting for its variety of foods and local color.

Culture and Society. More than 85 percent of the population belongs to the Fang ethnic group, a Bantu population, and more than 90 percent of the population is Christian, mostly Roman Catholic. The official language is Spanish, but Fang and other indigenous languages are more commonly spoken. The exploitation of oil has increased opportunities in Malabo and sparked a large migration. The city is badly underdeveloped, however, and the wealth from oil is in very few hands. Many citizens of Equatorial Guinea have emigrated to neighboring African nations or to Spain. Antimalaria spraying programs have brought the debilitating tropical disease under control in much of Equatorial Guinea.

Further Reading

Creus, Jacit. "Never Again: The Mission Against the City," *Scientific Journal of Humanistic Studies* 3, no. 5 (2011): 76–84.

Roberts, Adam. *The Wonga Coup: Guns, Thugs, and Ruthless Determination to Create Mayhem in an Oil-Rich Corner of Africa.* New York: Public Affairs (Perseus Book Group), 2006.

Seoane, Nora Salas. "'Welcome to the Eccentric Circus': Youth, Rap Music and the Appropriation of Power in Malabo (Equatorial Guinea)," *Scientific Journal of Humanistic Studies* 3, no. 5 (2011): 12–21.

Ugarte, Michael. *Africans in Europe: The Culture of Exile and Emigration from Equatorial Guinea to Spain.* Urbana: University of Illinois Press, 2010.

MALÉ

Malé is the capital and largest city in the Republic of the Maldives, a nation of 1,192 small coral islands grouped in a double north–south chain of 26 atolls southwest of India in the Laccadive Sea of the Indian Ocean. The city is located near the center of the country on the southern edge of North Malé Atoll (Kaafu Atoll), one of atolls of the easternmost chain. The city's population is about 104,000, nearly one-third that of the Maldives as a whole. In addition to being the administrative center of the country, Malé is the nation's leading business and cultural center. Malé's airport on the neighboring island of Hulhule is the principal point of entry for foreign tourists who head to resorts on other islands for beach and scuba vacations.

Historical Overview. Dravidian people from the Indian subcontinent settled the Maldives many centuries ago. The first written record dates to about the fifth-century BC with the arrival of Sinhalese people from Sri Lanka. In the second-century BC, Buddhism arrived on the islands from the Indian subcontinent and remained the principal religion until the conversion of the Maldives to Islam in the 12th century. From 1153 to 1968 when the country became a republic, the Maldives were governed as an independent Islamic Sultanate; and from 1887 to until independence was achieved on July 26, 1965, they were a British Protectorate. The history of Malé began in the 16th century when it was founded by Portuguese navigators as a trading post. The islands traded with the African coast, the Arabian Peninsula, and India. Before the modern era, the main exports were cowry shells that the islanders cultivated on floating branches of coconut palms in the sea, and coir, a fiber made from dried coconut husk that was used to make saltwater-resistant ropes and cordage. The tourism industry began to develop in the 1970s and now accounts for 28 percent of the national GDP.

The Maldives Islands have often been unstable politically after independence and have a history of repeated coups, attempted coups, and other conflict. The most recent coup resulted in the ouster of President Mohamed Nasheed on February 7, 2012. On December 26, 2004, the Maldives were devastated by the Indian

Ocean tsunami that followed the earthquake on that date centered off the Indonesian islands of Sumatra. Because the islands are extremely low-lying they are at great risk from global warming and rising sea levels. The Maldives government has argued for reductions by industrialized countries in carbon emissions to reduce global warming and is considering options for its population should the islands need to be evacuated because of flooding.

Major Landmarks. The symbolic center of Malé is Independence Square (Jumhooree Maidhaan) located on the city's north shore. A high flagpole with a large flag of the Maldives dominates the square. Just south of the Square is the golden-domed Islamic Center, Malé's largest mosque. The Friday Mosque, built in 1656, is another major landmark. Sultan Park is the site of the former palace. The lone surviving building on the site now houses the Maldivian National Museum. Other attractions are the Malé Market and the fish market.

Culture and Society. The Maldives is a devoutly religious Islamic state, and the open practice of any other religion on the islands is specifically forbidden. In recent years, fundamentalist Maldivians have destroyed artifacts in the National Museum that dated back to the island's early Buddhist history. In order to protect local culture, the Maldives segregates international tourists for most of the local population, with tourist resorts being permitted on certain islands only and tourists are discouraged from visiting other islands. Most Maldivians engage in fishing and subsistence farming. The official language of the Maldives is Dhivehi, an Indo-Aryan language that derives from early settlers from India and Sri Lanka. The language employs its own written script called Thaana.

Further Reading

Gee, John. "Unsettled Maldives," *Washington Report on Middle East Affairs* 31, no. 2 (2012): 36–37.

Neville, Adrian. *Malé: Capital City of the Maldives.* Malé: Novelty Printers and Publishers, 1995.

Sovacool, Benjamin. "Expert Views on Climate Adaptation in the Maldives," *Climatic Change* 114, no. 2 (2012): 295–300.

MANAGUA

Managua is the capital and largest city in Nicaragua, a country on the Central American isthmus that is bordered by the Caribbean Sea to the east and the Pacific Ocean to the west, and by Honduras to the north and Costa Rica to the South. The city is in the west center of the country, between its two large fresh water lakes, Lake Managua and Lake Nicaragua, and is on the southern shore of Lake Managua. Lake Managua is also known as Lake Xolotlán. The city is located on a seismic fault and experiences frequent earthquakes. The population of the Managua metropolitan area is about 2.4 million (2010), accounting for nearly one-half of the national population. Managua is the second-largest city in Central America after Guatemala City, the capital of Guatemala.

Historical Overview. Managua was founded in about 1819 and was originally a lakeshore fishing village. After Nicaragua established its independence in 1823, two older cities, León and Granada, competed to be capital and alternated in that role until Managua was selected as a compromise in 1852 because it was located between the two rivals. When Granada was destroyed in 1857 by a U.S. mercenary army led by William Walker, Managua rose to undisputed preeminence. It developed rapidly in the late 19th and early 20th centuries and became one of the three leading cities of Central America. In 1931, the city suffered a catastrophic earthquake, and in 1936 there was a fire that destroyed large parts of the city. It was rebuilt and continued to develop as a center of national government, industry, and trade. From 1936 to 1979 Nicaragua was ruled by the hereditary dictatorship of Anastasio Somoza Garciá and his sons. The Somoza dynasty was overthrown in the Nicaraguan Civil War of 1979, after which the country suffered from 11 years of war against the successor government of the Sandinista National Liberation Front (FSLN) by the so-called Contra rebels. The Contra War was supported financially and with arms by the United States. As these conflicts waged, Managua was still trying to recover from a second disastrous earthquake that struck the city on December 23, 1972. About 90 percent of the city had been destroyed and more than 19,000 citizens were killed. In 1998, the rains and winds of Hurricane Mitch brought more misery. The storm struck neighboring Honduras harder, but caused widespread damage in Nicaragua as well, including extensive flooding of Lake Managua. Rural poverty and environmental degradation, overpopulation, warfare, and the natural disasters have all combined to drive Nicaraguans from towns and villages of the countryside to the capital city where many live in poor conditions in peripheral squatter settlements.

Major Landmarks. The center of Managua is the Plaza de la Revolución, formerly known as Plaza de la República. It was badly damaged in the 1972 earthquake and is not yet fully restored. The plaza contains various monuments, including the tomb of Carlos Fonseca, the founder of the FSLN, and a monument to Nicaragua's unknown guerilla soldier. Catedral de Santiago, the old Roman Catholic Cathedral of Nicaragua that was built in the 1920s, remains in damaged condition from the earthquake and awaits renovation. A new cathedral, Catedral de la Concepción, has been built in another part of the city in 1993. Other landmarks are the National Palace, which is now a museum, the Museum of Acahualinca which features 6,000-year-old human footprints in volcanic ash, and the Rubén Dario National Theater. The Orange House is the official residence of the president of Nicaragua, although the present leader Daniel Ortega has chosen to not live there because he espouses a simpler lifestyle. A monument to Nicaraguan revolutionary leader Augusto Sandino stands in the Tiscapa Lagoon Natural Reserve.

Culture and Society. Most Nicaraguans are either *mestizo* (mixed Amerindian and European; 69%) or of European origin (17%), while the rest include African-Nicaraguans (9%) and Amerindians (5%). Roman Catholicism has been the major religion, but it has been declining recently with the growth of evangelical Protestantism and the Mormon faith. Baseball and football (soccer) are the most popular sports in Nicaragua.

Further Reading

Massey, Doreen. *Nicaragua*. Milton Keynes, UK: Open University Press, 1987.

Morris, Kenneth Earl. *Unfinished Revolution: Daniel Ortega and Nicaragua's Struggle for Liberation.* Chicago: Lawrence Hill Books, 2010.
Wall, David L. "Managua," *Cities: The International Journal of Urban Policy and Planning* 13, no. 1 (1996): 45–52.

MANAMA

Manama is the capital and largest city in the Kingdom of Bahrain, a small nation of 33 islands, including one called Bahrain Island that is much larger than all the others, off the western shore of the Persian Gulf. The city is on a peninsula in the far north of Bahrain Island, and has about 160,000 people. Both the city and the island nation as a whole are quite wealthy from oil. The dramatic skyline of newly built high office towers and hotels in Manama reflects this wealth. Bahrain is a constitutional monarchy, and since 1783 it has been ruled by the Al Khalifa dynasty.

Historical Overview. The north of Bahrain Island has been settled for as long as 5,000 years. In ancient times, it may have been the center of the Dilmun civilization. From the sixth- to the third-century BC, the island was ruled by the Persian Empire. The conversion of Bahrain to Islam took place in the late seventh century. In the tenth century, Bahrain was a center of the Qarmatian Republic, an idealistic and conservative Islamic movement that sacked Mecca and Medina and took the sacred Black Stone to Bahrain until it was returned 22 years later. The Qarmatians were overthrown by the Arab Uyunid dynasty from the oasis region of al-Hasa who then ruled the island until 1235. In 1253, the Bedouin Usfurids ruled Bahrain. Manama itself was first mentioned in Islamic chronicles in 1345. It was captured by the Portuguese in 1521, and then in 1602 by the Persians. The Al Khalifa dynasty took control in 1783. A series of treaties in 1863–1914 placed Bahrain under the increasing protection of Great Britain. Full independence was achieved in 1971.

For a long time the economy of Manama was based on fishing, pearling, boat building, and trading. Oil was discovered in 1932, resulting in a reorientation of the economy and restricting of the city, its port and other ports on Bahrain, and transportation networks more generally. The survey work for the 16 mile-long (25 km) King Fahd Causeway that links Bahrain with Saudi Arabia began in 1968, and the project was completed in 1986. There are plans to construct an even longer causeway (25 miles; 40 km) to link Bahrain with Qatar, but construction has not yet started.

Major Landmarks. The al-Fateh Mosque, which accommodates more than 7,000 worshippers, is one of the largest mosques in the world and a leading symbol of both Manama and Bahrain. The National Library of Bahrain is part of the mosque complex. The al-Fateh Corniche is a pleasant seaside park with views of Manama's impressive array of new skyscrapers such as the Bahrain Financial Harbour project and the 787-ft-high (240 m) twin towers of the Bahrain World Trade Center. Important Museums are the Bahrain National Museum, Museum of Pearl Diving, and at Bahrain Fort the Bahrain Fort Museum. The Barbar Temple complex is an archaeological site located west of Manama that is linked to ancient Dilmun culture and dates back as far as 3,000 BC. South of Manama is the mysterious Tree of Life, a lonesome tree in the desert with no apparent source of water to sustain it.

Culture and Society. Manama is a mix of about 50 percent Bahraini citizens and 50 percent foreign residents. The former are mostly of Arabic descent, while the foreigners are from Iran, India, Pakistan, the Philippines, Great Britain, and the United States, each in various parts of the economy and various occupations. Men from Pakistan and the Philippines, for instance, are often employed in the construction industry, while Britons and Americans are more often engaged in finance or in the oil industry as engineers and other specialists. The official language of Bahrain is Arabic, but English is widely used as well. Some Bahrainis from Persian origin speak a dialect from southern Iran. Urdu and Hindi are also widely used in Bahrain by foreign workers. Almost all citizens of Bahrain are Muslims. There is freedom for foreign residents to practice other religions. The most popular sports in Bahrain are football (soccer), and horse and camel racing.

Further Reading

Fuccaro, Nelida. *Histories of City and State in the Persian Gulf: Manama since 1800*. Cambridge, UK: Cambridge University Press, 2012.

Hamouche, Mustapha Ben. "The Changing Morphology of the Gulf Cities in the Age of Globalization: The Case of Bahrain," *Habitat International* 28, no. 4 (December 2004): 521–40.

MANILA

Manila is the capital and second-largest city of the Republic of the Philippines, an archipelago country of 7,107 islands in Southeast Asia in the western Pacific Ocean, and the center of one of the largest urban regions in the world. The city is located on the western shores of the island of Luzon, the country's largest island, at the head of Manila Bay. The Pasig River, which is technically a tidal estuary and only 16 miles long (25 km), cuts through the heart of the city. The population of the City of Manila itself is about 1.7 million, second in the Philippines only to Manila's adjacent neighbor Quezon City (2.7 million). The Manila metropolitan area has about 16.3 million inhabitants and ranks approximately 11th in the world in population, while a more widely defined urban area has about 20.7 million people and ranks fifth in the world. By some definitions, the City of Manila may be the most densely settled city in the world. The word Manila is believed to come from the local Tagalog language word *nila*, a flowering mangrove plant that grew locally.

Historical Overview. Manila's history dates at least as far back as the 10th century, when it is known to have had trading ties with what is today Java in Indonesia, China, and Japan. The first known record of the city is the Laguna Copperplate Inscription that was found in the sands of a river outside Manila in 1989. By the time the Spanish arrived in Manila in 1570, the city was under the rule of the Bolkiah dynasty of Brunei and had been Islamized. In 1571, conquistador Miguel Lopez de Legazpi founded a Spanish city at Manila in what is today the walled city of Intramuros. Spain controlled the Philippine Islands from 1565 to 1898, with Manila as capital after 1571. In the early part of that period, Manila was the Asian port of Spain's trans-Pacific trade with its partner port across the ocean, Acapulco. In 1898, Spain ceded the Philippines to the United States as a result of the Spanish-American War and the Philippine-American War. Self-government commenced in 1934 and independence from the United States came on July 4, 1946. During World War II, the Philippine Islands were occupied by Japan, and Manila was the scene of a bloody battle against the Japanese by Allied and Philippines troops combined. The city was almost totally destroyed in the fighting. From 1965 to 1986, the Republic of the Philippines was ruled by Ferdinand Marcos. He was democratically elected at first, but then established a dictatorship and a kleptocracy in which he, his wife Imelda Marcos, other family members, and many cronies collectively impoverished the country by robbing it of resources and wealth. The Philippines is still recovering from a long legacy of inequality that preceded the Marcos regime but worsened when he was in power. The city has also suffered from destructive tropical storms such as Ketsana (also known as Ondoy) in 2009.

Major Landmarks. Malacañang Palace, built by the Spanish in 1750, has been the residence of the leadership of the Philippines ever since and is now the official residence and principal workplace of the country's president. Intramuros is the old Spanish walled city. Within the walls are the Manila Cathedral, St. Agustin Church, and the remains of historic Fort Santiago. Major museums include the National Museum of the Philippines and the Metropolitan Museum of Manila. Other significant places are the Manila Hotel, an elegant vestige of the American era in Manila, the bay walk along the city's waterfront, Manila's Chinatown, the tourist neighborhood of Ermita, and Makati, Manila's upscale district with modern office towers, shopping malls, and expensive residential areas. The American Cemetery with graves of more than 17,000 U.S. World War II dead is nearby. There are many monuments and statues in Manila such as those honoring freedom fighter Jose Rizal, first president Manuel Roxas, and Benigno (Ninoy) Aquino Jr., who was assassinated while campaigning against Ferdinand Marcos. Other monuments honor People Power (the grassroots force that eventually toppled the Marcos regime and elected Ninoy's widow Corazon as president), the Philippine water jar, and the water buffalo.

Culture and Society. The national languages of the Philippines are Filipino, which is based on Tagalog, and English. There is also a large Chinese minority in the city. Roman Catholicism is the predominant faith, but there are also many Protestants. Although the city has a sizable middle class, there are also a great many very poor people and crowded squatter settlements. The Tondo district of Manila is one such area. Many citizens of the Philippines work abroad and send remittances home. Common occupations are home and health care, domestic service, entertainment, and construction work.

Further Reading

Mason, Paul. "Dirty Secret of the Modern Mega-City," *New Statesman* 140, no. 5065 (2011): 24–28.
Porio, Emma. "Shifting Spaces of Power in Metro Manila," *City* 13, no. 1 (2009): 110–19.

Porio, Emma. "Decentralization, Power, and Network Governance Practices in Metro Manila," *Space and Polity* 16, no. 1 (2012): 7–27.

Reed, Robert, R. *Colonial Manila: the Context of Hispanic Urbanism and Process of Morphogenesis.* Berkeley: University of California Press, 1978.

Tyner, James A. *The Philippines: Mobilities, Identities, Globalization.* New York: Routledge, 2009.

MAPUTO

Maputo is the capital and largest city of Mozambique, a country on the Indian Ocean in southern Africa. It is located on Maputo Bay in the southern part of the country. Before Mozambique obtained its independence from Portugal in 1975, the city was known as Lourenço Marques after a 16th-century Portuguese explorer and settler. The population is nearly 1.8. million (2007). The city's busy port exports chromite and various local farm products, including sugar, sisal, and cotton. The industrial port city of Matola (population 675,000; Mozambique's second-largest city) is only 12 km away and is part of Maputo's metropolitan area.

Historical Overview. The origins of the city are traced to a fortress established by Portuguese colonialists in the middle of the 16th century. It was subsequently destroyed in conflicts with the native population. A new fort was built nearby in 1878, and the first parts of the contemporary city were laid out beside it in 1850. The city was made capital of the Portuguese colony in 1898. Railroads were built about the same time, particularly one to Pretoria in South Africa, and the port became the mainstay of the local economy. The port is still the closest port to the rich industrial heartland of South Africa. The coming of independence was a joyous moment for the local population and resulted in the exodus to Portugal of some 250,000 residents of Portuguese origin. There was political and economic chaos after independence, and civil war between 1977 and 1992. Mozambique's economy began a recovery after peace in 1992, and there has since been increasing foreign investments in the Maputo-Matola urban area and in the countryside.

Major Landmarks. Among the major landmarks of Maputo are the 19th century Portuguese fortress, the distinctive Central Railway Station that was built between 1913 and 1916, and the gleaming white Roman Catholic cathedral. Portuguese architect Pancho Guedes, a long-time resident of Maputo, designed many of the city's notable buildings that were erected during the colonial period. Independence Square is a major focal point of the city's downtown, while the Museum of Natural History and the Museum of the Revolution, which tells about Mozambique's struggle against colonialism, are two of the most important visitor attractions in the city.

Culture and Society. As a port city, Maputo has a diverse society that blends cultural influences from across the Indian Ocean basin with those of native Bantu peoples and Portuguese colonialists. Seafood is an important part of the diet. The population is mostly poor and the majority of residents live in shanty town neighborhoods without access to clean water and sanitary sewage disposal. Thousands of Maputo residents live at the city's garbage dump at Huléne near the airport where they make a living picking through trash. The city ranks as one of the dirtiest in the world. In 2011, Maputo played host to some 5,000 athletes from 53 African nations in the All-Africa Games.

Further Reading

Brooks, Andrew. "Riches from Rags or Persistent Poverty? The Working Lives of Second-hand Clothing Venders in Maputo, Mozambique," *Textile: The Journal of Cloth Culture* 10, no. 2 (2012): 222–37.

Jenkins, Paul. "Maputo," *Cities: The International Journal of Urban Policy and Planning* 17, no. 3 (2000): 207–18.

Söderbaum, Fredrik and Ian Taylor, eds. *Regionalism and Uneven Development in Southern Africa: The Case of the Maputo Development Corridor*. Burlington, VT: Ashgate, 2003.

MASERU

Maseru is the capital and largest city of Lesotho (officially the Kingdom of Lesotho), a small landlocked country that is an enclave surrounded by the Republic of South Africa at the southern tip of the African continent. Maseru is located on the Caledon River in the west of the country, one of Lesotho's borders with South Africa. The river is called the Mohokare in the local Sesotho language. The Maseru Bridge crosses the river and connects Maseru to the small South African town of Ladybrand on the other side. The population of Maseru is approximately 228,000. It is the only sizable city in Lesotho. The name Maseru means "place of the red sandstone" in Sesotho.

Historical Overview. Maseru was founded in 1869 as a British police camp after the Free State-Basotho Wars and was made the administrative capital of the British protectorate Basutoland. The capital was moved to Cape Colony in 1871 and then back to Maseru in 1884. During the Gun War of 1881, many buildings were burned in Maseru.

Basutoland achieved its independence from Great Britain in 1996 and the name was changed to the Kingdom of Lesotho. Maseru then became a national capital. It grew quickly after independence because of migration from the Lesotho countryside. In 1998, in the wake of widespread fraud in an election in Lesotho, there was severe rioting in Maseru and most of the city was burned and looted. Troops from South Africa intervened in what was codenamed Operation Boleas to restore order. The city has still not been fully rebuilt after that episode.

Major Landmarks. Maseru is not noted for landmark architecture or significant visitor attractions. Its main buildings are the Royal Palace, the Parliament, and the State House, as well as the Roman Catholic Cathedral of Our Lady of Victories and St. John's Anglican Church. Setsoto Stadium is the national stadium of the country and is used mostly for football (soccer) matches. The Basotho Hat Shop is a souvenir store shaped like a hat.

Culture and Society. Both Lesotho and its capital city Maseru are very homogenous in ethnic makeup, with more than 99 percent of people being Basotho. Sesotho is the dominant language and is the official national language along with English. Indeed, the name of the country means "land of people who speak Sesotho." Approximately

90 percent of the population is Christian, about evenly divided between Protestants and Roman Catholics. The country has one of the highest literacy rates in Africa. Poverty rates are high and much of the economy depends on remittance income from migrant laborers to South Africa. Life expectancy is low, 41.18 years for men and 39.54 years for women. This is largely because of an extremely high rate (23.2 percent) of HIV/AIDS afflictions in the country.

Further Reading

Murray, Colin. *Families Divided: The Impact of Migrant Labor in Lesotho.* Cambridge, UK: Cambridge University Press, 1981, reissued in 2009.
Rosenberg, Scott, Richard F. Weisenfelder, and Michelle Frisbie-Fulton. *Historical Dictionary of Lesotho.* Lanham, MD: Scarecrow Press, 2004.

MBABANE

Mbabane is the administrative capital and largest city of the Kingdom of Swaziland, commonly referred to as Swaziland, a small land-locked country in the south of Africa that is almost entirely encircled by the Republic of South Africa except for a much shorter border on the east with Mozambique. The city is in the west-center of the country on the Mbabane River in the Mdimba Mountains. It addition to being the national capital, Mbabane is the capital of Hhohho, Swaziland's northwestern district. The population of Mbabane is about 95,000 (2003).

Even though Mbabane is the administrative capital, the legislative capital and the royal capital of Swaziland are both in Lobamba, a much smaller city also in Hhohho only about 10 miles (16 km) to the south of Mbabane. The population of Lobamba is about 5,800 (2003).

Historical Overview. The history of Swaziland as a separate entity goes back many centuries and was recognized as such by the British Crown during the time of colonialism in southern Africa in the 19th and 20th centuries. The country gained its independence from the United Kingdom in 1968. It is a unitary parliamentary democracy with a constitution, as well as an absolute monarchy. The present King is Mswati III, the head of the Swazi Royal Family. Mbabane has been capital of Swaziland since 1902 when the capital was moved from Bremersdorp, a city now called Manzini in the center of the country. The name Mbabane comes from the name of a local Chief, Mbabane Kunene.

Major Landmarks. The main attraction of Mbabane is the Swazi Market located in the center of the city on the banks of the Mbabane River. In nearby Lobamba, the main landmarks include the Embo State Palace, the Royal Kraal, the Swazi National Museum, and the Parliament of Swaziland.

Culture and Society. Most residents of Swaziland are ethnic Swazis. There are also Zulus and a small minority of white Africans, mostly of British origin and Afrikaner. The main languages are SiSwati (also known as Swazi) and English, as well as Zulu. More than 80 percent of the population is Christian. The country as a whole consists mostly of subsistence farmers and herders, but in cities such as Mbabane there are people engaged in commerce, handicrafts, and other urban occupations. Some Swazi work in the mining industry in South Africa. For the most part, Swaziland is a poor country. The HIV/AIDS epidemic has ravaged Swaziland since the first reported incidence in 1986 and is a significant impediment to national economic and social progress. Each year in Lobamba, Swazi people celebrate the Reed Dance in honor of the Queen Mother and Incwala in honor of the King. These are joyful ceremonies with singing and dancing, and the wearing of traditional attire.

Further Reading

Grotpeter, John J. and Alan R. Booth. *Historical Dictionary of Swaziland*. Lanham, MD: Scarecrow Press, 2000.
Rose, Laurel L. *The Politics of Harmony: Land Dispute Strategies in Swaziland*. Cambridge, UK: Cambridge University Press, 2006.

MELEKEOK

Melekeok is the capital of Palau, officially the Republic of Palau, an island nation in the Pacific Ocean to the east of the Philippines, south of Japan and north of the island of New Guinea. Sometimes the name of the country is spelled Belau. The archipelago has also been known as the Black Islands. The entire country has a population of 20,056 (2005). Melekeok is located on the east-central coast of Babeldaob Island, the country's largest, and has a population of 271. The entire state of Melekeok has a population total of 391. Hence, the capital is but a settlement or a village, and not a city. Melekeok is the world's smallest national capital by population.

Historical Overview. Palau was first settled some 3,000–4,500years ago by migrants from the Philippines or the Sunda Islands in Southeast Asia. British traders visited the islands in the 18th century, and in the 19th century the islands came under Spanish control. In 899, Spain sold its claim in the islands to Germany. After Germany's loss in World War I, the islands were administered by Japan. They were taken by the United States in 1944 and served as a strategic American base in its war against the Japanese Empire. In 1947, the United States assumed formal control of the islands as part of the United Nations Trust Territory of the Pacific Islands. In 1979, four of the districts of the Trust Territory ratified a constitution to become the Federated States of Micronesia, but Palau and the Marshall Islands opted to not join and became independent countries. Palau's independent status became official in 1994. Palau depends heavily on foreign aid from the United States and has voted with the United States in the United Nations, even when almost all member countries oppose the American position.

Melekeok has been the official capital of Palau since 2006 when the capital was transferred there from Koror, the island group's largest settlement, located on Koror Island to the south. The actual offices of government are in Ngerulmud, an even smaller settlement on Melekeok State than Melekeok, located about a little more than 1 mile (2 km) inland.

Major Landmarks. The capitol building in Ngerulmud reflects the architectural influence of the U.S. capitol building in Washington, DC. The Odalmelech Stone Faces are images of deities carved from stone that have been dated to about 895 AD. There are sand beaches along the shoreline of Melekeok.

The capitol building of Palau. (Peter Robert)

Culture and Society. Palauans are a mix of Melanesian, Micronesian, and Austonesian people. There are also citizens of mixed Palauan and Asian ancestry. The official languages are Palauan and English. About one-half of the population is Roman Catholic, 21 percent is Protestant, and 9 percent Modekngei, a blend of Christianity and traditional Palauan religion and fortunetelling. There is also a small Jewish community that in 2009 sent three athletes to the Maccabiah Games in Israel. Baseball has been a popular sport in Palau since being introduced to the islands by the Japanese in the 1920s. Traditional society in Palau is strongly matrilineal.

Further Reading

Cook, Ben et al. *Federated States of Micronesia and Republic of Palau.* (No City Listed): Other Places Publishing, 2010.

Leibowitz, Arnold H. *Embattled Island: Palau's Struggle for Independence.* New York: Praeger Publishers, 1996.

MEXICO CITY

Mexico City is the capital and largest city of the Republic of Mexico, the largest country in Central America and southern neighbor of the United States. The city's formal name is *Ciudad de México* (City of Mexico) or *México Distrito Federal* (Mexico Federal District); it is sometimes referred to simply as *México DF* or just DF Mexico City is located in the center of the country in the Valley of Mexico and is surrounded by active volcanic mountain peaks.

The population of Mexico City is about 8.84 million, while that of the metropolitan area as a whole is approximately 21 million. By either count, Mexico City is the largest urban place in the Western Hemisphere and one of the largest in the world.

The city's minimum elevation is 7,217 ft (2,200 m), while the volcanoes reach 16,400 ft (5,000 m). The city is built on the dry bed of a former lake, Lake Texcoco, and has enormous problems of subsidence as a result. There is also a shortage of water at high elevation, with requirements for extraordinary engineering works to pipe water from other regions. Other aspects of Mexico City's site are hazards from volcanic eruptions and earthquakes. The valley location concentrates the city's air pollutants, making Mexico City one of the world's foulest cities in terms of air quality.

Historical Overview. Mexico City was built in the 1520s by Spanish conquistadors led by Hernan Cortés on the site of Tenochtitlan, the capital of the Aztec empire and one of the greatest cities that the world had ever known. The Spanish conquered it in 1521 and destroyed it to rubble. They used forced Aztec labor to build their city, and erected their churches over the sites of demolished Aztec temples. The Spanish had conquered the Aztecs with a bit of luck and a bit of guile, and completely destroyed Tenochtitlan because it was evidence of a superior native culture in the Americas. Their new capital, which they had at first named *México Tenochtitlán*, was laid out according to instructions from the King of Spain for all colonial towns: a grid pattern of streets and a large central plaza along which is the town's main church or cathedral and the main government buildings. The name *La Ciudad de México* was given to the city in 1585. It served as capital of Spanish colonies in Middle America until Mexican independence was achieved in 1824. Since then, the city has been capital of Mexico.

Mexico City was occupied by American troops near the end of the Mexican-American War of 1846–1848. On September 13, 1947, the Americans stormed Chapultepec Castle, then a military school, as part of their assault on the city. Among those who defended the site to their death were six teenaged cadets who are known collectively as the Niños Héroes (child heroes) of Mexico. Strong-armed dictator Porfirio Díaz ruled Mexico for most of the years between 1876 and 1911 and gave Mexico City an impressive European look with beautiful boulevards, parks, museums, an opera house, and other grand buildings. He did not do much for the poor, however, and repressed dissent. He resigned in 1911 because of strong opposition, and Mexico was plunged into revolution from 1910 until about 1929. Great social inequality continues to be a defining characteristic of Mexican society.

Mexico City grew very rapidly in the 20th century, especially in the decades after World War II. As the leading industrial and business center of Mexico, it was an irresistible magnet for migrants from the Mexican countryside who poured into the capital in search of jobs and opportunity. The global status of Mexico City was enhanced with its hosting of the 1968 Summer Olympics, the first time the Olympics were in a developing country. There was much construction in the city in preparation. On September 19, 1985, a powerful earthquake brought considerable damage and loss of life to Mexico City. As a result of Olympics construction and reconstruction after the earthquake, archaeologists were able to learn many

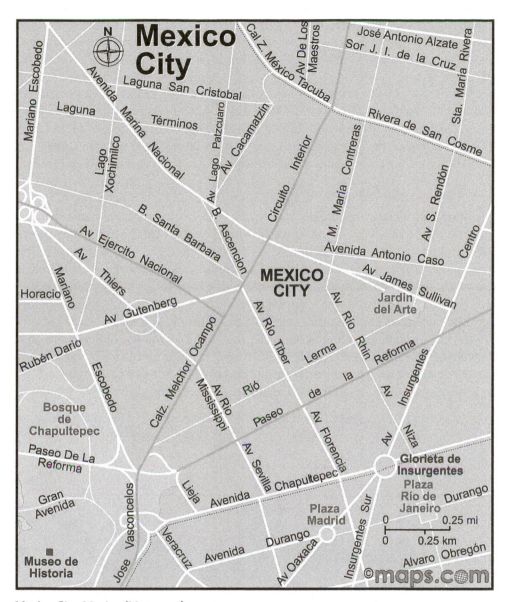

Mexico City, Mexico (Maps.com)

more details about the city of Tenochtitlan whose ruins are scattered beneath modern Mexico City.

Major Landmarks. The center of Mexico City is the Zócalo or Plaza de la Constitución that dates back to the original Spanish construction of the city. Facing it is the Metropolitan Cathedral and the National Palace. The ruins of the Aztec Templo Mayor (Major Temple) are on display nearby. The Palacio de Bellas Artes, now an art museum is also in the city center. It is a beautifully ornate building from the Díaz era that is so heavy from its marble that it has been sinking into the soft much on which Mexico City is built. The most symbolic avenue is Paseo de la Reforma modeled after the Champs-Élysées in Paris. It runs from the city center to Chapultepec Castle, another major landmark, and is the address for part of Mexico City's financial district. Also on Paseo de la Reforma is a high-column independence monument with the iconic gold Angel of Independence at the top. In Chapultepec Park itself is another attraction. Within it are many museums, monuments, and a zoo. The Nation Museum of Anthropology is one of the best museums of its kind in the world. Other important sites are the shrine to Our Lady of Guadalupe and the park and gardens at Xochimilico. The Coyoacán district of Mexico City has museums and landmarks related to Mexican artists Diego Rivera and Frida Kahlo. The Museo Soumaya is a unique post-modern construction that holds the private art collections of Carlos Slim, Mexico's richest man.

Culture and Society. Mexico City is marked by great social and economic contrasts, with upscale districts such as the Santa Fe business district being worlds apart from crowded squatter communities located at the city's margins. The largest squatter community in the world is Nezahualcóyotl, or simply Neza, a sprawling development of more than 1 million inhabitants on the old bed of Lake Texcoco. It is now formally integrated into Mexico City, but is known for poverty, high crime, and gangs of young toughs known as *cholos*. Drug trafficking is a major business in Mexico, and Mexico City has been a battleground between rival gangs and between law enforcement and traffickers. Wealthy residents of Mexico City and many businesses rely on heavily armed private guards for protection against crime. The tourist areas of Mexico City are very aggressively patrolled by police and are generally safe.

Historic Metropolitan Cathedral in Mexico City with ruins of Aztec Tenochtitlan in the foreground. (Photo courtesy of Roman Cybriwsky)

More than 90 percent of the Mexico City population is Roman Catholic. Spanish is the principal language, but languages such as Náhuathl, Otomi, and Mixtec are spoken as well by migrants from indigenous ethnic groups. English is widely understood in the elite sections of the city and in tourist areas such as the urban center and the Zona Rosa hotel and nightlife district.

Further Reading

Canclini, Néstor García. "Mexico City, 2010: Improvising Globalization," in Andreas Huyssen, ed., *Other Cities, Other Worlds: Urban Imaginaries in a Globalizing Age*, 79–98. Durham, N: Duke University Press, 2008.
Gallo, Ruben, ed. *The Mexico City Reader*. Madison: University Wisconsin Press, 2004.
Hernandez, Daniel. *Down and Delirious in Mexico City: The Aztec Metropolis in the Twenty-First Century*. New York: Scribner, 2011.
Lida, David. *First Stop in the New World: Mexico City, The Capital of the 21st Century*. New York: Riverhead Books, 2008.

MINSK

Minsk is the capital and largest city of Belarus, a country in Eastern Europe that is between Poland and the western border of Russia, and north of Ukraine. Lithuania and Latvia border Belarus to the northwest. From 1919 to 1991 when the Soviet Union began to come apart, Belarus was one of the 15 Soviet republics, the Belorussian Soviet Socialist Republic, and Minsk was its capital. The city is in the approximate center of the country on the Svislach and Namiga Rivers. In 2009, the city had a population counted at 1,836,808.

Historical Overview. The officially recognized founding date of Minsk is September 2, 1067, although it is probable that the settlement had existed for some time earlier. The *Primary Chronicle of Ancient Rus'* (also known in English as the *Tale of Bygone Years*), which was written in 1113, mentions the city in the context of a landmark battle on the Namiga River in 1067. From 1129 to 1242 Minsk was ruled from Kyiv, the Ukrainian city that was the capital of a principality named Rus'. After 1242 it was part of the Grand Duchy of Lithuania. In 1413, Lithuania entered into a union with Poland, and Minsk became part of a Polish-Lithuanian Commonwealth. In 1499, the city received town privileges under the Magdeburg Law. In the 17th and 18th centuries, as a result of various wars Minsk was ruled alternatively by Russia, Poland, and Sweden. In 1793, Minsk was annexed by Russia into its growing empire. Except for the World War II years, 1941–1944, when the city was occupied by Nazi forces, Minsk remained under Russian or Soviet control until Belarusian independence in 1991. Minsk suffered greatly during World War II, with much destruction to the city and many civilian deaths. The Jewish residents of Minsk and other cities of Belarus were herded into a large ghetto in Minsk that had as many as 100,000 inhabitants, the majority of whom perished in the Holocaust.

After the war, Minsk was reconstructed according to principles of Soviet city planning, with large government buildings and monumental plazas in the center, and big factories and expansive estates of apartment blocks for workers in the periphery. The city's population grew rapidly in the 1960s and 1970s because of industrialization. The city's biggest factories were the Minsk Tractor factory, the Minsk Automobile Factory, and the Minsk Refrigerator Factory. In 1974, Minsk was honored by Soviet Union authorities as a Hero City because of its fight against Nazi invaders. Because of the authoritarian rule of President Alexander Lukashenka (since 1994), Minsk retains more of a Soviet look and feel than most other cities of the former Soviet Union. The population is no longer growing rapidly because of economic difficulties and emigration.

Major Landmarks. Independence Square is in the Center of Minsk. The huge Soviet-era House of Government is nearby, and a high statue of Soviet Communist leader Vladimir Lenin stands in front. Victory Square has an obelisk monument to victory in World War II. The former site of the Minsk Ghetto has a haunting memorial to Holocaust victims called "The Pit." Not far from the center of Minsk is a small Old Town district with churches and other buildings, mostly from the 19th century, that survived the destruction of World War II. The new National Library of Belarus completed in 2006 in an outlying residential district is a truly spectacular modern structure. There is a rooftop deck that offers a panoramic view of the city and the forests at its edges.

Culture and Society. Belarusians are the main ethnic group in Minsk's population, accounting for about 80 percent of the total. Most of the rest are Russians (16%) and Ukrainians (2%). The Belarusian language has enjoyed a revival since independence, but Russian is the dominant language for everyday use, even among ethnic Belarusians.

A stern Belarus government building in Minsk with a statue of Russian revolutionary leader Vladimir Lenin in front. (Photo courtesy of Roman Cybriwsky)

During the time of Polish rule, Polish was the official language. Before the killing of Jews in the Holocaust, Minsk had a large Jewish minority, and even a majority at the turn of the 20th century before mass emigration. With few Jews remaining in the city, the main religious affiliations are Orthodox and Roman Catholic.

Further Reading

Barykina, Natalia. "Architecture and Spatial Practices in Post-Communist Minsk: Urban Space under Authoritarian Control," *Spaces of Identity* 8, no. 2 (2008).

Epstein, Barbara. *The Minsk Ghetto 1941–1943: Jewish Resistance and Soviet Internationalism.* Los Angeles: University of California Press, 2008.

Liskovets, Irina. "Trasjanka: A Code of Rural Migrants in Minsk," *International Journal of Bilinguialism* 13, no. 3 (2009): 396–412.

MOGADISHU

Mogadishu is the capital and largest city in the Republic of Somalia, a country at the Horn of Africa where the continent borders the Gulf of Aden and the Indian Ocean. The country's recent history has been tragic, and Somalia is currently classified as a failed state with no effective central government, considerable violence, and great poverty. The city is located on the Indian Ocean coast in the Benadir region in the southern part of the country. Accurate population counts are not possible given recent strife in Somalia and the many refugees, but the population of Mogadishu is estimated to be about 2 million in round numbers. The city is popularly known as Xamar.

Historical Overview. Mogadishu has a long history. The famous Moroccan traveler Ibn Battuta visited the city in the 14th century and described it as a large center with prosperous merchants and high-quality fabrics for trade. Later, the city resisted European attempts at colonization, particularly by the Portuguese. In the 19th century, Mogadishu was ruled by the sultanate in Zanzibar and was leased in 1892 to Italy. In 1905, the city was purchased by Italy that made it the capital of its colony Italian Somaliland. Many Italians settled in the city in the early 20th century and invested in industrialization. Other parts of Somalia were colonized by the British and by the French. Somalia gained its independence in 1960. Civil war broke out in Somalia in 1990 and Mogadishu has lived without peace since despite attempts by the United Nations, the United States, and countries from the African Union to intervene. In 1993, during what is called the Battle of Mogadishu, the bodies of U.S. soldiers were dragged through the city by Somali militants. By 2008, it was reported that the city had emptied of much of its population and that many neighborhoods had been totally devastated by the warfare.

Major Landmarks. The Abra-Rucum Mosque (the mosque of the four pillars) is the main landmark of Mogadishu. It dates to 1269 and still stands because warring factions in Somalia consider this to be a holy place. Other notable places are the Hamarwein old town district, the Bakaara Market, and Gezira Beach, which was once a tourist attraction and considered to be one of the world's most beautiful beaches.

Culture and Society. Mogadishu has long been a multiethnic city comprised of Bushmen migrants from the surrounding countryside, and Arabs and Persians who entered the city via the port. During the colonial period the city also had a substantial Italian population. Since the civil war, the city has both attracted refugees from areas of fighting in the country and created refugees when the Mogadishu was itself under siege. The population of Mogadishu is mostly Islamic. The major languages are Somali and Arabic. Many Somalis have emigrated to begin new lives abroad to other Arabic-speaking countries, and to the United States, Canada, and Italy, among other destinations. The city of Mogadishu is currently considered to be the most lawless city on earth, and faces a major crisis of too many refugees and insufficient food, potable water, and medical assistance.

Further Reading

Bowden, Mark. *Black Hawk Down: A Story of Modern War*. New York: Grove Press, 1999.

Grünewald, François. "Aid in a City at War: The Case of Mogadishu, Somalia," *Disasters* 36, supplement (July 2012): S105–25.

Kapteijns, Lidwien and Maryan Muuse Bogor. "Memories of a Mogadishu Childhood," *International Journal of African Historical Studies* 42, no. 1 (2009): 105–26.

MONACO

Monaco, officially the Principality of Monaco, is a sovereign city-state and, therefore, is itself a capital. The principality has a population of 35,986 and measures only 0.76 square miles (1.98 sq km), making it the second-smallest country in the world (after Vatican City) and the most densely settled. It is located on the French Riviera on the Mediterranean Sea and borders only France and the sea. The nearest border with Italy is only less than 10 miles (16 km) away. Monaco is divided into four traditional quarters: Monaco-Ville, La Condamine, Monte Carlo, and Fontvieille. A fifth quarter called Monghetti is a "modern" quarter of the city located inland from shoreline. Because the palace of the Albert II, Sovereign Prince of Monaco, is located in Monaco-Ville, one could argue that this district of the city-state is Monaco's capital.

Historical Overview. Monaco was founded in 1215 as a colony of the Italian port and city-state Genoa. Since 1297, it has been ruled more or less continuously by the House of Grimaldi. After the Grimaldis purchased Monaco in 1419 from the King of Aragon, they became undisputed rulers of the territory. From 1642 until the French revolution in 1793 Monaco was a protectorate of France, and from 1793 to 1814 it was under direct French control. From 1815 to 1860 it was a protectorate of the Kingdom of Sardinia. In 1861, Monaco became a sovereign state. The Prince of Monaco held absolute authority until the promulgation of a constitution in 1911. In 1918, Monaco and France entered into an agreement that aligned common military, political, and economic interests. A new constitution promulgated in 1972 that abolished capital punishment and provided for female suffrage. In 1993, Monaco joined the United Nations. Ranier III was Prince of Monaco from 1949 until his death in 2005, a rule of nearly 56 years. He married the American actress Grace Kelly in 1956. Their son Albert II is the current Prince of Monaco.

Major Landmarks. The Grand Casino (Monte Carlo) is an iconic landmark of Monaco. It has a high admission fee and a strict dress code, and is known as a haunt of the rich, famous, and high-roller gamblers. Other landmarks are Palais Princier (the Prince's Palace), Monaco Cathedral, and the Monaco Opera House (Salle Garnier). The Oceanographic Museum and Aquarium is another popular attraction. The city also has beautiful and picturesque neighborhoods: Monaco-Ville, also known as "le rocher" or "the rock," is an unspoiled medieval urban center; while La Condamine is a beautiful area alongside the yachts on the waterfront.

Culture and Society. Monaco is a very prosperous society. The economy is based on tourism, commerce, and finance. Because there is no income tax on individuals, the country has become a tax haven for wealthy Europeans and others who move there as tax refugees. The official language of Monaco is French, although English, Italian, and Monegasque (a blend of French and Italian that is unique to Monaco) are also spoken. The Roman Catholic faith is the official religion and most Monegasque citizens are at least nominally Catholics, but the constitution guarantees freedom for all religious worships. The Formula One Monaco Grand Prix and the Monte Carlo Rally are two automobile racing events that attract many visitors to Monaco and help to underpin the local economy.

Further Reading

Eccardt, Thomas M. *Secrets of the Seven Smallest States of Europe: Andorra, Liechtenstein, Luxembourg, Malta, Monaco, San Marino, and Vatican City.* New York: Hippocrene Books, 2005.

Edwards, Anne. *The Grimaldis of Monaco,* New York: Morrow, 1992.

Glatt, John. *The Royal House of Monaco: Dynasty of Glamour, Tragedy and Scandal.* New York: St. Martin's Press, 2000.

MONROVIA

Monrovia is the capital and largest city of Liberia, a West African country on the coast of the Atlantic Ocean. It is located on the coast in the northwest of the country at Cape Mesurado, where the Mesurado River empties into the sea. Monrovia is a port city as well as the national capital. The population of Monrovia is approximately 1 million (2008), nearly 30 percent of the population of Liberia. The coastal area of Liberia along which Monrovia is located is sometimes referred to as the Pepper Coast or the Grain Coast because of the commodities that were once shipped from there.

Historical Overview. The area of Monrovia has been inhabited for many centuries. Portuguese navigators arrived in the 1560s and named it Cape Mesurado. Liberia's modern history is traced to 1821 with the arrival of emancipated African slaves from the United States under the sponsorship of the American Colonization Society. In 1822, a second ship of freed slaves landed at Cape Mesurado and established Christopolis. It was renamed Monrovia in 1824 after then president of the United States James Monroe, a supporter of the return of emancipated American slaves to Africa. In 1845, the independent Republic of Liberia was formed and Monrovia became a national capital. The city had two parts: (1) Monrovia proper, where most of the city's Amero-Liberians are settled and (2) Krutown, a section for native African ethnic groups, most prominently the

Krus. Since 1980, when a military coup led by Samuel Doe ousted the government of President William R. Tolbert, Liberia has been marked by political instability, dictatorial government, and violence. From 1989 until about 2003, two civil wars raged in Liberia. Monrovia was badly damaged during a siege of the city by Liberians United for Reconciliation and Democracy (LURD) rebels against the repressive government of President Charles Taylor between July 18 and August 14, 2003, in which an estimated 1,000 civilians were killed. The conflict ended when a West African peace force supported by American troops entered the city, sending Taylor into exile. Earlier, Liberian women had demonstrated en masse for peace in a fish market in Monrovia. The situation in Liberia is now much improved since the election in 2003 of Ellen Johnson Sirleaf as president, the first female to be elected head of state of an African nation.

Major Landmarks. Monrovia has few major landmarks other than the waterside market, the National Museum of Liberia, and the country's various government buildings. In suburban Paynesville is Blo Degbo a natural rock formation shaped like a human face. From 1976 until it was taken down in 2011, Paynesville also had the Paynesville Omega Transmitter, a 1,368-ft-high (417 m) radio antenna that was the highest structure in Africa.

Culture and Society. The population of Liberia is comprised of 16 indigenous ethnic groups, plus other various foreign ethnic minorities. Descendants of Americo-Liberians make up about 2.5 percent of the population. The foreigner population includes Lebanese, South Asians, and migrants from various West African countries. More than 30 languages are spoken in Liberia as first languages. English is the common and the country's official language. The dialect that is spoken in Liberia is called Liberian English. About 85 percent of the population is Christian, with Muslims, mainly from the Mandingo ethnic group, making up most of the rest of the population. The 1968 novel *Murder in the Cassava Patch* by Bai T. Moore is regarded as Liberia's best-known novel. It is read by every school student in the country. Liberia is one of the three countries in the world, along with Myanmar and the United States, that does not officially use the International System of Units, the modern metric system.

Further Reading

Dunn, Elwood D., Amos J. Beyan, and Carl Patrick Burrowes. *Historical Dictionary of Liberia.* Lanham, MD: Scarecrow Press, 2000.

Waugh, Colin M. *Charles Taylor and Liberia: Ambition and Atrocity in Africa's Lone Star State.* London: Zed Books, 2011.

Williams, Gabriel I. H. *Liberia: The Heart of Darkness.* Trafford: Victoria, BC, 2002.

MONTEVIDEO

Montevideo is the capital and largest city of Uruguay, formally the Oriental Republic of Uruguay, a country in South America on the coast of the South Atlantic Ocean and the north shores of the estuary of the Rio del la Plata. Argentina lies across the estuary, while to the north Uruguay is bordered by Brazil. Montevideo is located in southernmost Uruguay on the Rio de la Plata and is the country's principal port as well as the chief administrative and business center. The population of Montevideo is about 1.4 million, with about 1.8 million in the metropolitan area as a whole. Uruguay as a whole has a population of 3.3 million, so the metropolitan area of Montevideo makes up about 55 percent of the national population.

Historical Overview. Montevideo was founded by Spanish settlers led by Bruno Mauricio de Zabala, who arrived from Spanish Buenos Aires across the Rio de la Plata in 1726 specifically to establish a base for Spain that would counter the Portuguese settlement of Colonia del Sacramento that had been founded in 1680 by colonists from Portugal. Other early settlers arrived from the Spanish Canary Islands. From the beginning, Montevideo was laid out as a fortress town in order to assert Spanish claims to both sides of the Rio de la Plata in the face of the Portuguese presence in Brazil. In the late 18th and 19th centuries, Montevideo and Buenos vied for commercial and supremacy in the region. For a time in 1807, Montevideo was occupied by British troops as a result of the Battle of Montevideo of the Napoleonic Wars, but it was recaptured by Spain within the same year. The struggle for independence from Spain took full root at about the same time, but so did Portuguese ambitions to annex what is today Uruguay to Brazil. The Portuguese captured Montevideo in 1817 and annexed Uruguay four years later. Brazil broke from Portugal in 1822, and in 1825, a revolutionary group called the Thirty-Three Orientals (after the province of Banda Oriental) declared Uruguay's independence. A 500-day war with Argentina followed, resulting in Uruguay becoming an independent nation in 1828 via the British-mediated Treaty of Montevideo. The country has prospered since from a strong agricultural economy, particularly exports through the port of Montevideo of wool, hides, and meat. The agricultural and industrial economies attracted many immigrants from Spain, Italy, and Central Europe to Montevideo in the late 19th and early 20th centuries, causing the city to grow quickly in population and become ethnically diverse and cosmopolitan.

The origin of the word "Montevideo" is uncertain and disputed, but it probably did not originate with the most commonly repeated thesis that it came from the words of a lookout on a 1520 Spanish ship captained by Magellan, who supposedly uttered *Monte vide eu*, "I see a mountain," when he sighted land.

Major Landmarks. The symbolic center of Montevideo is Plaza de Independencia (Independence Plaza). Within the plaza is an equestrian statue of national hero José Gervasio Artigas and his mausoleum. Many important buildings stand nearby, most conspicuously the eclectic 328-ft-high (100 m) Palacio Salvo, completed in 1928, once South America's tallest building and still an iconic landmark of Montevideo. Plaza Zabala is in Ciudad Vieja, Montevideo's Old Town, and focuses on an equestrian statue of the city's founder. Plaza de la Constitución, also in Ciudad Vieja, houses the Cabildo, the seat of colonial government, and the Roman Catholic Montevideo Metropolitan Cathedral. Major museums include the National History Museum, the Natural History Museum, and Museo Torres Garcia which is dedicated to the works of the famous Uruguayan artist. The Solis Theater is Uruguay's oldest theater. Barrio Reus is a small historic neighborhood with picturesque streets and colorful houses. Montevideo has a popular beach on the Rio de la Plata.

Culture and Society. The large majority of residents of Uruguay is of European descent, most commonly from Spanish and secondarily Italian origin. More than 90 percent of the population is Roman Catholic. Montevideo is a city rich in cultural life, including theater, art, and musical performances, as well as dance, particularly tango. The city also has a tradition of great literature, thanks to great Uruguayan writers such as José Enrique Rodó and Eduardo Galeano, the latter the author of the 1971 *Open Veins of Latin America.* Football (soccer) is the most popular sport. Montevideo's Estadio Centenario hosted the first World Cup in 1930, which was won by Uruguay.

Further Reading

Canel, Eduardo. *Barrio Democracy in Latin America: Participatory Decentralization and Community Activism in Montevideo.* University Park: The Pennsylvania State University Press, 2010.

Garcia Ferrari, Maria Soledad. "Montevideo," *Cities: The International Journal of Urban Policy and Planning* 23, no. 5 (2006): 282–299.

Willis, Jean L. *Historical Dictionary of Uruguay.* Metuchen, NJ: Scarecrow Press, 1974.

MORONI

Moroni is the capital and largest city of the Union of the Comoros, usually referred to as the Comoros, a small island nation in the Indian Ocean to the east coast of Africa and northwest of Madagascar. The country consists of three main islands and several tiny ones, and excludes Mayotte, an island in the Comoros archipelago that voted to be administered as an overseas department of France rather than join with the other islands when they became independent. Moroni is located on the western coast of the island called Grande Comore, the largest and northwestern-most island of the Comoros group. Grande Comore is also called Ngazidja. The population of Moroni was about 60,000 in 2003.

Historical Overview. The Comoros have been settled for at least 1,500 years, probably initially by people who arrived by boat from the coast of Africa. In the seventh century the residents of the islands converted to Islam. In the Middle Ages, Comorans navigated the Indian Ocean and traded with the coasts of Africa and the Arabian Peninsula. The first Europeans to arrive were the Portuguese in 1505. The bloodthirsty Alfonse de Albuquerque sacked the islands in 1514. Afterward, the islands were attacked by pirates from Madagascar who took slaves from the Comoros. Colonial rule by France began in 1841. The French established sugar plantations, and exported coconuts, tortoise shells, vanilla, coffee, sisal, and ylang-ylang, a tree product used in the production of perfume. On July 6, 1975, the Comoros became independent and Ahmed Abdallah became the first president. There has been considerable strife among the islands of the Comoros, and from 1997 to 2002, two of the islands, Anjouan and Moheli, had seceded from the national union until a new constitution was put into effect. The Comoros was one of the countries that received damage from the Indian Ocean tsunami that followed the December 26, 2004, earthquake centered off the coast of Sumatra Island, Indonesia.

Moroni was founded in the 10th-century AD by Arab traders. It became capital of the Comoros in 1958 replacing Dzaoudzi. Its name means "in the heart of the fire" in the Comorian language, referring to its location near a large volcano.

Major Landmarks. The Old Friday Mosque (Ancienne Mosquée du Vendredi) is a prominent landmark in Moroni that stands on the shoreline of the city. Outside Moroni there are various beaches. Mount Karthala (7,746 ft; 2,361 m) is an extremely active volcano on Grande Comore Island located to the southeast of Moroni.

Culture and Society. The Comoros is a small and very crowded country, and is also very poor. Birthrates are high and a large percentage of the population is young. The leading sectors of the economy are fishing and agriculture. The population is mostly of African-Arab origin. The dominant religion is Sunni Islam. The common language is Comorian, a language that is related to Swahili. French and Arabic are also commonly heard. Comorian, French, and Arabic are the three official languages of the Comoros.

Further Reading

Newitt, M. D. D. *The Comoro islands: Struggle against Dependency in the Indian Ocean.* Boulder, CO: Westview Press, 1984.

Ottenheimer, Martin and Harriett Ottenheimer. *Historical Dictionary of the Comoros.* Lanham, MD: Scarecrow Press, 1994.

MOSCOW

Moscow is the capital and largest city in the Russian Federation, the world's largest country in size that stretches from a European exclave on the Baltic Sea and adjacent to Poland, through the vastness of Siberia and to the Pacific Coast of North Asia and Pacific islands within sight of Japan. There are 83 federal subjects in the Russian Federation. The city is located in the west of this vast empire, within the eastern reaches of Europe and within Russian ethnic territory, but its governance extends across many other ethnic territories that were acquired by Russia and the Soviet Union during more than three centuries of aggressive colonial expansion. The main geographical features of Moscow are the Moscow River

Kazan: Tatar Metropolis

Kazan is the capital of the Republic of Tatarstan, an oil-rich region within the Russian Federation near the eastern margins of European Russia. The city is at the confluence of the Volga and Kazanka Rivers and centers on a prominent fortress (*kremlin*) with landmarks of both Moslem-Tatar heritage and Russian Orthodoxy. The Kazan Kremlin is a UNESCO World Heritage Site. The Russian tsar Ivan the Terrible conquered the city in 1552 and massacred most of the indigenous population. The Annunciation Cathedral of Kazan was built soon afterward on the site of a mosque that the Russians had destroyed. After the 1991 fall of the Soviet Union, the Qolşärif Mosque was built in Kazan's Kremlin near the Orthodox church as a landmark of new religious freedoms. It is acknowledged to be one of the most beautiful places of worship in the world, and is the largest mosque in Europe outside Istanbul. Because of enormous oil deposits in Tatarstan, Kazan is a pleasant and prosperous city with much new construction and modern new facilities. It enjoys a rebirth of Tatar cultural life and Islamic faith.

that flows through the city, the heavily forested East European Plain that surrounds it, and location at 55°45'N latitude, making the city the world's north-most giant metropolis. The population of Moscow is about 11.5 million (2010), by some definitions the largest city in Europe.

Historical Overview. The history of Moscow dates to the mid-12th century when a prince named Yuri Dolgorukiy ordered the construction of a *kremlin* (fortress) at a location deep in the northern forests that would be safe from invaders on horseback. It evolved into the capital of the Russian Empire except for the years 1703–1918 when the newly built city of St. Petersburg was capital, and was capital of the Union of Soviet Socialist Republics from its founding in 1922 to its disintegration in 1991. During that time, there were "great" leaders who ruled from the city such as Ivan the Great (Ivan III; ruled 1462–1505) and Peter the Great (ruled 1662–1725; from 1703 in St. Petersburg); and leaders who are said to be "terrible" because of the terrors they inspired and deaths they caused: Ivan the Terrible (ruled 1533–1584) and Soviet leader Joseph Stalin (various high offices 1922–1953). Russia is now ruled from Moscow by President Vladimir Putin. In 1956, Moscow was awarded the title Hero City by Soviet authorities because of its heroic defense in the war against Nazi Germany. In 1979, the Soviet Union invaded Afghanistan, and the following year Moscow hosted the summer Olympics games that were in turn boycotted by many nations from around the world because of that invasion. In the late 1980 to 1991, the city was the focus of much political activity and popular demonstrations that led to the fall of the Soviet Union and declarations of independence by the former republics of the Soviet Union that had previously been under Moscow's control. On March 29, 2010, Moscow was the victim of two suicide bombings in the subways system that took at least 40 lives and injured an additional 100 individuals. This terrorist attack is believed to be related to Russia's conflict with Chechnya, an ethnic region in the south of Moscow that wants to break away from the Russian federation.

Major Landmarks. The Kremlin fortress has been rebuilt many times since its original construction and is now the iconic center of the city and symbol of government authority. It overlooks the Moscow River and features the ornate St. Basil's Cathedral (secularized since 1929) and Red Square within which lies the body the Soviet revolutionary leader Vladimir Lenin (1870–1924), among other

---

Grozny: Chechnya in Crisis

Grozny is the capital city of the Chechen Republic, a federal subject of the Russian Federation located north of the Caucuses Mountains. The population is overwhelmingly ethnically Chechen and the main religion is Islam. There are strong separatist feelings in Chechnya, but attempts to break away from Russia have been brutally suppressed. The city was greatly damaged in wars against Russia in 1994–1995 and in 1999–2000, and in a very costly battle for control of the city in August, 1996, with great losses of life. The word Grozny means "fearsome" in Russian and refers to a Russian fortress that was built there during expansion of the imperial frontier in the 19th century. Chechen separatists often refer to the city as Dzhokhar after Dzhokhar Dudaev, one of their political leaders. Most Russian residents have moved from Grozny because they did not feel welcome or safe, while Chechens in Moscow and other parts of the Russia are constant victims of ethnic harassment and intimidation by police.

landmarks. Lenin's mausoleum faces an historic large department store that has been converted since Soviet times into upscale shopping mall. A towering statue crafted of Peter the Great as navigator along the Moscow River is another interesting landmark. The statue was crafted by favored sculptor Zurab Tsereteli, and was originally of Christopher Columbus, but the head was changed to that of the Russian czar allegedly after Tsereteli could find no purchasers for his Columbus statue. Nearby is the reconstructed Cathedral of Christ the Savior. The original cathedral was consecrated in 1883 and stood until it was dynamited by Soviet authorities in 1931. The site then housed a large swimming pool. The new cathedral was built after Soviet times and was consecrated in 2000. It is gaudily adorned with dozens of statues of saints and other figures made by Tsereteli. The workshop-museum of Zurab Tsereteli is an interesting landmark off the beaten path. It is near the famous Moscow Zoo and the doorway to the workshop is decorated with figures of clowns. Other landmarks are the famous Bolshoi Theater, Ostakino Tower, television and

Moscow, Russia (Maps.com)

Controversial statue of Czar Peter the Great as navigator in the Moscow River and the Orthodox Christ the Savior Cathedral in the background. (Photo courtesy of Roman Cybriwsky)

communications tower that at 1,772 ft (540 m) was the world's tallest structure from 1967 to 1976 and is still the tallest in Europe and the Lubyanka Building, the former headquarters and prison building of the Soviet KGB and now the offices of Russian security agencies and the site of a museum about KGB repressions.

Culture and Society. Moscow has long been a major center for Russia's rich arts and cultural activities, and is known for ballet, concert halls, theater, and museums. It is also a center of sport, with popular followings for ice hockey, football (soccer), basketball, and other sports. The city also has a rich intellectual life, and was the focus of considerable dissident activity during the latter part of the Soviet period. Political fervor continues in the city with frequent protest activity in support of freedoms of the press, democracy, and honesty in elections. The city is reported to have more billionaires than any city in the world and is one of the most expensive in terms of cost of living. There is a great contrast within the population between those who have become excessively wealthy with the emergence of capitalist society and the many Russians and other ethnics in the city who are desperately poor. There is a serious problem of alcoholism in Moscow and low life expectancy by European standards, particularly among males.

Further Reading

Brooke, Carolin. *Moscow: A Cultural History.* New York: Oxford University Press, 2009.
Gessen, Masha. *The Man without a Face: The Unlikely Rise of Vladimir Putin.* New York: Riverhead Books, 2012.

Schlögel, Karl. *Moscow*. London: Reaktion Books, 1984.
Shevchenko, Olga. *Crisis and the Everyday in Postsocialist Moscow*. Bloomington: Indiana
    University Press, 2009.

MUSCAT

Muscat is the capital and largest city of Oman, officially called the Sultanate of Oman,
a country in the Middle East on the southeast coast of the Arabian Peninsula. Muscat
is located on the north coast of the country on the Arabian Sea's Gulf of Oman. The
population is 734, 697 (2010). The city has long been an important trading port.
Since the 1960s, the oil industry has also been an important element in Muscat's
economy.

Historical Overview. Because of its sheltered natural harbor and strategic
location, Muscat has been a city of seafarers for many centuries. Archaeological
evidence indicates that the site has been inhabited since at least the sixth-
millennium BC, and that had been engaged in trade centuries ago with the Indian
subcontinent. It was converted to Islam in the seventh century. Unification
of disparate tribes in Oman into a single state began in the ninth century.
In 1507, the Portuguese admiral Alfonso de Albuquerque attacked the city,
killing almost all the inhabitants and beginning nearly a century of intermittent
Portuguese control. At various times the city was under Persian rule, Ottoman
rule, and rule by local Omanis. In 1749, the al Said dynasty gained control of
Oman and has ruled ever since. In 1840, Said bin Sultan al Said took control
of Zanzibar off the Indian Ocean coast of Africa and designated it as his capital
until the island was lost to the al-Said rulers in until 1856. From the late 19th
century, Muscat has had to contend with many decades of uprisings by tribes
from Oman's interior, such as the Dhofar Rebellion of 1962. Qaboos bin Said al
Said has been Sultan of Oman since 1970 and has ruled as absolute monarch.
In 2011, Muscat was the scene of protests by citizens against government
corruption and other issues as part of the wider Arab Spring movement across
the Middle East.

Major Landmarks. Qasr al Alam is the royal palace of Oman. On either side
are the old Portuguese forts, al Jalali Fort and al Mirani Fort. Both are now
museums. Other museums include the Museum of Omani Heritage and the
National Museum of Oman. Major mosques in Muscat are the Sultan Qaboos
Grand Mosque, Ruwi Mosque, and Zawai Mosque. The Mutrah Souk is a
traditional marketplace in the city center. A major new landmark is the Royal
Opera House of Muscat.

Culture and Society. Islam is the predominate religion in Muscat and the
main language is Arabic. As a port city, however, the city has a wide mix of
ethnic groups, languages, and religious beliefs. Omani law permits freedoms
of religious practice, but non-Moslems are not permitted to proselytize. Many
of the expatriates in Muscat work in oil-related occupations or in trade.
The Corniche area at the inner waterfront of Muscat is a popular place for
holiday and evening strolls by visitors and residents alike, and for a variety of
inexpensive restaurants.

Further Reading

Allen, Rory Patrick. *Oman: Under Arabian Skies*. New York: Vanguard, 2010.
Critchfield, Lois M. *Oman Emerges: An American Company in an Ancient Kingdom*. Vista, CA: Selwa Press, 2010.
Hawley, Donald. *Oman*. London: Stacey International, 2005.
"Oman: Into the Future," *Middle East* 346 (June 2007): 49–55.

NAIROBI

Nairobi is the capital and largest city in Kenya, a country in East Africa on the India Ocean coast of the continent. The city itself is in the south-central part of the country, approximately midway between the Kenyan port city of Mombasa and Kampala, the capital city of Uganda, closer to the continent's interior. The city is near the equator at latitude 1°17′ S, but has a moderate climate because of relatively high elevation (5,450 ft; 1,661m). The name Nairobi comes from a local Masai language phrase that means "the place of cool waters," referring to the Nairobi River nearby.

The population of Nairobi is more than 3.1 million (2009), making the city the most populous in East Africa and one of the largest cities on the African continent in general. The city is disproportionately prominent in the economy and political affairs of Africa because it is the site of the headquarters of many international companies that do business in Africa, the location of United Nations offices for Africa and the Middle East, and host for many international NGOs and foreign aid associations. For these reasons, Nairobi is also venue for many international conferences. In addition, the city is an important hub for international tourism, being a gateway to national parks, safari tourism, and Mt. Kilimanjaro, Africa's highest peak located just across the border from Kenya in neighboring Tanzania.

Historical Overview. Nairobi was founded in 1899 as a supply depot for the Uganda Railway about midway between its two terminals in Mombasa and Kampala. It then became the headquarters for the railway company and, in 1905, the administrative center for the British East African Protectorate, as the British colony was known at the time. Local tribes, particularly the Nandi, resisted construction of the railway through their lands and many Africans viewed Nairobi as a foreign intrusion. Under British colonial administration, Nairobi was divided into three ethnic zones: an area for British and other Europeans on hillier terrain to the west of the city center; residential districts for Africans in the low-lying eastern flatlands; and a district for Indian residents in between. The Indians had originally been imported as railway construction workers, but they and their descendants eventually became prominent in local business. There were grand hotels in the center of Nairobi that catered to foreign game hunters. In 1920, the British East Africa Protectorate became the colony of Kenya, with Nairobi as its capital. The British High Commission for East Africa was later headquartered in Nairobi, so the city also administered the British colonies of Tanganyika, Uganda, and Zanzibar, in addition to Kenya.

Kenyan independence from Great Britain came on December 12, 1963, and the Republic of Kenya was proclaimed on December 12, 1964. Jomo Kenyatta was the first president and is regarded as the founder of independent Kenya. Independence was achieved after a long period of armed rebellion against the British, including the so-called Mau Mau Uprising or the Kenya Emergency of 1952–1960 by Kikuyu tribesmen. Periods of political instability continued in Kenya after independence, but in general the country has evolved into a peaceful, democratic society, and Nairobi is regarded as one of Africa's most progressive and most comfortable cities. On August 7, 1998, an Al Qaeda terrorist bomb was set off in a truck parked outside the U.S. Embassy building in Nairobi, killing 212 people.

Major Landmarks. The Kenyatta International Conference Center has an observation tower with a panoramic view of Nairobi. Other landmarks are a memorial to the victims of the 1998 bombing at the U.S. Embassy, Uhuru Gardens, a memorial to the struggle for independence, the Jomo Kenyatta monument in Nairobi's Central Park, and the Dedan Kimathi statue, a monument honoring one of the leaders of the Mau Mau Uprising. The Nairobi National Museum displays artifacts from Kenya's history and rich cultural heritage. The Times Tower, or the New Central Bank Tower as it is also called, in downtown Nairobi is the tallest skyscraper in East Africa (460 ft; 140 m). The Nairobi National Park borders the city. It is a game preserve and a very popular tourist attraction.

Culture and Society. The population of Nairobi is diverse and includes various Kenyan tribal groups, as well as minorities of Indians and Europeans. Swahili is the national language, along with English. While many Nairobians live comfortably, many others are poor and live in crowded slum neighborhoods, most notably Kibera and Mathare, two of the largest slum neighborhoods of Africa. Because of the poverty, Nairobi has a high crime rate. Westlands is a Nairobi district with an active nightlife, perhaps the best in all Africa. There is an active music scene in Nairobi, including African genres called *benga* and *soukous*, as well as Kenyan hip-hop.

Further Reading

Charton-Bigot, Helene and Deyssi Rodriguez-Torres. *Nairobi Today: The Paradoxes of a Fragmented City*. Dar es Salaam, Tanzania: Mkuki na Nyota Publishers, 2010.

Gathanju, Denis. "Nairobi Redraws its Planning Strategy," *Planning* 75, no. 7 (2009): 28–31.

Hendricks, Bob. *Urban Livelihoods, Institutions and Inclusive Governance in Nairobi: 'Spaces' and Their Impacts on Quality of Life, Influence, and Political Rights*. Amsterdam: Vossiuspers UvA, 2010.

NASSAU

Nassau is the capital and largest city of the Commonwealth of the Bahamas, a country of 29 islands and 661 cays, plus more than 2,000 tiny uninhabited islets, in the North Atlantic Ocean east of Florida. The city is on an island named New Providence, which is in the north-center of the island group, and has a population of 353,658 (2010 census). The city's population is about 70 percent of the national population. In addition to being the government center, Nassau is the main business center of The Bahamas and a popular tourism destination. The entire island of New Providence is now urbanized.

Historical Overview. The Bahamas were originally inhabited by Taino (Lucayan) people. Their population was decimated by diseases soon after the arrival of Europeans (Columbus's 1492 landing in the New World was on one of the islands of the Bahamas). The first European settlers may have been English Puritans who arrived in 1648 from Bermuda. The first name of Nassau was Charles Town, after English King Charles II. The town was burned to the ground in 1684 by Spanish forces led by Juan de Alcon in a conflict between Spain and England about control of the North Atlantic Ocean for shipping. It was rebuilt in 1695 and renamed Nassau, a Dutch name associated with the family lineage of England's King William III. At the start of the 18th century, Nassau was a haven for pirates who preyed on Atlantic and Caribbean shipping, particularly on Spanish ships laden with booty from New Spain, and on ports along the coasts. For a time, pirates proclaimed Nassau to be a pirate republic. The English eventually regained control of Nassau in 1718, but had to contend with an attack by Spanish forces in 1720. In 1776, during the American War of Independence, Nassau was briefly occupied by American rebels against the English crown. Spain captured Nassau in 1782, only to lose it in 1783. Nassau's Fort Charlotte was built soon thereafter by the British in order to better protect Nassau and their island colony. Several thousand English loyalists from former English colonies in North America settled New Providence and other islands of the Bahamas after the Americans won their war of independence. They became the dominant population on the islands, established a plantation-based economy, and exploited African slaves. Slavery was abolished in England's colonies in 1834. Bahamian independence was achieved on July 10, 1973.

Major Landmarks. Downtown Nassau, centered on Bay Street near the port, is a popular shopping area that is often crowded with tourists. Nearby are the Old Town district and the pretty, pink Parliament Building of the Bahamas. There is a fine monument to England's Queen Victoria in front. The main museums in Nassau are the National Art Gallery of the Bahamas and tourist-oriented Pirates of Nassau Museum. The ruins of the historic Fort Fincastle from 1793 stand on a low hill overlooking Nassau. Nearby Paradise Island, with its enormous Atlantis Resort, is a major destination for foreign tourists, primarily from the United States and Canada. The main hotel and ocean beach district of Nassau is called Cable Beach.

Culture and Society. Approximately 85 percent of the population of the Bahamas is of African origin and about 12 percent is of European origin, mostly the descendants of English Puritans and loyalist refugees from the United States. There are also wealthy retirees from North America and the United Kingdom who have settled in and near Nassau. Many of the outer islands have lost population as young Bahamians migrate to New Providence and Paradise Islands to work in the tourism industry. The outer islands provide locally made handicrafts for sale in tourist shops. Junkaroo is a traditional street festival that takes place on various holidays. It is of African origin and features a happy parade with music, dance, and exotic costumes.

A view of Parliament Square in downtown Nassau. (Photo courtesy of Julie Dunbar)

Further Reading

Cleare, A. B. *History of Tourism in the Bahamas: A Global Perspective.* Bloomington, IN: Xlibris Corporation, 2007.

Martin, Nona Patara and Virgil Henry Storr. "Whose Bay Street? Competing Narratives of Nassau's City Center," *Island Studies Journal* 4, no. 1 (2009): 25–42.

NAYPYIDAW

Naypyidaw is the new official capital of Burma, a country in Southeast Asia that is also called Myanmar (official name: Republic of the Union of Myanmar). The city is currently being built from the ground up since 2005 for the specific purpose of replacing Yangon (also called Rangoon) as capital. The word Naypyidaw means "royal capital." The city is situated about 200 miles (320 km) north of Yangon near the center of the country, a better location in the opinions of the government officials who decided on the change away from Yangon which is near the coast and is perhaps more vulnerable to foreign invasion. A more central location is also thought to be advantageous for administration. The population of Naypyidaw is said to be about 925,000, which if true, would represent an extraordinarily remarkable growth over a short period from essentially nothing. That would make Naypyidaw Burma's third-largest city after Yangon and Mandalay.

Historical Overview. The decision to move Burma's capital to the site of Naypyidaw was made early in the 2000s, and construction commenced in 2002. As many as 25 construction companies from various countries were hired to build the new city. The moving of government offices from Yangon to Naypyidaw

The Pyidaungsu Hluttaw complex, or the Assembly of the Union, the large new legislature building in Myanmar's new capital. (Soe Than Win/AFP/Getty Images)

began precisely at 6:37 a.m. on November 6, 2005, a time that was considered to be auspicious from an astrological standpoint. Then five days later, there was more moving of the government, on the 11th day of the 11th month and 11:00 a.m. by 1,100 military trucks carrying 11 military battalions and 11 government ministries. The first public event held in Naypyidaw was a large military parade on March 27, 2006, the anniversary of an uprising in 1945 against the Japanese occupation of Burma. Construction of Naypyidaw continues and the population of the city is growing extremely rapidly.

Major Landmarks. The largest and most impressive building in Naypyidaw is the Pyidaungsu Hluttaw Complex, the building that houses Burma's bicameral parliament. The Uppatasanti Pagoda, also called the "Peace Pagoda," was completed in 2009 is a same-sized replica of Yangon's Shwedagon pagoda and houses a sacred tooth relic from the Buddha. Other landmarks are the Naypyidaw Water Fountain Garden, the Naypyidaw Gems Museum, the National Herbal Park, and Naypyidaw Zoo.

Culture and Society. Naypyidaw is being populated from the ground up and may have gained as many as 925,000 residents in less than a decade. It must be a fascinating story of urban migration, settlement, and orientation in a large city in which everyone is a new resident. That story, however, has not yet been written. As the capital of a nation with a diverse ethnic population, we assume that Naypyidaw is ethnically diverse. Like Burma as a whole, the main language of the population is Burmese, with English being widely understood by the educated class, and the main faith is Buddhism. The government of Burma has long been repressive, but there are recent signs that a shift toward democracy might be in the making for the future.

See also: Yangon.

Further Reading

Preecharushh, Dulyapak. *Naypyidaw: The New Capital of Burma*. Bangkok, Thailand: White Lotus Books, 2009.
Seekins, Donald M. "'Runaway Chickens' and Myanmar Identity: Relocating Burma's Capital," *City: Analysis of Urban Trends, Culture, Theory, Policy, Action* 13, no. 1 (2009): 63–70.

N'DJAMENA

N'Djamena is the capital and largest city of Chad, a poverty-stricken, and strife-torn, country in central Africa. The city is located at the confluence of the Chari and Logone Rivers at the country's western border with Cameroon. The Cameroonian town of Kousséri is directly across the Chari River from N'Djamena. N'Djamena has been important locally as a hub of trade. The population of the city is about 1 million. Many residents are refugees from civil wars and environmental disaster in Chad and other parts of central Africa; poverty and substandard living conditions are widespread. The population jump was especially large in the 1970s and 1980s.

Historical Overview. N'Djamena is a product of French colonialism in Africa. The city was founded in 1900 by French commander Émile Gentil, leader of military missions to conquer African territory for France, and was named Fort-Lamy after one of his officers who had been killed that year in the Battle of Kousséri. The French territorial possession of Chad was established in that year, and Fort-Lamy was its administrative center. In 1920, France incorporated the colony into the federation of French Equatorial Africa. It was administered from a colonial capital in Brazzaville, now the capital of the Republic of the Congo. Fort-Lamy was an important military base for the French in World War II.

Chad achieved independence in 1960 and Fort-Lamy became a national capital. The city's name was changed in 1973 to N'Djamena, the name of a nearby village that means "place of rest" in Arabic. The name change was part of authoritarian President Françoise Tombalbaye's aggressive campaign of Africanization. Chad became engulfed in civil war in 1970s, which resulted in Tombalbaye's overthrow and death in 1975, and has been in turmoil ever since. The city was destroyed in fighting in 1979 and again in 1980, and most of it residents fled for a time to Cameroon. In 2006, it was the scene of violent unrest in the Battle of N'Djamena when rebels attempted to overthrow the government. Another attack on the city took place in February 2, 2008. Despite the conflicts, the population of the city has risen, because of refugees from the countryside and from conflicts in the Darfur region of neighboring Sudan.

Major Landmarks. N'Djamena is not known for tourism or important landmarks. There is, however, a presidential palace, many mosques in this Islamic capital city, and a Roman Catholic cathedral. There is also a Chad National Museum.

Culture and Society. The Chadian population is a mix of ethnic groups. In N'Djamena the main ethnic populations are Ngambaye, Chadian Arabs, Hadherai, and Daza. The predominant religion is Islam. The city continues to be torn by violence and unrest, and people find it difficult to lead a normal life in the city.

Further Reading

Decalo, Samuel. *Historical Dictionary of Chad*, Lanham, MD: Scarecrow Press, 1997.
Hammer, Joshua. "Heartbreak, CHAOS, Mayhem," *Outside* 34, no. 12 (2009): 100–32.
    https://www.cia.gov/library/publications/the-world-factbook/geos/cd.html

NEW DELHI

New Delhi is the capital of India, the largest country of South Asia and, after China, the second-most populous country in the world. The city is located in north-central India within the Indo-Gangetic Plain and is just west of the Yamuna River. It is within the metropolitan area of Delhi, India's second-largest city, and was built in the early 20th century as a planned, new capital adjacent to the historic city of "Old Delhi." The distinctions between Delhi and New Delhi are somewhat blurred, although it is only New Delhi that is officially the capital. The core of New Delhi is unambiguously a national capital city, distinct in architecture, land use, and urban design from the rest of the metropolitan area. The population of New Delhi is approximately 250,000 (2011), while that of Delhi exceeds 11 million. The metropolitan area of Delhi, of which New Delhi is a small part, has more than 16 million residents.

Historical Overview. New Delhi was built to be the capital of the British Indian Empire and became the capital of India when the country achieved its independence in 1947. The previous British capital had been Calcutta (now Kolkata) near the Bay of Bengal in the east of the country. Delhi had been capital of various empires on Indian land over the centuries, most notably the Mughal Empire from 1649 to 1857, and it was decided that the British Raj could better administer its most important colony from a central location rather than the coast. On December 12, 1911, during a mass gathering that is called the Delhi Durbar, King George V, then the Emperor of India, along with his Consort Queen Mary, announced that henceforth the capital of India would be in Delhi, and laid a foundation stone for the construction of a new city. Construction of New Delhi was started in 1912 at a site about 3 miles (5 km) south of the center of Delhi, but was soon slowed until 1917 by World War I. The new capital was formally inaugurated on February 13, 1931. The city was planned by Edwin Lutyens and Herbert Baker, and construction was supervised by Sobha Singh. New Delhi is sometimes referred to as Lutyens' Delhi. The design features straight and diagonal, wide tree-lined streets, open vistas, and large green spaces, all in sharp contrast to the winding narrow streets and crowded neighborhoods of Old Delhi. The city also was designed with monumental government buildings and landmarks.

On January 30, 1948, Indian leader Mohandas Gandhi was assassinated in New Delhi by a Hindu nationalist. The site of his death is marked by the "Martyr's Column" located on the grounds of the Gandhi Smriti, a museum dedicated to Gandhi that was once the home of a wealthy Indian family. In the 1950s, the first major extension of Lutyens' Delhi was undertaken by construction of the diplomatic enclave of Chanakyapuri, an enclave of foreign embassies, ambassadors' residences, chanceries, and other foreign diplomatic institutions. Later in the 1950s, additional farmland was appropriated to construct new residential areas for New Delhi. In 1982, New Delhi hosted the Asian Games, and in 2010 it was host to the Commonwealth Games. On December 12, 2011, the city celebrated its centennial as capital.

Major Landmarks. India Gate is located in the heart of New Delhi and is considered to be the national monument of India. It was designed by Lutyens as an Indian version of the Arc de Triomphe in Paris and was completed in 1931.

It was originally called the All India War Memorial and once had a statue of King George V beneath the now-vacant canopy. Rashtrapati Bhavan (Presidential House) is the official residence of India's Head of State. It is in the center of the city atop Raisina Hill. Construction started in 1911 and was completed in 1930. The iconic Jaipur Column stands in front of the building while in back are the Mughal Gardens. The Secretariat Building houses various ministries of the Cabinet of India. Major museums include the Indira Gandhi Memorial Museum, the National Gallery of Modern Art, the National Museum of Natural History, the National Rail Museum, and Nehru Planetarium, in addition to Gandhi Smriti. Because of India's diverse society, the city's main houses of worship represent various faiths. Important religious landmarks include the Hindu Laxminarayan Temple, the Sikh Gurudwara Bangla Sahib, the Bahá'í Lotus Temple, and Sacred Heart Roman Catholic Cathedral. Connaught Place, also called CP and officially Rajiv Chowk, is a large, circular financial, commercial, and office district in New Delhi that was laid out by Lutyens.

Culture and Society. New Delhi is a cross-section of India's population. The main religion is Hinduism, accounting for about 87 percent of the city's population. Other religions include Muslims (6%), Sikhs (2.4%), Jains (1%), and Christians (1%). There are also small numbers of Parsis, Buddhists, and Jews. The official languages are Hindi and English. Hindi is the main spoken language, while English is the main written language. The many other languages of India are also represented, including Punjabi, Urdu, Bihari, Bengali, Sindhi, Tamil, Telugu, Kannada, Malayalam, Marathi, Gujarati, and others. More than any other

Rashtrapati Bhavan, the official home of the president of India. (William Perry/Dreamstime. com)

city in India, New Delhi has a high proportion of middle class, highly educated population. This, of course, is because the city has many civil servants and other government workers. In many ways, New Delhi is thought of as India's greenest and most comfortable city.

Further Reading

Johnson, David A. "Land Acquisition, Landlessness, and the Building of New Delhi," *Radical History Review* 108 (Fall 2010): 91–116.

Legg, Stephen. *Spaces of Colonialism: Delhi's Urban Governmentalities.* Malden, MA: Blackwell, 2007.

Singh, B. P. and Pavan K. Varma, eds. *The Millennium Book on New Delhi.* New Delhi: Oxford University Press, 2001.

NIAMEY

Niamey is the capital city and most populous urban center in Niger, a large landlocked country in west-central Africa. The city is on the Niger River in the far southwestern portion of the country. Much of the distant rest of the country is the Sahara Desert and its edges. The city was formed on the east (left) bank of the Niger River, but it has grown as well on the west bank. These two zones of Niamey are connected by the John F. Kennedy Bridge and by passenger ferry services. The population of Niamey is reported to be about 775,000, but that is probably an underestimate because recent years have seen considerable migration from the countryside in response to drought conditions. The population grows rapidly as well because of high birthrates.

Historical Overview. Niamey's origins are said to date back to the 18th century, but its urban character did not begin to develop until construction of a fort in the 1890s by French colonialists. It grew steadily as a result. In 1905, the city was designated as the capital of the French Military Territory of Niger. In 1911, the capital was transferred to Zinder in the south-center of Niger, but it was returned to Niamey in 1928. From 1895 to 1958, Niger was one of the eight constituent colonies of French West Africa, with administration centered in Saint-Louis and then in Dakar, both in Senegal. Independence from France was declared on August 3, 1960. Drought and civil unrest have propelled Niger's rural citizens to resettle in their national capital. As a result, there are many squatter settlements in Niamey, especially at the edges of the city. The years 2005–2006 and 2010 were especially tragic with respect to drought, locusts, and starvation. During a hard-line military regime that ruled the country from 1974 until 1987, there were sporadic attempts to round-up migrants and send them back to home villages.

Major Landmarks. The Grande Marché in the center of Niamey is the main market in the city. It has more than 5,000 stalls and is an excellent place to buy fabrics and other local products. The city's most spectacular building is the Grande Mosquée on Avenue de l'Islam, the main religious center for the city's Muslim worshippers. Another notable landmark is the Roman Catholic Cathedral de Maorey.

Culture and Society. More than 90 percent of the population of Niger is Muslim. There are also Protestants and Roman Catholics, but in small numbers. French is the national language of the country, but for most citizens it is a second language, as local African languages are spoken at home. About one-half of the population of Niger speaks Hausa, but the Zarma language (second in the country) has been especially numerous in Niamey as the city is in the traditional heartland of Zarma settlement. Niger has the world's highest fertility rate at 7.2 births per woman. As a result, nearly one-half (49%) of the population of Niger is under 15 years of age.

Further Reading

Abdourahmane, Idrissa and Samuel Decalo. *Historical Dictionary of Niger*. Lanham, MD: Scarecrow Press, 2012.
Fuglestad, Finn. *A History of Niger*. Cambridge, UK: Cambridge University Press, 1983.
Göpfert, Mirco. Security in Niamey: An Anthropological Perspective on Policing and an Act of Terrorism in Niger," *Journal of Modern African Studies* 50, no. 1 (2012): 53–74.

NICOSIA

Nicosia is the capital and largest city of the Republic of Cyprus, a country on the island of Cyprus in the eastern Mediterranean Seas. The country is the third-largest island in the Mediterranean, and is in the east of Greece, south of Turkey, and west of Syria. The northern part of the island has been occupied by Turkey since an invasion in 1974 and is the self-proclaimed Turkish Republic of Northern Cyprus, a breakaway region that only Turkey recognizes. Nicosia is actually capital of both, as the Turkish side considers the city to be its capital just as it is the capital of the Republic of Cyprus. A "Green Line" divides the capital between the Turkish-occupied districts in the north from the rest of the city. Nicosia is located near the center of the island on the Mesaoria Plain between the Kyrenia Mountains to the north and the Troodos highlands to the south. It is on the Pedieos River. Greeks know the city as Lefkosía, while Turks call it Lefkoşa. More than 400,000 people live in the city.

Historical Overview. The area of Nicosia has been settled for more than 4,000 years. In about 1000 BC, it was a city-state known as Ledra. In the fourth-century AD, the town became a bishopric under Byzantine rule. In approximately 965, it became the capital of Cyprus largely because its inland location was safer from enemy raids by sea. During the Third Crusade in 1187, the town was defeated by Richard I of England (Richard the Lionhart) and was then sold to the Knights Templar who controlled the city until a revolt by Nicoseans in 1192. Afterward, Nicosia was ruled by Lusignan kings until 1489, Venetians from 1489 to 1571, the Ottoman Empire from 1571 to 1878, and the British from 1878 until 1960. Independence for Cyprus was achieved in 1960 after about five years of struggle against British rule by EOKA, the Greek-language acronym for the National Organization of Cypriot Struggle, and after an agreement by the island's two main communities, Greek and Turkish Cypriots, to live peacefully and share the island. However, conflict continued from the start, in Nicosia was divided into Greek and Turkish ethnic zones by drawing a green line on a map of the city, thus formalizing Nicosia's Green Line. In 1974, there was a coup attempt in which a Greek military junta sought to overthrow the government of Cyprus and annex the island to Greece. Turkey invaded in response and took control of the northern part of the island and northern Nicosia. In 1975, the Turkish Cypriot community proclaimed the Turkish Republic of Northern Cyprus that only Turkey recognizes. Nicosia is the world's only divided capital. There are separate municipal governments for the two parts of the city. Despite this situation, Nicosia has developed into an attractive city with a high quality of living and a strong economy based on international business and banking.

Major Landmarks. The historic center of Nicosia is a walled city that is quite well preserved, although it is divided between a northern half that is occupied by Turkey and the south that falls under the legitimate government of Cyprus. There are rows of barbed wire and guard towers at the border, and crossing is controlled. The walls of the city themselves are in place, as are the three historic gates: the Kyrenia Gate, the Famagusta Gate, and the Paphos Gate. The moats that surrounded the complex have long since been made into a circle of parkland. In the south, the main landmarks include St. John's Cathedral, the Archbishop's Palace, Phaneroméni Church, Omeriye Mosque, Freedom Monument, and the National Struggle Museum. Other museums in the south are the Folk Art Museum, the Ethnology Museum, the Byzantine Art Museum, and the Leventis Municipal Museum. Outside the walls is the Cyprus Archaeological Museum. In the north, the major landmark is the former Orthodox Cathedral of St. Sophia, now the Selimiye Mosque. It was so named in 1954 in honor of the Ottoman sultan who had conquered Cyprus. Construction of this structure started in 1209 and was completed in 1325. It was made into the chief mosque of Cyprus in 1572. The hub of Turkish Nicosia is Atatürk Square, also called Saray Square after the Lusignian Palace (saray) that once stood there. In 1821, the Ottomans beheaded or hung 486 Greek Cypriots in this square, including four bishops and other clergy, accusing them of conspiracy against the government.

Culture and Society. About 80 percent of the population of Cyprus is Greek Cypriot and 18 percent is Turkish Cypriot. Although Turkey occupies the north, both Greeks and Turks live there, just as both Greeks and Turks live in the rest of Cyprus. The same is true for Nicosia itself: both Greeks and Turks are found in both the north and the south of the divided city. Most Greek residents are members of the Greek Orthodox Church of Cyprus, while most Turkish residents are Sunni Muslims. Greek and Turkish are the two official languages of the country.

Further Reading

Mallinson, William. *Cyprus: A Modern History*. London: I. B. Tauris, 2005.

Papadakis, Yiannis, Nicos Peristianis, and Gisela Welz, eds. *Divided Cyprus: Modernity, History, and an Island in Conflict*. Bloomington: Indiana University Press, 2006.

Zetter, Roger. "Nicosia," *Cities: The International Journal of Urban Policy and Planning* 2, no. 1 (1985): 24–33.

NOUAKCHOTT

Nouakchott is the capital and largest city of Mauritania, a country on the North Atlantic coast of the Maghreb region of West Africa. The city is in the west of the country on the coast, about midpoint along the country's Atlantic shore. In addition to its role as capital, it is an important port, handling especially imports to Mauritania. The city is growing very rapidly, having had a population of only several thousand in the late 1950s and now numbering as a many as 2 million inhabitants. The increase has come largely because of migration to Nouakchott from the countryside of Mauritania, where social unrest and the spread of deserts has displaced much of the population in recent decades.

Historical Overview. Nouakchott was but a small village (*ksar*) before the late 1950s when Mauritania began preparations for independence from France and started to consider a possible site for a capital. Before, there was no capital city on Mauritanian territory, and the French administered it from Saint-Louis in neighboring Senegal. The site was chosen by Moktar Ould Daddah, the leading political figure of Mauritania and future first president, because of its central location along the coast and because it seemed like a compromise between the territories inhabited by the country's two fractious main ethnic groups, Arabs or Moors and blacks. Mauritania achieved independence from France on November 28, 1960, and Nouakchott began to grow. The city was laid out in grid plan, but a constant flow of economic and environmental refugees from hardship zones in West Africa has resulted in considerable unplanned construction and many illegal squatter settlements with little or no infrastructure or amenities. Currently, Mauritania is receiving considerable investment from the People's Republic of China, including funds for enlargement of the port in Nouakchott and the development of Mauritania fledgling oil industry. China has also invested in improvements for water and sewer networks in Nouakchott's residential neighborhoods, road and rail construction, and many other projects.

Major Landmarks. Nouakchott's most impressive structure is perhaps its Central Mosque with two high minarets. Other landmarks are the St. Joseph Roman Catholic Cathedral, the Nouakchott Stadium (used mainly for football-soccer), the Nouakchott Museum, and the National Library, and National Archives. The University of Mauritania is also in Nouakchott. The city has fine natural beaches which are enjoyed by its citizens.

Culture and Society. More than one-half and perhaps as many as two-thirds of the residents of Mauritania live in or near Nouakchott. The country is mostly poor and most parts of the city are lacking in basic amenities and public services. Because of high birth rates and low life expectancy, the majority of population is quite young in demographic profile. Islam is the religion of almost all Mauritanians. The Roman Catholic Cathedral mentioned above serves only several thousand Catholics. Arabic is the official language, although French is spoken as well. Mauritania is one of the two countries in the world (the other being Madagascar) to not use a decimal-based currency system. Its basic unit of currency, the *ouguiya*, is divided into five *khoums*.

Further Reading

Chenal, Jérôme and Vincent Kaufmann. "Nouakchott," *Cities: The International Journal of Urban Policy and Planning* 25 (2008): 163–75.

Gerteiny, Alfred and Anthony G. Pazzanita. *Historical Dictionary of Mauritania.* Lanham, MD: Scarecrow Press, 2008.

NUKU'ALOFA

Nuku'alofa is the capital and largest city of the Kingdom of Tonga, a sovereign state comprised of 176 islands (52 of which are inhabited) scattered over 270,000 square miles (700,000 sq km) in the South Pacific Ocean, about one-third of the distance from New Zealand to Hawaii. The city is on the north coast of the island of Tongatapu, the country's largest island and main population center, which is located in Tonga's southernmost island group. In addition to being the administrative center of Tonga, Nuku'alofa is the country's main port and main commercial center. The country's main airport, Fua'amotu International Airport, is on Tongatapu about 22 miles (35 km) southeast of the capital. The population of Nuku'alofa is about 24,500, nearly one-quarter that of the Tonga as a whole. The word Nuku'alofa means "abode of love," and originates with Tongan historical mythology.

Historical Overview. The Tongan Islands have been inhabited for more than 3,000 years. Little is known about their early history, although there is evidence that there may have been a Tongan Empire in the Pacific in the 12th century. The first Europeans to arrive were Dutch explorers in 1616. They were followed by their countryman Abel Tasman in 1643, and then the British Captain James Cook in 1773, 1774, and 1777. Because of a friendly reception on his first visit, Cook named the islands the Friendly Islands. The first foreign missionaries arrived in 1797. In 1845, Tonga was united into a kingdom by Tāufa'āhau, who had been baptized into Christianity in 1831 as Jiaoji (George). With the help of British missionary Shirley Waldemar Baker, Tonga became a constitutional monarchy with a code of enlightened laws. In 1900, Tonga became a British Protectorate. It was administered from Fiji as part of the British Western Pacific Territories, although local rule was maintained by the Tongan monarchy. Independence was achieved on June 4, 1970. In 1999, the country was admitted into the United Nations. Nuku'alofa was the scene of rioting, arson fires, and looting in late 2006 that spun from demonstrations by Tongan citizens who had gathered in the city center to protest for increased democracy. In 2010, Tonga held its first fully representative elections and elected Siale'ataongo Tu'ivakanō as its first prime minister.

Major Landmarks. Important landmarks in Nuku'alofa are the Royal Palace of Tonga, the Tongan Parliament Building, the 1885 Free Church of Tonga (originally the Wesleyan Free Church of Tonga), and Talamahu Market. To the east of Nuku'alofa is Mu'a, the second-largest town on the island and historic capital where there are the tombs of Tongan kings.

Culture and Society. Most residents of Tonga are ethnic Polynesians. There are small minorities of Europeans, mostly of British origins, and Chinese. The arson that accompanied the 2006 riots targeted mostly Chinese businesses, as a result of which the majority of Tonga's Chinese have emigrated. The official language of the islands is Tongan, along with English. Most islanders are Christian. The main denominations are Free Wesleyans/Methodists (37%), Mormons (17%), Roman Catholics (16%), and Free Church of Tonga (11%). Many Tongans live abroad, especially in Australia, New Zealand, and the United States, and send remittances to family members on the islands. Some 90 percent of the population of Tonga is overweight, mostly because of diet, with 60 percent classified as obese. The national sport of Tonga is rugby.

Further Reading

Ferdon, Edwin N. *Early Tonga as the Explorers Saw It: 1616–1810.* Tucson: University of Arizona Press, 1988.

Rutherford, Noel. *Shirley Baker and the King of Tonga.* Honolulu: University of Hawaii Press, 1996.

Theroux, Paul. *The Happy Isles of Oceania: Paddling the Pacific.* New York: Mariner Books, 2006.

OSLO

Oslo is the capital and largest city in Norway, a Scandinavian country in northern Europe that faces the North Atlantic Ocean. The city is the southern part of Norway at the head of Oslo Fjord and is an important port. The city is also rimmed by hills and mountains. There are many lakes in the city as well as islands in Oslo Fjord. The population of Oslo is 613,285 for the city proper, 925,242 for the urban area, and 1,442,318 for a more widely defined metropolitan area (2012). Population is increasing rapidly because of foreign immigration.

In addition to being the government center for Norway, Oslo is also the country's leading business and cultural center, as well as the city from which the Nobel Peace Prize is awarded each year. The city has a very high standard of living and ranks high in terms of sustainability and green urban design. It is, however, extremely expensive and often shows up as number one or number two in the world in terms of high cost of living.

Historic Overview. Oslo's origins go back to about 1000 AD. It was designated as a city in 1050 and became the capital of Norway in 1299. Akershus Castle was built in the 1290s in order to defend the city. In 1350, the Black Death befell Oslo and killed about three-quarters of the population. In the Middle Ages there were many large fires that destroyed major parts of the city. In 1624, King Christian IV decided to rebuild the city in brick and stone. The city was renamed Christiania in his honor in 1624 until it reverted back to Oslo by popular vote in 1925. The Oslo Cathedral was completed in 1697. The Royal Palace was constructed between 1825 and 1849, while the Stortinget, the Parliament of Norway, was completed in 1866. Oslo City Hall opened in 1950 and the Oslo Opera House in 2008.

Major Landmarks. Main landmarks in Oslo include the Akershus Fortress, Storinget, the Royal Palace, the University of Oslo, Oslo Cathedral, City Hall, and the Oslo Opera House. The Nobel Peace Centers has exhibits about past winners of the Nobel Peace Prize. The main museums are the National Gallery and the Munch Museum which features paintings by the great Norwegian painter Edvard Munch. The Henrik Ibsen Museum celebrates the life and work of the famed 19th-century Norwegian playwright and poet. The Norwegian Maritime Museum is in the city's harbor area.

Culture and Society. Oslo has many parks, and opportunities abound for outdoor recreation in all seasons. It is also a cultural center with a wide array of theatrical performances, concerts (e.g., the Oslo Philharmonic), and gallery exhibitions. There are also several districts noted for nightlife.

The Oslo Opera House. (Photo courtesy of Roman Cybriwsky)

About 28 percent of Oslo's residents are from ethnic backgrounds other than Norwegian. There is considerable foreign immigration, and the Norwegian percentage of the population has been declining. The main foreign ethnic groups are Pakistanis, Somalis, Swedes, and Poles, in descending population order. Because of the Pakistani and Somali communities, about 11 percent of the population of Oslo is Muslim. There is racist backlash against immigration and so-called white flight from the city to suburban neighborhoods with fewer immigrants.

Further Reading

Isaksen, Arne and Heidi Wiig Aslesen. "Oslo: In What Way an Innovative City?" *European Planning Studies* 9, no. 7 (2001): 871–87.

Naess, Peter and Teresa Naess. "Oslo's Farewell to Urban Sprawl," *European Planning Studies* 19, no. 1 (2011): 113–39.

Wessel, Terje. "Losing Control? Inequality and Social Division in Oslo," *European Planning Studies* 9, no. 7 (2001): 889–906.

OTTAWA

Ottawa is the capital of Canada, the world's second-largest country in size. It is located on the right (south) bank of the Ottawa River in the province of Ontario. Across the river are the province of Quebec and the city of Gatineau. The population of Ottawa is about 812,000 (2006 data), the fourth-largest total in Canada, while that of the Ottawa-Gatineau Census Metropolitan Area is about 1,239,000 (official estimate, 2010), the fifth-largest in Canada. The name "Ottawa" is said to derive from an Algonquin word for trade, referring to the early fur trade along the

Ottawa River between European traders and indigenous people. The name was given to the city when it was incorporated in 1855; prior to then the settlement was called Bytown.

In addition to being the capital of Canada, Ottawa is also an important business, educational, and cultural center. There are many high-technology companies in the city and surroundings, with the result that Ottawa is sometimes referred to as the "Silicon Valley of the North."

Historical Overview. Ottawa was founded in 1826 by John By, a Colonel in the British Engineers, who was put in charge of constructing a canal from where the Rideau River empties into the Ottawa River to Kingston, a city on Lake Ontario to the south. Called the Rideau Canal, the waterway had significant impact in the past on Ottawa's trade and commerce, and is still in use today, especially for recreation. The main stimulus for Ottawa's subsequent growth was the timber trade; that business had been started in the area as early as 1800 in Hull, now a part of the city of Gatineau across the Ottawa River. Large lumber mills opened in Ottawa in the 1850s, and from 1854 began shipping products by rail to other Canadian cities. On December 31, 1857, when Canada was still a British colony, Queen Victoria designated Ottawa to be the capital of the Province of Canada. The choice was made because Ottawa was further from the border with the United States than other candidate cities, and because the Ottawa River offered easy transportation. It also turned out that the city was astride the boundary between Canada's French speakers and speakers of English. Canada achieved independence from Great Britain in 1867, and Ottawa became a national capital. In 1900, an enormous fire engulfed the city of Hull and spread across the river to burn sections of Ottawa as well. Another fire in 1916 destroyed part of the Parliament Building. Reconstruction was completed in 1922, and featured a high Gothic revival tower that is now an icon of both Canada and its capital city. In the 1940s, Ottawa underwent many capital improvements and urban planning projects under the direction of French urban designer Jacques Gréber.

Major Landmarks. The most famous landmark in Ottawa is Parliament Hill, the beautiful main government building of Canada. Its high clock tower is called Peace Tower (302 ft; 92.2 m). Nearby are statues of Canada's founders and leading politicians in history, as well as of Queens Victoria and Elizabeth II of England, and the National War Memorial. The Rideau Canal is beside Parliament Hill and is a UNESCO World Heritage Site. Major museums include the Bytown Museum at the canal, the National Gallery of Canada designed by the famous architect Moshe Safdie, and the Museum of Civilization in Gatineau. Outside the National Gallery is Maman, a famous large (30-ft high; 9.1 m) sculpture by Louise Bourgeois of a spider with a sac with eggs. In Lower Town is Byward Market, the oldest farmers' market in Canada. Carleton University and the University of Ottawa are the city's two major universities.

Culture and Society. Ottawa ranks as one of the most livable cities in the world, known especially for its many beautiful parks, its opportunities for outdoor recreation across four seasons, an excellent public transit system, and high standards of cleanliness and public safety. There are many bicycle paths and bicycling is a way for many Ottawans to commute to work of school. In winter,

The landmark Parliament Building in Ottawa as seen from across the Ottawa River. (Corel)

Ottawans ice skate on the Rideau Canal, some as commuters to Parliament Hill. Ice hockey is a popular sport for spectators and participants. Ottawa has a lively art and culture scene, as well as a variety of festivals and holiday celebrations. The streets near Byward Market are popular for nightlife.

The population of Ottawa is ethnically diverse, like that of Canada as a whole, with immigrants and their descendants from all corners of the globe. As national capital, it also has residents from all parts of Canada. English is the mother tongue for 62.8 percent of the population and French for 14.9 percent. Other residents list a great many other languages from around the world as their mother tongue. Ottawa is Canada's largest truly bilingual city, with 37 percent of the population speaking both English and French. Ottawa's population is also highly educated, reflecting its strong economic base in government services, high technology, many museums, and two large universities, among other areas of employment.

Further Reading

Keshen, Jeff and Nicole St-Onge. *Ottawa: Making a Capital*. Ottawa: University of Ottawa Press, 2001.

McLennan, Rob. *Ottawa: The Unknown City*. Vancouver, Canada: Arsenal Pulp Press, 2008.

Taylor, John H. *Ottawa: An Illustrated History*. Ottawa: Canadian Museum of Civilization, 1986.
    http://www.ottawa.com/

OUAGADOUGOU

Ouagadougou (pronounced "wa ga doo goo") is the capital and largest city of Burkina Faso, a poor land-locked country in West Africa. Its name is shortened in colloquial usage to Ouaga and its residents are called *ouagalais*, a word from the French language. The city is located in the province of Kadiogo in the very center of Burkina Faso. Before 1984, the name of the country was the Republic of Upper Volta, a name referring to a major West African river with branches that flow through the area. The name Burkina Faso means "land of the upright people" in local languages. The population of Ouagadougou is approximately 1.7 million (2012), about 10 percent of the national total.

Historical Overview. The history of Ouagadougou goes back to 1441 when, after a decisive military battle, it became the capital of the third kingdom of the local Mossi people, a kingdom named Ouagadougou (or Wogodogo in a spelling from English) that corresponds approximately to the geographical limits of Burkina Faso today. From 1681, the city became the permanent residence of the Moro-Naba, the kingdom's supreme ruler. The French defeated the Mossi people in 1919 and colonized them. They called their colony Haute Volta and retained Ouagadougou as the colonial capital. In English the colony was referred to as French Upper Volta. Independence from France came in 1960, at which time the name Republic of Upper Volta was put into use. Ouagadougou's growth was stimulated by the opening of a rail connection to the neighboring French colony of Côte d'Ivoire in 1954, giving Haute Volta a link to a sea port.

Major Landmarks. The most distinctive landmark in Ouagadougou is the National Heroes Memorial, a high monument that was erected in 2000 at a busy traffic circle. At a different traffic crossroad is a large globe sculpture that honors the United Nations. Other landmarks are the Grand Mosque of Ouagadougou and the city's Roman Catholic Cathedral. Bangre-Weoogo is an urban park and nature preserve in the city. At the central railway station is Naba Koom, a statue of a woman with a calabash with water to welcome visitors to the city. The central market of Ouagadougou burned in 2003, but has recovered since and is a place to buy cloth.

Culture and Society. The Mossi people are the main ethnic group of Ouagadougou. They are descendants from warriors who migrated north centuries ago from Ghana. The Moro-Naba is still considered to be the symbolic leader of the people and a ritual ceremony takes place every Friday to affirm this leadership. The main religions are Islam and Christianity, particularly Roman Catholicism, and many people practice a blend of both. French is the official national language but most *ouagalais* speak local languages at home. Most residents of Ouagadougou trace roots to the countryside of Burkina Faso, where farming and livestock raising predominate.

Further Reading

Engberg-Perderson, Lars. *Endangering Development: Politics, Projects, and Environment in Burkina Faso.* New York: Praeger, 2003.

Lindén, J., J. Boman, B. Holmer, S. Thorrson, and I. Eliasson. "Intra-Urban Air Pollution in a Rapidly Growing Sahelian City," *Environment International* 40, no. 2 (2012): 51–62.

McFarland, Daniel Miles and Lawrence Rupley. *Historical Dictionary of Burkina Faso (Former Upper Volta).* Lanham, MD: Scarecrow Press, 1998.

P

PALIKIR

Palikir is the capital of the Federated States of Micronesia (FSM), a sovereign nation of approximately 607 small, tropical islands north of the equator in the Western Pacific Ocean. The islands are part of the larger Caroline Islands group of Micronesia. The total land area of the country is small, being only 271 square miles (702 km sq), but the territory of FSM covers more than 1,000,000 square miles (2.6 million km sq) of ocean. There are four states in FSM: Yap, Chuuk, Pohnpei, and Kosrae, going west to east. Palikir is located in the northern part of Pohnpei Island in Pohnpei state. Pohnpei Island is one of the largest, highest, and most populous of the islands of FSM. The population of Palikir is about 4,645 (2011). It is the third-largest settlement in FSM, after Weno in Chuuk and Kolonia, Palikir's neighbor on Pohnpei about 6 miles (10 km) to the east. Palikir's population is quite spread out geographically and lacks the form of a nucleated community. Kolonia is more like a town. It is the former capital of FSM and is still capital of Pohnpei state, as well as the island's major business center.

Historical Overview. Palikir became capital in 1989 when a complex of new buildings for the national government was built there in place of older facilities in Kolonia.

Micronesia has been settled for more than 4,000 years. From about 1100 to 1628, the island of Pohnpei was the ceremonial and political base of the Saudeleur Dynasty, thought to be an autocratic and highly centralized government by foreigners who migrated to the island. The Saudeleur capital was at Nan Madol on the opposite side of Pohnpei from Palikir. The Portuguese and then the Spanish were the first Europeans to reach the Caroline Islands. Spain established sovereignty, calling the islands Islas Carolinas, and then sold them in 1899 to Germany that administered them as Karolinen, an extension of German New Guinea. Japan administered the islands after a League of Nations mandate in 1914, and used Truk Lagoon (Chuuk) as a major navy base in World War II. After the War, the islands became United States Trust Territories until achieving independence in 1986.

Major Landmarks. The FSM government center is located off the main road of Pohnpei Island. The site had previously been a Japanese airfield. The campus of the College of Micronesia is nearby. The stone ruins of the historic Saudeleur capital of Nan Madol probably comprise the most striking of the island's landmarks. The site is sometimes referred to as the Venice of the Pacific because it consists of a series of small offshore artificial islands linked by a network of canals. In Kolonia are remnants of island history such as the Old Spanish Wall, the German Bell Tower, the Old German Cemetery, and Japanese tanks on display from World War II.

Culture and Society. The people of FSM belong to many different Micronesian ethnic and linguistic groups. On Pohnpei the principal ethnic and linguistic group is Pohnpeian. There is also a small minority of immigrants from the Philippines. English is the official language of government throughout FSM, although local languages are used at meetings of local governments. English is also the language of secondary education and the College of Micronesia. In terms of religion, the population of Pohnpei Island is overwhelmingly Christian, about equally divided between Protestants and Roman Catholics.

Further Reading

Cook, Ben. *The Federated States of Micronesia and the Republic of Palau.* (No City Listed): Other Places Publishers, 2010.

East, Andrew J. and Les A. Dawes. "Home Gardening as a Panacea: A Case Study of South Tarawa," *Asia Pacific Viewpoint* 50, no. 3 (2009): 338–52.

Warren, Mary. "Federated State of Micronesia," *New Internationalist* 448 (December, 2011): 32-33.

PANAMA CITY

Panama City, or simply Panama, is the capital and largest city in the Republic of Panama, the Central American country closest to the South American continent. The city is located on the Pacific Ocean side of Panama at the head of the Gulf of Panama, and is near the Pacific entrance to the Panama Canal, an important passage for ship traveling between the Pacific and Atlantic Oceans. In addition to the national government of Panama, the city's economy derives from trade businesses related to the canal, as well as from international banking and finance, and from tourism. The city has the look and feel of a very modern and prosperous city. The population of Panama City is about 881,000 and about 1.3 million in the metropolitan area (2010).

Historical Overview. Panama City was founded in 1519 by the Spanish conquistador Pedro Arias Dávila as a base for Spanish expeditions down the Pacific Coast to the Inca Empire in what is today Peru. Much of the gold and silver that the Spanish extracted from South America passed through Panama City. On January 28, 1671, the city was set ablaze during an attack by the Welsh pirate Henry Morgan and was completely destroyed. It was rebuilt soon thereafter about 5 miles (8 km) from the original site. The old city, known as Panama Viejo (Old Panama), is still in ruins and is a tourist attraction. In 1821, Panama achieved independence from Spain and joined into a union called the Republic of Gran Colombia with northern South American territories (today's Colombia, Venezuela, and Ecuador). That union was partly dissolved in 1830. Then in 1903, by virtue of the Hay–Bunau-Varilla Treaty and pressure from the United States, Panama broke away from Colombia and established its own republic with Panama City as the capital. A strip of land roughly 10 miles (16 km) wide and 50 miles (80 km) long was ceded to the United States "in perpetuity" as the Panama Canal Zone for the purposes of building, maintaining, and defending a canal. It did not include Panama City and Colón, the Panamanian city at the Atlantic end of the canal.

The Panama Canal was constructed between 1904 and 1914, and it has been a central aspect of the national economy and Panama City ever since. Because of pressure from Panamanian citizens, the Canal Zone was disestablished in 1979 and the territory was returned to the republic of Panama. In the 1970s and 1980s, Panama suffered from corrupt and repressive leadership, and from economic sanctions imposed by the United States. In December 1989, on orders from U.S. President George H. W. Bush, the United States invaded Panama, ostensibly in order to depose the corrupt and repressive military government, protect the lives of American citizens in the country, and combat drug trafficking. The United States termed the invasion as Operation Just Cause, but there was much more international outrage than approval. Panama City was a main battleground, with casualties on both sides and considerable property damage. As a result of the invasion, the United States was able to strengthen its influence over the country.

Major Landmarks. In addition to the historical ruins of Panama Viejo, the historic core of the second Panama City that was built after 1671 is also a major visitor attraction. The area is known as Casco Viejo and was designated as a World Heritage Site in 2003. Ancon Hill offers a terrific panoramic view of Panama City and its waterfront. Other landmarks are the Roman Catholic cathedral in the center of Panama City, other historic churches, Teatro Nacional (the National Theater), and the Interoceanic Canal Museum. The Bridge of the Americas, built in 1962, crosses the Panama Canal near its Pacific entrance, while the newer (2004) Centennial Bridge crosses it 9 miles (15 km) to the north.

Culture and Society. Panama has a diverse society that includes *mestizos* (people of mixed European and indigenous American origins), native Amerindians,

The booming skyline of Panama City rising from the Pacific Ocean shores. (Cienpies Design & Communication/Dreamstime.com)

Afro-Panamanians, and Chinese. The latter group originally came to Panama to help build the canal, but there have been migration flows from China since, including recently. About three-quarters of the population is Roman Catholic, with the majority of the rest belonging to evangelical Christian denominations. The national sport of Panama is baseball. There are local leagues as well as a national baseball team, and many players from Panama play in the U.S. Major Leagues.

Further Reading

Friar, William. *Panama City and the Panama Canal.* Berkeley, CA: Avalon Travel Publishing, 2009.
Holston, Mark. "Panama City on the Rise," *Americas* 62, no. 5 (2010): 12–19.
Uribe, Alvaro. "Integration of the Former Panama Canal Zone into Metro Panama City," *Ekistics* 69, no. 415–417 (2002): 202–208.

PARAMARIBO

Paramaribo is the capital and largest city of the Republic of Suriname, a country on the north coast of South America. The country was previously a colony of England and The Netherlands, and was once called Dutch Guiana. It achieved independence in 1975, and is now the smallest sovereign state in South America. Paramaribo is located on the banks of the Suriname River about 9 miles (15 km) from Suriname's Atlantic Ocean coast. The latitude is equatorial—5°52′N—and the climate is hot and muggy.

The population of Paramaribo is about 250,000 (2011), about one-half of that of Suriname. Since 2002, the core of Paramaribo has been a UNESCO World Heritage Site. The city's nickname is Parbo.

Historical Overview. Paramaribo was founded as a Dutch trading post and alternated in history between being under the Dutch or English flag. The English made it a colonial capital in 1650, but in 1667 the Dutch gained control of the colony via the Treaty of Breda and Paramaribo became a Dutch capital. Major fires in 1821 and 1832 destroyed large parts of the original settlement. Both English and Dutch settlers established a plantation economy in Suriname and exploited slaves imported from Africa as labor. The crops that were grown in the valley of the Suriname River were cotton, sugarcane, coffee, and cocoa. Slavery was abolished in 1863 and slaves were released from plantations in 1873 after a 10-year transition period. Many freed slaves then settled in Paramaribo.

In the period leading up to independence in 1975, as well as in the decades since, many Surinamese have migrated to the Netherlands. For more than two decades after independence, Suriname was heavily reliant on Dutch foreign aid. The country is now stronger economically because of exports of bauxite and the discovery of oil and gold. Paramaribo is the gateway city for exploitation of Suriname's resources.

Major Landmarks. The major government buildings in Paramaribo are the Presidential Palace, the National Assembly, and the Court of Justice. The most significant religious landmarks are the Roman Catholic Cathedral of St. Peter and St. Paul, the Suriname Mosque, and Neveh Shalom synagogue. Another notable landmark is the Helstone Monument erected in honor of Johannes Nicolaas Helstone (1853–1927), who was born a slave and became an accomplished Surinamese pianist and composer. Waterkant is a riverfront street in Paramaribo with historic architecture. The Jules Wijdenbosch Bridge, completed in 2000, is nearly a mile long (4,934 ft; 1,504 m) and spans the Suriname River.

Culture and Society. The population of Paramaribo comes from a great many different ethnic origins and ethnic admixtures. Significant groups include Creoles (mixed descendants of West African slaves and Europeans); descendants of 19th-century contract laborers from India, China, and Java, then a part of the Dutch East Indies; Amerindians (native South Americans); Maroons (mixed descendants of escaped African slaves and Amerindians); Jews (descendants of emigrants from Portugal in the 17th century), and Surinamese of European origin, mostly Dutch. Dutch is the official language of Surinam and the first language of about two-thirds of the residents of Paramaribo. Many other residents of Paramaribo speak Sranan Tongo, a local creole language.

Further Reading

Bruinje, A. "Ethnic Residential Patterns in Paramaribo: Spatial Segregation or Blending," in R. Jaffe, ed., *The Caribbean City*, 162–88. Kingston, Jamaica: Ian Randle, 2008.

Potter, R. B. *The Urban Caribbean in an Era of Global Change*. Aldershot, UK: Ashgate, 2000.

Verrest, Hebe. *Home-Based Economic Activities and Caribbean Urban Livelihood: Vulnerability, Ambition and Impact in Paramaribo and Port-of-Spain*. Amsterdam: Amsterdam University Press, 2007.

Verrest, Hebe. "Paramaribo," *Cities: The International Journal of Urban Policy and Planning* 27, no. 1 (2010): 50–60.

PARIS

Paris is the capital and largest city in France, a historic country in western Europe, and one of the world's most famous and beloved cities. The city is on the Seine River in the north-central part of the country, and is centered on an island in the river called Île de la Cité that once defined the medieval walled city. The Seine divides the city into a Left Bank and a Right Bank. Since 1860 the core of the city has been divided into 20 numbered districts called *arrondissements* that are arranged geographically in a clockwise spiral beginning with number one near the center of the city. Among the many nicknames for the city of Paris is "City of Light." The French often refer to the Paris region as Île de France, the Island of France.

The population of Paris is about 2.2 million, while that of its much larger metropolitan area is about 12.1 million (2009). The city is one of the largest urban centers of Europe. For about 1,000 years, Paris was the largest city in the Western world and from the 16th to the 19th centuries it may have been the largest city in the entire world. In addition to being the capital city of France, Paris is one of the world's leading financial, corporate, and industrial centers, as well a major center of higher education, research and technology, and fashion. With as many as 42 million visitors annually from all over the globe, Paris is a leading center for tourism and is considered to be the most visited city in the world.

Historical Overview. Archaeological evidence indicates that the site of Paris was inhabited as far back as 4,200 BC. In around 250 BC the site was inhabited by a tribe called the Parisii, after whom the city is named. Paris was conquered by the Romans in 52 BC, and became an outpost of the Roman Empire. The Romans named it Lutetia or Lutetia Parisorum. The collapse of the Roman Empire in the fifth century opened the way for invasions by Germanic tribes and sent the city into a period of decline. By 508, however, the first king of the Merovingian dynasty, Clovis the Frank, had selected Paris to be his capital. In the late eighth century, the Carolingian dynasty moved the capital to Aachen, leaving Paris vulnerable to attacks by Viking raiders. A fortress was built on Île de la Cité to defend the city. Paris became capital again in 987 with the coronation of Hugh Capet as King of France. Warfare, particularly with England such as the 1337–1453 Hundred Years War, and major outbreaks of disease such as the Black Death that began in 1348 plagued Paris throughout the Middle Ages. The French Wars of religion between Catholics and Protestants (French Huguenots) consumed the city between 1562 and 1598. On August 23, 1572, Paris was the scene of a targeted series of religion-based assassinations by Catholics against Huguenot leaders that is known as the St. Bartholomew's Day Massacre. The Siege of Paris by Henry of Navarre took place between May and September, 1590. Civil war caused the royal family to flee Paris in 1648, and in 1682 King Louis XIV moved the royal court to Versailles, a lavish estate outside the city. The French Revolution took place from 1789 to 1799, highlighted by the storming of the Bastille on July 14, 1789, and the overthrow of the monarchy in 1792.

The 19th century began with more war and the imperial ambitions of Napoleon Bonaparte, defeat in 1814 and occupation of Paris by Russian and Allied forces, restoration of the monarchy, more revolution, and more epidemics of disease. The Industrial Revolution began to transform Paris in the 1840s. From 1852 up to about 1870, much of Paris was rebuilt under the direction of Baron Georges-Eugène Haussmann, a planner who was commissioned by Napoleon III to modernize the city. He leveled old slums and put in grand boulevards and beautiful buildings that became the foundation of the beautiful city that Paris is today. In 1889, Paris hosted a world's fair called the Exposition Universelle that celebrated the city as a center for innovative architecture and technology. The Eiffel Tower was opened in that year as the centerpiece for the fair. Another world's fair in Paris in 1900 was highlighted by the opening of the Paris Metro, the city's subway system.

During the 20th century, Paris evolved even more into a global center of intellectual life, arts, fashion, and decadence. The city was occupied by Nazi forces from 1940 to 1944, but was spared the kind of wartime destruction that befell other European cities at the time. Adolf Hitler had ordered that the city's monuments be destroyed, but his general in the city Dietrich von Choltitz disobeyed that command as the Germans retreated from the city before Allied advances. After World War II, Paris grew with construction of large housing estates near the margins of the urban area, and the development of a major new commercial center known as La Défense. The population mix of the city changed with large-scale immigration from abroad, especially from France's former colonies in North, Central and West Africa.

Major Landmarks. Paris has too many landmarks to mention. There are as many as 3,800 historical monuments in and near the city, plus countless museums, theaters, concert halls, churches, and other attractions. The most famous landmarks are the Eiffel Tower, the Cathedral of Notre Dame, the Napoleonic Arc de Triomphe, and the Louvre Museum with its spectacular glass pyramid, new entry designed by architect I. M. Pei. Other famous places are Montmartre and its hilltop Basilique du Sacré Coeur, Place de la Bastille, the Tuileries Gardens, the Champs Élysée, the shopping area Les Halles, the Montparnasse district on the left

> ### Baron Haussmann and Paris
>
> Georges-Eugène Haussmann (1809–1891), commonly referred to as Baron Haussmann, was a Paris-born surveyor and city planner who is most famous for rebuilding the center of his home city in the 1860s. He converted the grimy industrial city of Paris into one of spectacular architecture, boulevards, and monuments, as well as modern infrastructure such as a sewer system and new bridges. He was charged with the task by Emperor Napoleon III who wanted Paris to become one of the world's most beautiful cities. The city that resulted is spectacular indeed with monuments such as the Champs-Élysées, the Arc de Triomphe, and the Opera Garnier, but critics lamented the destruction of the medieval city and the heavy hand of undemocratic redevelopment. The word "Haussmannization" has entered the urban literature to refer to the rebuilding of a city for the sake of public image without regard to previous historic fabric or the needs of poor people in old neighborhoods.

bank known for artist studious and café life, the Latin Quarter and its Sorbonne University, Opéra Garnier, Opéra Bastille, the banks of the Seine River and its rows of book and art sellers, and the La Défense. Must-see museums in addition to the Louvre include Musée Cluny, Musée d'Orsay, and the Centre Georges Pompidou. The old dance halls Le Lido and Moulin Rouge are now popular tourist attractions. The palace and gardens at Versailles outside the city are major visitor attractions. Paris Disneyland outside the city also attracts millions of visitors each year.

Culture and Society. The central areas of Paris are generally prosperous and prestigious places to live and do business, while much of the periphery of the urban region is industrial or formerly industrial and are often concentrations of poor immigrant groups. It is a "tale of two cities," with one being the core and the other isolated at the margins. The suburbs especially have many immigrants from North Africa and a corresponding increase in the Muslim population of Paris. A major challenge for France is to integrate these newcomers into French society while allowing them the freedoms to openly display their faiths and to speak their own languages. There have been great social and economic tensions in recent times in Paris and other French cities, and angry riots by unemployed immigrant youth. To improve the situation, the French government has introduced French language and job training programs for immigrants, and invested in transportation infrastructure in order to break down the barrier of distance between outlying suburbs and employment and cultural integration opportunities closer to the urban center.

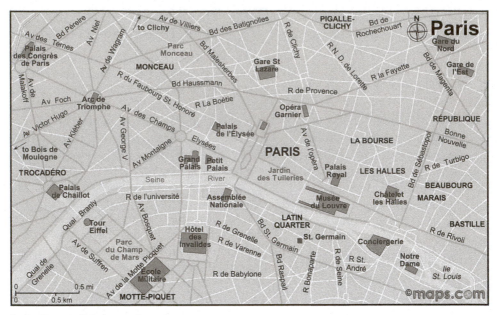

Paris, France (Maps.com)

Further Reading

Harvey, Davis. *Paris, Capital of Modernity*. New York and London: Routledge, 2003.

Higonnet, Patrice. *Paris: Capital of the World* (translated by Arthur Goldhammer). Cambridge, MA: The Belknap Press of Harvard University Press, 2002.

Horne, Alistair. *Seven Ages of Paris*. New York: Vintage Books, 2002.

Jones, Colin. *Paris: The Biography of a City*. New York: Penguin Books, 2005.

PHNOM PENH

Phnom Penh is the capital and largest city in Cambodia, a compact country of more than 14 million people in Southeast Asia that borders Thailand, Laos, Vietnam, and the Gulf of Thailand. The city is on flat land in the south-center of the country on the banks of the Mekong River and its tributaries rivers, the Bassac and Tonlé Sap. In addition to being capital, Phnom Penh is Cambodia's leading economic, cultural, and education center, and a popular stop for foreign tourists. The population of Phnom Penh is about 2.3 million (2012). The name "Phnom Penh" means "Penh's hill," originating with the legend that in 1372 a certain Madam Penh found four bronze and one stone Buddha statues in a hole in a dead tree that she saw floating in the river. She was thus inspired to order workers to shape a hill near the river atop which Wat Phnom (the "hill temple"), still a major landmark in the city, was constructed to enshrine the statues. The city's original name was Krong Chaktomuk, meaning "city of four faces" and referring to an 'X' shape that was made by the confluence of the various rivers where the early settlement grew up.

   Historical Overview. Phnom gained prominence from the temple built by Madam Penh and then in 1432 from the decision by King Ponhea Yat to move the capital of what was left of the Khmer Empire to the site after the original capital of Angkor Thom was destroyed by invaders from Siam (Thailand). It remained the royal capital until 1505, after which the city languished for more than three centuries because of political infighting and decisions by subsequent kings to move the capital to a succession of other cities. The city was restored in 1866 when King Norodom I chose Phnom Penh as his capital and began construction of his spectacular royal palace. In the 1870s, Phnom Penh began to be transformed by French colonial interests in Southeast Asia, and by 1920 it was being called the "Pearl of Asia" by Europeans because of beautiful boulevards, tree-lined residential streets, sumptuous mansions, and outstanding cultural institutions. Cambodian independence from France was achieved in 1953, after which there was a period of construction of great monuments and public buildings in the Khmer Revival style that was advocated by nationalist architect Vann Molyvann. The Vietnam War brought great damage and hundreds of thousands of poor refugees from the war-ravaged countryside to the city, swelling the population to some 3 million in a short time. In 1975, the Khmer Rouge took control of the city and emptied it of its population by sending people to work camps in the countryside and to their deaths in the Cambodian genocide. The Vietnamese liberated Phnom Penh in 1979, and a period of restoration was begun. The population has grown with new residents, and the city is once again a bustling hub of government and commerce. Cambodia is still very poor and receives considerable foreign aid, including aid in the form of construction projects to improve Phnom Penh's infrastructure and competitiveness in a global economy.

Major Landmarks. In addition to the very lively Wat Phnom and its grounds and the Royal Palace, major landmarks in Phnom Penh include Independence Monument, the statue of Lady Penh, the Tuol Sleng Genocide Museum in a school that had been converted by the Khmer Rouge into a prison and interrogation-torture site, the Buddhist Institute, the National Museum, and the beautiful new Cambodian National Assembly. The Naga World Hotel and Casino on the city's riverfront is a new landmark too, although one that is roundly criticized as an architectural monstrosity and a form of foreign investment that does more harm to the poor people of the city than good.

Culture and Society. Phnom Penh differs from some of the other capitals of Southeast Asia in that it is not yet a city of high-rise office towers and modern shopping malls, although such change is coming in the guise of modernization and economic development. For the most part, the city is still one of poor people's neighborhoods and enterprise, and an assortment of development and aid projects by foreign and Cambodian NGOs that focus on local needs. Unfortunately, corruption is rampant in Cambodia and often impedes true progress. The city has long had a problem with sex and drugs tourists from abroad, and has been the scene of sex trafficking and child prostitution. Some parts of Phnom Penh's outskirts are becoming industrial zones that employ low-wage female labor in garment and electronics factories for foreign manufacturers. About 90 percent of the population is ethnically Khmer (Cambodian). There is also a significant Chinese minority that is disproportionately engaged in business, and a community

Sculpture in Phnom Penh made from melted weapons. (Photo courtesy of Roman Cybriwsky)

of Moslem Cham people who had migrated from the countryside of Cambodia and Vietnam. The main language of Phnom Penh is Khmer, and other than Islam among the Cham, the dominant religion is Buddhism.

Further Reading

Cybriwsky, Roman. "Phnom Penh at a Crossroads," *FOCUS (American Geographical Society)* 52, no. 3–4 (Winter 2009): 9–15.

Gilboa, Amit. *Off the Rails in Phnom Penh: Into the Dark Heart of Guns, Girls, and Ganja.* Bangkok, Thailand: Asia Books, 1998.

Osborne, Milton. *Phnom Penh: A Cultural and Literary History.* Oxford, UK: Signal Books, 2008.

PODGORICA

Podgorica is the capital and largest city of Montenegro, a small country on the Adriatic Sea in southeastern Europe. It is located in the south center of the country where the Ribnica and Moraça rivers come together in the fertile Zeta Plain. It is at the crossroads of several historically important trade routes. The population of Podgorica is approximately 156,000 (2011), about 30 percent of the nation's total. The metropolitan area numbers about 186,000 people. The meaning of the word Podgorica is "under the small hill."

Historical Overview. The history of Podgorica is traced to a Roman Empire trade center called Birziminium, although about 2 miles (3 km) to the northwest of the city are ruins of an even earlier settlement from Greek civilization called Doclea. In the early Middle Ages the city was a local administrative center named Ribnica. The name Podgorica came into use in about 1326. It was conquered by the Turks in 1474 and remained part of the Ottoman Empire until 1878. The Ottomans built a fortress in the city and made it into a strategic regional stronghold. After 1878, the city became capital of the Kingdom of Montenegro, and then from 1916 to 1918, it was occupied by Austria-Hungary. In 1918, Montenegro was annexed into Serbia. It was occupied by the Italians in 1941 and then Germans in 1943. The city was very heavily damaged in World War II. Montenegro became part of Yugoslavia after the war. From 1946 to 1992, the city was named Titograd in honor of Josip Broz Tito, a leader of Yugoslavia. The Socialist Federal Republic of Yugoslavia broke apart in 1992, and Montenegro became an independent country with Podgorica as capital under its restored name.

Major Landmarks. World War II destroyed much of Podgorica's historical fabric, although there is still an old town district (Stara Varoš) with narrow windings streets from the Ottoman era and a landmark Ottoman Clock Tower. St. George's Church predates the Ottoman period and is on the slopes of Gorica, the "small hill" after which the city was named. The Cathedral of the Resurrection of Christ is a new religious landmark that was constructed beginning in 1993. Other landmarks are Trg Republike, the central square of the city, Hercegovacka, a popular pedestrian street nearby, and Millennium Bridge, a striking new bridge over the Moraça River. Important cultural institutions include the Montenegrin National Theater, the Podgorica City Museum, the Museum of Natural History. The University of Montenegro was founded in 1974 and is the country's leading university.

Culture and Society. About 57 percent of the population of Podgorica belongs to the Montenegrin ethnic group, with Serbians, Bosniaks, Roma, and Albanians being the main ethnic minorities. About 80 percent of the population is Eastern Orthodox, and a little more than 10 percent is Muslim. The national language is Montenegrin. Industry is an important part of Podgorica's economy, with smelting of locally mined aluminum ores being the largest single enterprise. Foreign tourism is increasing in Montenegro, both on the Adriatic Sea coast and in mountains with ski resorts, with Podgorica's airport as an important hub.

Further Reading

Dragicevich, Peter, William Gourlay and Vesna Maric. *Lonely Planet's Montenegro.* Footscray, Australia: Lonely Planet, 2009.

Morrison, Kenneth. *Montenegro: A Modern History.* London: I. B. Tauris & Company, 2009.

Roberts, Elizabeth. *Realm of the Black Mountain: A History of Montenegro.* Ithaca: Cornell University Press, 2007.

PORT-AU-PRINCE

Port-au-Prince is the capital and largest city in the Republic of Haiti, a country in the West Indies that occupies the western half of the island Hispaniola which it shares with the Dominican Republic. The city is at the apex of the Gulf of Gonâve and is protected by the open sea by the offshore island of La Gonâve. Despite the coastal location, Port-au-Prince is more or less centrally located in Haiti because the country has an irregular shape. The population of the city is about 900,000, while that of the metropolitan area, which includes Pétionville, Carrefour, and other nearby cities, exceeds 2.5 million (2009). The name Port-au-Prince is derived from a French ship named *Le Prince* that docked there in 1706 to protect the city from possible incursion by the English.

Historical Overview. The original inhabitants of Hispaniola were Taino Amerindians. The Spanish began arriving in 1492 and claimed the island as theirs, leading to decimation of native population by conflict and disease. In about 1650, French buccaneers called *flibustiers* began to use the western part of the island as a base. They established a hospital at one of their ports, calling the area Hôpital. The Spanish challenged the French presence on Hispaniola but were defeated militarily, and ceded the western part of the island in 1697 via the Treaty of Ryswick. The French began settling the area and established a colony based on sugar plantations and slave labor imported from Africa. The colony was named Saint-Domingue. Port-au-Prince was founded in 1749. It became capital in 1770, replacing Cap-Français (today's Cap-Haïtien). In 1793, the city was renamed Port-Républicain in recognition of the founding of the French Republic. Following a slave revolt against French planters, Haiti achieved its independence in 1804, becoming the second independent nation in the western hemisphere after the United States. The country was given the name Haiti and the city was once again named Port-au-Prince. For a time, it was renamed Port-aux-Crimes (Port of Crimes) because of the assassination of Haiti's emperor Jacques I. When Haiti was divided between a kingdom in the north and a republic, Port-au-Prince was the republican capital.

Haiti was occupied by American troops from 1915 to 1934, during which time there was American investment in infrastructure and orientation of the economy to favor American business. This was followed by a conflict between Haiti and its neighbor the Dominican Republic, and dictatorial rule in both countries. Haiti was greatly impoverished by the dictatorship of the Duvalier family in 1957–1986, and suffered a great brain drain during this time. Although there have been steps toward better government and democracy in Haiti since, the country remains very poor, and Port-au-Prince is a city with many impoverished shantytown neighborhoods. Remittance income from Haitians living abroad props up the economy.

Port-au-Prince was devastated by earthquakes in 1751 and 1770, and then again in January 2010 that killed some 316,000 people in the city and its surroundings, injured approximately 300,000 more, and made more than 1 million people homeless. The historic center of the city was almost totally destroyed. Port-au-Prince is still recovering from this disaster, with reconstruction underway of basic infrastructure, public buildings, and residential neighborhoods. Hundreds of thousands of city residents are still in temporary shelters, and the country still depends on international relief efforts for daily sustenance.

Major Landmarks. The most important landmarks in Port-au-Prince, the National Palace and the Roman Catholic Cathedral of the Assumption, that collapsed during the 2010 earthquake are still in disrepair, as is much of the rest of the city center. The historic Iron Market, however, has been rebuilt and is now a symbol of the resiliency of Port-au-Prince. The Hotel Oloffson is a grand old hotel that is also an icon of the city. It too is again in operation. The Musée du Panthéon National Haïtien is another important city landmark. The suburb of Pétionville is the best part of town and has upscale shops, restaurants, and nightclubs.

Culture and Society. The majority of Haitian society is descendant from African slaves. About 95 percent of the population is classified as black, with the rest being either mulatto, Haiti's term for lighter-skinned people of mixed race, or white. The majority of Port-au-Princes mulatto population resides in Pétionville. In contrast to this better suburb are many slum neighborhoods such as Cité Soleil, one of the largest and poorest communities in the western hemisphere. The official languages of Haiti are French and Haitian Creole. About 80 percent of the population is Roman Catholic and 16 percent Protestant. Many Haitians combine their religious practices with elements of voodoo. Many Haitians live abroad, most notably in the Dominican Republic, in the United States (New York City), in Canada (Montreal), and in France.

Further Reading

Dubois, Laurent. *Haiti: The Aftershocks of History*. New York: Metropolitan Books, 2012.
Farmer, Paul. *The Uses of Haiti*. Monroe, ME: Common Courage Press, 2005.
Munro, Martin, ed. *Haiti Rising: Haitian History, Culture, and the Earthquake of 2010*. Kingston: University of the West Indies Press, 2011.

PORT LOUIS

Port Louis is the capital and largest city of the Republic of Mauritius, a small island country in the southwest Indian Ocean approximately 560 miles (900 km) east of the coast of Madagascar. The country consists of a main island Mauritius and several smaller islands, and is generally categorized as one of the countries of Africa. Port Louis is on the northwest coast of the main island. It is one of Africa's most prosperous cities, with an economy based on a modern port, manufacturing of clothing and textiles, chemicals, plastics, and pharmaceuticals. The city is also an important financial services center and tourism destination. Port Louis has the highest per capita income of any African city. The population of Port Louis is 128,483 for the administrative district of the city and 148,416 for the wider urban area (2010).

Historical Overview. Mauritius was visited by Arab sailors in the Middle Ages. The first European visitor was probably Portuguese navigator Diogo Fernandes Pereira in 1511. Dutch sailors arrived in 1598 and named the island in honor of Prince Maurits van Nassau. The Dutch settled the island in the 17th century but left by 1710. France took control in 1715 and renamed the island Ile de France (Island of France). Port Louis was named after French King Louis XV and became the French administrative center and the main port for French ships travelling across the Indian Ocean. The French developed sugar production on the island and exploited slaves imported from the African continent and Madagascar. British administration began in 1810 and slavery was abolished in 1835. The island was renamed Mauritius. Indentured laborers were brought from India to work the sugar fields. With time, the Indian population became numerically dominant. Independence from the United Kingdom was achieved in 1968, and in 1992,

Mauritius became a republic. The economy of Mauritius has developed since independence from one based on sugar production to a diversified mix of industry, financial services, tourism, and agricultural exports.

Major Landmarks. The landmarks of Port Louis reflect the diversity of the city's cultures. Aapravasi Ghat was the landing point of indentured servants from India in the 19th century and is a UNESCO World Heritage Site. Other landmarks with ethnic significance are St. Louis Cathedral, the Jummah Mosque, and the Chinatown district. Museums in Port Louis are Mauritius Natural History Museum, the Blue Penny Museum, and the Stamp Museum. The Champ de Mars racetrack near the center of Port Louis was founded in 1812 and still races horses. The Caudan Waterfront is trendy district opened in 1996 that features many shops, restaurants, casinos, and many other businesses.

Culture and Society. More than three-quarters of the population of Port Louis is of Indian origin, descendant from laborers who were brought to Mauritius in the 19th century. Other ethnicities are Chinese, Eurasian (mostly of French origin), and Mauritian Creoles (mostly of African origin). The official language of government is English, although French is also permitted. Most citizens of Mauritius speak a French-based Mauritian Creole, English, and French. English and French are the languages of schooling in Mauritius. In terms of religion, Hindus comprise more than one-half of the population, Roman Catholics a little more than one-quarter, other Christian faiths about 9 percent, and Muslims about 17 percent. The coat-of-arms of Mauritius displays the dodo bird among other objects. The dodo was unique to Mauritius and became extinct in 1681 soon after the island became settled. It has since become a famous example of the extinction of a species and the word "dodo" has entered popular culture in many ways, including the expression "dead as a dodo."

Further Reading

Allen, Richard B. *Slaves, Freedmen and Indentured Laborers in Colonial Mauritius.* Cambridge, UK: Cambridge University Press, 1999.
Bowman, Larry W. *Mauritius: Democracy and Development in the Indian Ocean.* Boulder, CO: Westview Press, 1991.
Eisenlohr, Patrick. *Little India: Diaspora, Time, and Ethnolinguistic Belonging in Hindu Mauritius.* Berkeley: University of California Press, 2006.
Simmons, Adele. *Modern Mauritius: The Politics of Decolonization.* Bloomington: Indiana University Press, 1982.

PORT MORESBY

Port Moresby is the capital and largest city in Papua New Guinea (PNG), an island nation in Oceania north of Australia. The city is on the south coast of the main part of the PNG, where the country comprises the eastern one-half of the island of New Guinea, and faces the Gulf of Papua. The estimated population of Port Moresby is 308,000 (2009), although the number grows rapidly from in-migration from the country's mountainous provinces.

Port Moresby was named by English Captain John Moresby, who arrived on the island in 1873 as the first known European visitor, in honor of his father, Admiral

Sir Fairfax Moresby. It is commonly called Moresby. The native Motu-Koitabu people refer to the city as Pot Mosbi in Tok Pisin, as their "talk" is called.

Historical Overview. The Motu-Koitabu people have long inhabited Port Moresby's site and used it as a base for interisland trade. European colonization began about 10 years after John Moresby's 1873 landing, and the area was soon claimed for England as a territorial possession. In 1906, control of British New Guinea was transferred to the Commonwealth of Australia and the territory came to be called Papua. Port Moresby was its only notable European settlement and the administrative center. In the New Guinea Campaign (1942–1945) of World War II, Port Moresby was the object of bitter fighting between Japanese and Allied forces, as control of the area would provide an important military base between Australia and Southeast Asia. The Battle of the Coral Sea on April 4–8, 1942, was a landmark in that campaign in which an attempted Japanese invasion of Port Moresby was turned back. After Allied victory, German New Guinea was annexed to Papua, forming the Territory of PNG under Australian administration. PNG gained its independence from Australia in 1975 and Port Moresby became a national capital. Administration buildings for the new country were built in Waigani, a suburb of Port Moresby.

Major Landmarks. The Parliament Haus (as it is known) is the principal government building of PNG and a distinctive landmark for its design. Nearby is the PNG National Museum. The city's Botanical Gardens and Varirata National Park outside the city are also noteworthy. Touaguba Hill has a nice view of the city and Port Moresby's better homes.

Culture and Society. The majority of Port Moresby residents are indigenous people, among whom there is a very wide range of ethnic and tribal groups and spoken languages. The island of New Guinea in general is extraordinarily diverse with respect to cultural diversity, with at least 860 spoken languages, more than one-tenth of the world's total. There are three official languages used most commonly in Port Moresby: Tok Pisin, English, and Hiri Motu. Christianity is the main religion, although there are many admixtures of traditional animistic beliefs. Port Moresby has a small minority of foreign residents, mostly Australians. Among the foreigners, there are many missionaries and aid workers. There are also Chinese immigrants. There have recently been conflicts between indigenous Papua New Guineans and Chinese, including violence. Crime and violence are major problems across PNG and in Port Moresby in particular. The city has one of the highest crime rates in the world, which is compounded by police corruption and indifference, and by roving gangs called *raskals* (rascals).

Further Reading

Goddard, Michael. *The Unseen City: Anthropological Perspectives on Port Moresby, Papua New Guinea.* Canberra: Australian National University, 2005.

Goddard, Michael, ed. *Villagers and the City: Melanesia Experiences of Port Moresby, Papua New Guinea.* Wantage, UK: Sean Kingston Publishing, 2010.

May, R. J. *State and Society in Papua New Guinea: The First twenty-Five Years.* Canberra: Australia National University, 2001.

Stuart, Ian. *Port Moresby Yesterday and Today.* Sydney, Australia: Pacific Publications, 1970.

PORT-OF-SPAIN

Port-of-Spain, which is also spelled as Port of Spain, is the capital and third-largest city in Trinidad and Tobago, an island country in the southeastern West Indies off the north coast of South America's Venezuela. There are two main islands, a larger one named Trinidad which has about 96 percent of the nation's 1.2 million people, and Tobago to the northeast. There are also some islets. Port-of-Spain is located in the northwest of Trinidad on the Gulf of Paria, and is, of course, a port city. In addition, it is the economic hub of the country, an important banking center for the West Indies more generally, and its country's principal cultural and educational center. The population of Port-of-Spain is about 57,000. The northwest of Trinidad is densely built up, and includes the country's largest urban center Chaguanas, which is about 11 miles (16 km) south of Port-of-Spain and has a population of approximately 330,000 (2012). It is estimated that the urban area of Port-of-Spain has as many as 600,000 residents.

Historical Overview. The first explorer to sight Trinidad was Christopher Columbus in 1498. He gave the island its name because of the three hills that were seen from the water and the Holy Trinity. The island had been heavily settled by Carib Indians before arrival of the Spanish, but their population was then decimated by warfare and disease. Port-of-Spain was founded on the site of an Amerindian fishing village named Cumucurapo. The Spanish built a small mud-walled fort to protect their settlement. In 1757, the town became the administrative center for the island, replacing San José de Oruña about 7 miles (11 km) inland. In 1797, Trinidad was annexed to the British Empire after Port-of-Spain was taken by British forces led by General Sir Ralph Abercromby. Streets and places were renamed after British royalty and military leaders. Under British rule, Trinidad developed as an area of sugar plantations with slave labor from Africa. After emancipation of the slaves in 1838, British planters brought indentured laborers from China, Portugal, and India to work in the plantations. Indians became the largest group; their descendants now comprise more than one-half of the population of Trinidad and Tobago. From 1958 to 1962, Port-of-Spain was capital of the short-lived Federation of the West Indies. Trinidad and Tobago's independence from the United Kingdom was achieved in 1962, and in 1976, the country became a republic. Eric Williams was the first prime minister, serving from before independence in 1956 until his death in 1981. He is regarded as "The Father of The Nation." In July 1990, Port-of-Spain suffered major rioting and looting during a terrorist attack that held the prime minister and members of parliament hostage for five days. Since 2005, the city has been the seat of the Caribbean Court of Justice.

Major Landmarks. The Red House is an iconic building that is the seat of the Parliament of Trinidad and Tobago. Rosary Church is another historic building in the gothic style. There are beautiful historic mansions in the St. Clair neighborhood of Port-of-Spain. Other landmarks include the International Waterfront Centre, which is a modern commercial redevelopment project in the downtown, the National Academy of Performing Arts, the National Museum and Art Gallery, the Royal Botanic Gardens, and the statue of Christopher Columbus in Christopher Columbus Square. The port of Port-of-Spain is also a notable landmark.

Culture and Society. The population of Port-of-Spain is highly diverse ethnically. People of Indian background are most numerous, but there are also significant numbers of African-origin residents, English, French, Spanish, Portuguese, Venezuelans, Chinese, and Syrians, among others. Christianity is the religion of about two-thirds of the population, Hinduism of about one-quarter, and Islam of about 7 percent. The national language is English. The city has had two Nobel Prize winners for literature, Derek Walcott in 1992 and Sir Vidiadhar Naipaul in 2001. Cricket is a very popular sport. The pre-Lent Carnival of Port-of-Spain is one of the world's best.

Further Reading

Mason, Peter. *Bacchanal: The Carnival Culture of Trinidad.* Philadelphia: Temple University Press, 1999.

Pantin, Raoul A. *Days of Wrath: The 1990 Coup in Trinidad and Tobago.* Lincoln, NE: iUniverse, 2007.

Verrest, Hebe. *Home-Based Economic Activities and Caribbean urban Livelihood: Vulnerability, Ambition and Impact in Paramaribo and Port-of-Spain.* Amsterdam: Amsterdam University Press, 2007.

Red House, the beautiful parliament building of Trinidad and Tobago in downtown Port-of-Spain. (Robert Lerich/iStockPhoto.com)

PORT VILA

Port Vila is the capital and largest city of Vanuatu, a nation of approximately 82 small islands, 65 of which are inhabited, in the South Pacific Ocean about 1,090 miles (1,750 km) northeast of Australia. The city is located on the south coast of Efate, Vanuatu's third-largest and most populous island, and has a population that is estimated at a little more than 40,000. In addition to being the capital, Port Vila is the main business center of Vanuatu and the country's main port. One of the city's economic mainstays is a role as offshore tax haven.

Historical Overview. Melanesian people have occupied Efate Island for millennia. Europeans first learned about the islands in 1606 from the explorations of Portuguese navigator Pedro Fernandes de Queirós who sailed for the Spanish crown. The French admiral and explorer Louis Antoine de Bougainville rediscovered the islands in 1768, and then in 1774 the noted English explorer, Captain James Cook, arrived and named the island group New Hebrides. Catholic and Protestant missionaries began arriving in the 19th century. France and Great Britain jostled for control of the islands, and in 1887 declared them to be neutral territory and agreed to administer them jointly. Port Vila was established by French settlers in the 19th century and was initially named Franceville. Because of disputes about French versus British laws relating to marriage, Franceville declared independence in 1889 for about one year. The Independent Commune of Franceville, as it was known, is recognized in history as the first self-governing nation to practice universal suffrage (although only men could hold office). During World War II, Port Vila was an American and Australian air base. In 1980, the New Hebrides gained independence from France and Great Britain and the name of the islands was changed to Vanuatu, a word that means "independent homeland." At that point, Port Vila became a national capital. An earthquake and subsequent tsunami caused much damage to the city in 2002.

Major Landmarks. Important landmarks in Port Vila include the national parliament building, the major church buildings for the various Catholic and Protestant faiths of the city, and the Port Vila Market. The Mele Cascades are a short distance outside the city and offer a beautiful tropical waterfalls setting.

Culture and Society. The vast majority of residents of Port Vila are Melanesians, with small populations as well of Polynesians, Australians, Asians, and Europeans. More than 90 percent of the population is Christian, with Presbyterians being the most numerous denomination. Bislama, a creole language, is the official language of Vanuatu, along with English and French. There are many traditional cultural beliefs among the population, including considering pigs, particularly those with rounded tusks, to be a symbol of wealth. Traditional Vanuatu music has been adapted by contemporary island musicians into popular new sounds with commercial appeal.

Further Reading

Bolton, Lissant. "Women, Place, and Practice in Vanuatu: A View from Ambae," *Oceania* 70, no. 1 (1999): 43–56.

Rodman, Margaret, Lissant Bolton, and Jean Tarisese, eds. *House-Girls Remember: Domestic Workers in Vanuatu.* Honolulu: University of Hawai'i Press, 2007.

Troost, J. Maarten. *Getting Stoned Savages: A Trip through the Islands of Fiji and Vanuatu.* New York: Broadway Books, 2006.

PORTO-NOVO

Porto-Novo is the capital and second-largest city in the Republic of Benin, a West African country with a narrow coastline on the coast of wedged between Togo to the west and Nigeria to the east. The city is located near the coast on an inlet of the Gulf of Guinea. Although Port-Novo is the home of the national legislature and the national archives, and is formally the "official capital" of Benin, many of the other most important functions of a seat of national government such as diplomacy are carried out in Cotonou, Benin's largest city. Consequently, Cotonou is usually referred to as Benin's de facto capital. Cotonou is located on the coast as well, just to the west of Porto-Novo. The population of Porto-Novo is 224,000 (2002), while that of Cotonou is at least 761,000 (2006) and probably quite a bit more. Other names for Port-Novo are Hogbonou and Adjacé.

Historical Overview. Porto-Novo was founded by Portuguese slavers as one of their ports along the so-called Slave Coast of Africa. Spanish and French slave ships docked at the port as well. At the time, it was within the Yoruba kingdom named Oyo that was centered on what is today southwestern Nigeria. The Kingdom of Dahomey was nearby and ruled the southern part of today's Benin from about 1600 to 1900. The kings of Dahomey provided enemy captives from the interior for sale into slavery to the Europeans. The French increased their dominance over the region in the 19th century, and in 1892, the land that is now Benin was carved out of Africa across traditional boundaries between African states to become a French colonial possession called French Dahomey. In 1899, French Dahomey was incorporated into the larger colony French West Africa. Port-Novo became the capital of French Dahomey in 1900. Autonomy was granted to the French Dahomey in 1958 and in 1960, the Republic of Dahomey became an independent nation. It was renamed Benin in 1975 after the Bight of Benin, because the word Dahomey did not apply to all parts of the country's territory.

Major Landmarks. The Parliament Building of Benin is a distinctive modern structure and major landmark. The Museé Ethnographique de Porto Novo has displays about the kings of Porto-Novo, Yoruba masks, traditional musical instruments, and other artifacts about Benin's history and culture. Museé de Silva displays information about Benin's Afro-Brazilian population. The Great Mosque of Porto-Novo is a landmark building that has Islamic architectural features grafted on to the features of a church. The Araromi Mosque is another example of interesting mosque architecture. The Roman Catholic Cathedral of the Immaculate Conception is still another religious landmark in Porto-Novo.

Culture and Society. Cotonou is the principal business center of Benin and is much more dynamic and modern city than Porto-Novo, which is slower paced and less well developed. The main ethnic groups in Porto-Novo are Nagot, Goun, and Yoruba. The city also has a sizable minority of Afro-Brazilians who settled there after returning to Africa following the end of slavery in Brazil. Afro-Brazilians have added to Porto-Novo's distinctive architectural heritage and cuisine. The languages of the city include African ethnic languages and French, which is the official language of the country. There are approximately equal numbers of Muslims and Christians in the population of Benin as a whole. The literacy rate of the country is among the lowest in the world.

Further Reading

Decalo, Samuel. *Historical Dictionary of Benin.* Lanham, MD: Scarecrow Press, 1995.

Mandel, Jennifer L. "Mobility Matters: Women's Livelihood Strategies in Porto-Novo, Benin," *Gender, Place, and Culture: A Journal of Feminist Geography* 11, no. 2 (2004): 257–87.

Manning, Patrick. *Slavery, Colonialism and Economic Growth in Dahomey, 1640–1960.* Cambridge, UK: Cambridge University Press, 1982.

PRAGUE

Prague is the capital and largest city of the Czech Republic, a small country in central Europe that is bordered by Germany, Poland, Slovakia, and Austria. The city is near the center of the country on both banks of the Vlatava River. In Czech and other languages, the city is known as *Praha*. In addition to being a national capital, the city is an important regional business center, a major cultural center, and a very popular destination for foreign tourists. The population of Prague is about 1.3 million, while that of the metropolitan area is about 2 million (2011).

Historical Overview. The site of Prague has been settled for millennia. Legend has it that the city was founded by Libuše and her husband Přemysl. In the ninth century a fort, Vyšehrad, was built of a high promontory on the right bank of the Vlatava. This was the foundation of Prague Castle. In the 10th century, Prague's other fort, Hradćany, was erected on an equally commanding site on the left bank, a little downstream. The city was the seat of royalty of Bohemia and a busy trading center. It became a bishopric in 973 and an archbishopric in 1344. It has had a Jewish population since at least the 10th century. The Old New Synagogue was built in 1270 and still stands. Bohemian King and Holy Roman Emperor Charles IV, who reigned from 1346 to 1378, had enormous influence on the development of the city. In 1348, he established Charles University, the first in central Europe. Among other projects, he also oversaw construction of the planned New Town (Nové město) adjacent to the Old Town (Staré město), the construction of the Charles Bridge beginning in 1357, and the start of construction of the spectacular St. Vitus' Cathedral. During the reign of the son who succeeded him, Wenceslaus IV (r. 1378–1419), Prague experienced considerable turmoil, including a murderous ransacking of the Jewish ghetto in which almost all 3,000 inhabitants perished, religious conflicts related to the Protestant reformation, and the first of two notable episodes of defenestration of political opponents. The second Defenestration of Prague took place in 1618 when the governors of Bohemia were thrown from the windows of Hradćany, an event that helped start the Thirty Years War (1618–1648). The war plus outbreaks of plague led to a long period of decline for Prague. A great fire swept through the city in 1689 killing approximately 13,000 people.

The city recovered in the 18th century and grew to become once again a center of commerce and trade, culture and arts. Prague also became an industrial city based on the presence of coal and iron ore resources nearby. The Jewish population increased and comprised more than one-quarter of the city's total. The Austro-Hungarian Empire, of which Prague was a part, was dismantled after World War I and in 1918 Prague became the capital of the newly formed nation of Czechoslovakia. The first president, Tomáš Masaryk, governed from Prague Castle. Adolf Hitler's German Nazi forces entered Prague on March 15, 1939, and took control of Czechoslovakia. Most Jews were murdered. The city was heavily damaged in the war. After World War II, Prague was capital of a Communist Czechoslovakia, a satellite state of the Soviet Union. The "Prague Spring of 1968" was an attempt to liberalize Czechoslovakia but Soviet tanks entered the city and put down the popular movement for democratic reform. Communist was defeated and replaced with democratic institutions as a result of the popular Velvet Revolution of 1989 (also called the Quiet Revolution). In

1993, Czechoslovakia decided peacefully to divide into separate Czech and Slovak republics, with Prague becoming capital of the Czech Republic.

Major Landmarks. Prague has a great many important landmarks and historical buildings. Prague Castle is perhaps the most iconic structure in the city, occupying the highest point on the left bank. St. Vitus Cathedral and its lookout tower are nearby. The Charles Bridge connects Old Town with Lesser Town and is often jammed with tourists. In Old Town, important landmarks are the Astronomical Clock, the spectacular Gothic Týn Church, and the Jan Hus Monument to a famed priest and philosopher who was burned at the stake in Prague in 1415 for alleged heresy against the Catholic Church. In the Josefov district there are many buildings intact from the old Jewish ghetto including old synagogues. The main attraction in the New Town is Wenceslas Square. The National Museum is one of several prominent buildings that face the square. In Lesser Town there are many beautiful historic churches, including St. Nicholas Church and the Church of Our Lady Victorious in which is the famed statue of Jesus Christ known as the Holy Infant of Prague. Other landmarks of note include the Prague Giant Metronome, the unusual architectural structure by Frank Gehry called the Dancing House, and the Memorial to the 1989 Velvet Revolution. The Museum of Communism offers a very critical view of that period of history.

Culture and Society. Prague is a sophisticated and culturally diverse city with a great many theaters, concert halls, art galleries, and museums. Among the many natives of Prague who are known around the world for their contributions to arts and letters is the German-language writer Franz Kafka who was born in the city in 1883. Prague's population is mostly Czech in ethnic origin, but there are also Slovaks, Moravians, Poles, Germans, Roma, and others. At any given time, the city has many thousands of foreign tourists as well. The Czech people are among the least religious in the world, being described as tolerant of all faiths and interested in none. The main religion is Roman Catholicism, to which about 10 percent of the population adheres at least somewhat. The official language is Czech, but quite a few other central and eastern European languages are recognized as official minority languages. English is widely understood.

Further Reading

Lau, J. M. *Prague Then and Now.* San Diego: Thunder Bay Press, 2007.

Sugliano, Claudia. *Prague Past and Present.* New York: Metro Books, 2002.

Thomas, Alfred. *Prague Palimpsest: Writing, Memory, and the City.* Chicago: University of Chicago Press, 2010.

PRAIA

Praia is the capital and largest city of Cape Verde, an island nation in the North Atlantic Ocean about 355 miles (570 km) west of Senegal on the African continent. The country consists of two groups of islands, Ilhas de Barlavento (Windward Islands) and Ilhas de Sotavento (Leeward Islands). Praia is on the southern coast of Santiago Island, one of the Sotavento Islands. In addition to being the chief administrative center for the country, Praia is also Cape Verde's principal port. The main exports are coffee, sugar cane, and tropic fruits. There is also a fishing industry and tourism to nearby beaches. The word "Praia" itself means beach in the native Portuguese. The population of Praia is about 127,000, about one-quarter of the total population living on the Cape Verde Islands.

Historical Overview. The Cape Verde Islands were settled originally by Portuguese at the dawn of European maritime exploration. The first town that was founded was Ribera Grande on Santiago Island in 1462. Praia, originally called Praia de Santa Maria, was established in about 1615 as an alternative harbor to Ribera Grande in order to avoid paying the latter port's customs fees. Frequent pirate attacks against Ribera Grande eventually caused population to move to Praia for safety, and in 1770 the newer city was made the capital. Because of its position between Africa and the North and South American continents, the main business of the Cape Verde Islands was the slave trade. When the slave trade was ended in the 19th century, the islands went into economic crisis. There are few resources on the islands, little agricultural land, and a chronic shortage of fresh water. Cape Verdeans made a living by resupplying passing ships and from fishing. Many have immigrated to other countries. The relative poverty of Cape Verde hastened the call among locals for independence from Portugal. That was achieved in 1975. The present economy is supported mostly by the port in Praia and other island ports, light manufacturing, the fishing industry, and tourism. Remittances from Cape Verdeans living abroad are also very important to the islands' economy. There are more emigrants from Cape Verde in North America and in various countries of Europe and the African mainland than in the home country. The goal of energy independence is an important pillar for economic development in Cape Verde. Accordingly, the country derives much of its electricity from wind power, being one of the world leaders in that form of renewal energy.

Major Landmarks. Main landmarks in Praia are the Praia Roman Catholic Cathedral, the Palace of Justice, the Presidential Palace, and the Museum of Ethnography. Other notable landmarks are the statues of Portuguese navigator Diogo Gomes and Cape Verdean independence leader Amilcar Cabral.

Culture and Society. The people of Cape Verde are a mix of African and European ancestries. More than 95 percent of the population is Christian, with Roman Catholics being the predominant group. The official language of Cape Verde is Portuguese, but Cape Verdean Creole is the language that is spoken most on a daily basis by the majority of citizens. Birth rates are high and the population is generally youthful in demographic profile. Music is an important part of Cape Verdean culture, especially a form of song called *morna*. The singer Cesaria Evoria was a Cape Verdean whose distinctive voice was popularly known around the world.

Further Reading

Batalha, Luis. *Transnational Archipelago: Perspectives on Cape Verdean Migration and Diaspora.* Amsterdam: Amsterdam University Press, 2008.

Chabal, Patrick. *Amilcar Cabral: Revolutionary Leadership and People's War.* Trenton, NJ: Africa World Press, 2003.

Lobban, Richard A. *Cape Verde: Crioulo Colony to Independent Nation.* Boulder, CO: Westview Press, 1988.

PRETORIA

Pretoria is one of the three capital cities of the Republic of South Africa, the country at the southern tip of the African continent. It is the executive (administrative) capital and de facto national capital, with Cape Town being the legislative capital and Bloemfontein being the judicial capital. (See the separate entries for these two cities.) Pretoria is located in the northeast of the country at about 4,500 ft elevation (1,350 m) in a valley amidst the Magaliesberg Hills. It is in the northern part of Guateng Province, about 31 miles (50 km) north of Johannesburg, South Africa's largest city. The population of Pretoria is about 1.1 million (2009), but that of the City of Tshwane, the administrative jurisdiction of which it a part, is nearly 2.4 million. Guateng is itself a heavily urbanized region with a total population of over 11 million. Pretoria is sometimes referred to as Tshwane, a name that has been proposed as the official new name for the city. The city is sometimes called the Jacaranda City due to the large numbers of Jacaranda trees in the city's streets, parks, and gardens.

Historical Overview. Although the valley where Pretoria is situated has been occupied for a long time, the city itself was founded in 1855 by Marthinus Pretorius, a leader of Dutch-Afrikaner pioneers who left British-controlled Cape Colony to resettle in Transvaal, an independent Boer state. He named the city after his father, Andries Pretorius, also a prominent Voortrekker leader. The town's original name was actually Pretoria Philadelphia, Pretoria of Brotherly Love. On May 1, 1860, Pretoria became the capital of the South African Republic, abbreviated as ZAR from the Dutch *Zuid-Afrikaansche Republiek.* The republic was also called the Transvaal Republic. In 1910, the Boer Republics united with Cape Colony and Natal to become the Union of South Africa, and Pretoria became the administrative capital. South Africa became a republic in 1961 and Pretoria continued as the administrative capital. In Pretoria, in 1994, Nelson Mandela became the first president of post-apartheid South Africa.

Major Landmarks. Major landmarks in Pretoria include the beautiful sandstone Union Buildings, which are the main government buildings of South Africa and the office of the nation's president located atop Meintjieskop, a prominent hill in the city; Church Square (Kerkplein in Afrikaans) which marks the heart of the city and has a statue of ZAT President Paul Kruger in its center; an equestrian statue of Andries Pretorius; and the Voortrekker Monument that honors the early Boer settlers of the region. Other prominent landmarks are the main campus of the University of South Africa, the South African State Theater, Loftus Versfeld Stadium for both rugby and football (soccer), and the Transvaal Museum, also known as the Ditsong National Museum of Natural History. The African Window is a building that houses the Ditsong National Museum of Cultural History. Pretoria is also home to the National Zoological Gardens of South Africa and the Pretoria National Botanical Garden.

Culture and Society. For all of its history, Pretoria has been associated with Afrikaner settlement and a white majority population. Even after the end of apartheid in 1994, the city has maintained a mostly white population, and has the largest white population of any city on the African continent. There are about 400,000

The residence of the president of South Africa in Pretoria's Union Buildings complex, the main national government center. (Sean Nel/iStockPhoto.com)

Afrikaners living in the city and its surroundings. Increasingly, however, there is also a growing black middle class in the city, especially in the townships of Soshanguve and Atteridgeville. The main languages spoken in Pretoria are Pedi, Afrikaans, Tswana, Tsonga, Zulu, and English. The city is one of the most prosperous and comfortable cities in Africa.

Further Reading

Donaldson, Ronnie. "Challenges for Urban Conservation in the Historical Pretoria Suburb of Clydesdale," *Urban Forum* 12, no. 2 (2001): 225–47.

Donaldson, Ronnie. "Contesting the Proposed Rapid Rail Link in Guateng," *Urban Forum* 16, no. 1 (2005): 55–62.

Matloff, Judith. "South Africa's Sleepy Pretoria Awakened by Apartheid's End," *Christian Science Monitor* 88, no. 187 (August 21, 1996): 6.

PRISTINA

Pristina is the capital and largest city of the Republic of Kosovo, a sovereign, land-locked state in southeastern Europe. The country is locked in a dispute with neighboring Serbia, which does not recognize Kosovo's independence. Pristina is in the east center of the country near the Goljak Mountains. In addition to being the administrative center, Pristina is Kosovo's leading business, cultural, and educational center. The population of the city has been estimated to be about 200,000, but because of recent conflict, refugee flows, and resettlement processes, it may actually be as high as 500,000.

Historical Overview. Pristina is situated near the ruins of an old Roman town called Ulpiana (today's neighboring town of Lipljan), and grew as a trade center at a junction of a Balkan-region roads network. In the late 13th century, the city became the capital of the Kingdom of Serbia. The area came under Ottoman control in various stages after the June 15, 1389 Battle of Kosovo which was fought just west of Pristina. During the Ottoman period, the population of Serbians declined and that of Albanian-speakers increased, and the region came to be predominantly Moslem. In 1912, Serbia recaptured the city, killing many Albanian civilians. In 1918, Kosovo became part of the Kingdom of Serbs, Croats, and Slovenes, the precursor of Yugoslavia, and many thousands of Kosovar Albanians were forced to flee abroad. Serbian settlers replaced them in the population. Conflict between the mostly Orthodox Serbs and mostly Muslim Albanians continued throughout the history of Yugoslavia, including during World War II, and resulted in open warfare and bloody "ethnic cleansing" policies by both Serbia and its rival Kosovo Liberation Army after Yugoslavia broke up into independent countries. Again, a great many Albanian civilians had to flee the country. The conflict reached a peak in 1998–1999, and was finally quelled after intervention by NATO troops and peacekeeping forces from the United Nations (UN). With UN assistance, Kosovo became an independent country in 2008. Pristina suffered significant damage during the Kosovo conflict and is now recovering. The ethnic reshuffling has continued, as it is Serbians who have had to flee the city most recently, and the Albanian population has been on the rise.

Major Landmarks. The central landmark is a large sculpture in the city that reads NEWBORN in English in celebration of Kosovo's independence. It is the meeting place for many public rallies and other events in the city. Other landmarks are Rilindja Tower, Pristina's tallest building, the United Nations Mission Headquarters (UNMIK), the OSCE Building, and the Museum of Kosovo. Pristina's Clock Tower dates back to the 19th century. The Fatih Mosque is from the 15th century.

Culture and Society. Pristina has alternated between control by Orthodox Serbians and Muslim Albanian speakers. It is now populated overwhelmingly by the latter. About 90 percent of the Kosovo population speaks Albanian as the first language. The relations between Serbs and Albanians continue to be hostile. There is a small minority of Roma in the Kosovo population, as well as a minority of Albanians who are Roman Catholics.

Further Reading

Judah, Tim. *Kosovo: What Everyone Needs to Know*. New York: Oxford University Press, 2008.

Ker-Lindsay, James. *Kosovo: The Path to Contested Statehood in the Balkans*. London: I. B. Tauris and Company, 2009.

McKinna, Anita. "Kosovo: The International Community's European Project," *European Review* 20, no. 1 (2012): 10–22.

Tooland, David S. "Four Days in Kosovo," *America* 181, no. 21 (2003): 7013.

PYONGYANG

Pyongyang is the capital and largest city in the Democratic People's Republic of Korea (DPRK), the repressive Communist country that occupies the northern part of the Korean Peninsula and that is commonly called North Korea. The country is widely considered to be a totalitarian Stalinist dictatorship built on a cult of personality around the ruling Kim family. Its citizens are generally very poor and lack basic freedoms. There have been recent famines and reportedly several hundreds of thousands of deaths from hunger. The city is located on the Taedong River in a flat plain in the east center of the country about 30 miles (50 km) east of Korea Bay, an arm of the Yellow Sea. The name Pyongyang means "flat land." One of the city's historic names is Ryugyong, "capital of willows," because of the many willow trees that grew there. According to results presented in the 2008 census, the population of Pyongyang is 3,255,388.

Historical Overview. The site of Pyongyang has been settled since prehistory. The city evolved into a significant urban center during the Gojoseon Kingdom more than two millennia ago. In the early Three Kingdom period, Pyongyang was capital of the Nanglang Kingdom. In 427, it became capital of the powerful Gogureyo Dynasty. Silla captured the city in 627. During the Joseon Dynasty, the city was capital of Pyeongan Province. Pyongyang was severely damaged during the 1894–1895 war between Ming Dynasty China and Meiji Japan over control of the Korean Peninsula. It was later rebuilt and became an industrial center during Japanese rule in 1910–1945, during which time it was named Heijō. Pyongyang was destroyed again in the Korean War (1950–1953) and was for a time occupied by forces from South Korea. After the war, it was rebuilt with assistance from the Soviet Union, and took on much of the architectural form of a Soviet city: excessively wide streets, monumental government buildings and large government monuments and parade grounds, and blocks of identical apartment block housing for the masses of citizens. Kim Il-Sung was prime minister of the country from 1948 to 1972 and then president from 1972 until his death in 1994, a period of supreme rule that totaled about 46 years. He ruled from Pyongyang and had enormous impact on the city and North Korea's citizenry. After his death, the "Great Leader," as he is reverently referred to in North Korea, was succeeded by his son Kim Jong-Il, known as the "Dear Leader," who ran the country until his death in 2011. The current leader is Kim Jong-Un, a son of Kim Jong-Il. Pyongyang grew rapidly in the latter half of the 20th century with migration from the North Korean countryside, but it differs markedly from Seoul, the South Korean capital 121 miles (195 km) to the south. Seoul is a busting, crowded, and prosperous bright city, while Pyongyang is mostly dull, poor, and dark for want of electricity.

Major Landmarks. For the most part, the landmarks of Pyongyang celebrate Kim Jong-Il, his social philosophy, and his government. The Kumsusan Palace of the Sun, previously called the Kumsusan Memorial Palace and sometimes referred to as the Kim Il-Sung Mausoleum, is a large and lavish structure that was once Kim-Il-Sung's official residence and then after his death became his mausoleum. It is now also the mausoleum of Kim Jong-Il. On Mangyongdae Hill is the reputed birthplace of Kim

Il-Sung. The Juche Tower dominates the center of the city beside the Taedong River. It was completed in 1982 to commemorate Kim Il-Sung's 70th birthday, and is named after the Great Leader's espoused philosophy for North Korea of self-sufficiency, independence, nationalism, and Marxism-Leninism. The iconic structure is 560 ft high (170 m) and is topped with an illuminated metal torch. Pyongyang TV Tower is another high landmark. The Arch of Triumph was built in 1982 and resembles the similarly named landmark in Paris. It commemorates Korean resistance against Japanese colonial rule. The Arch of Reunification was constructed in 2001 to honor proposals by Kim Il-Sung for reunification of the two Koreas. It straddles Reunification Highway at the entrance to the city. Still another iconic monument honors the Workers' Party of Korea that Kim Il-Sung had headed. Perhaps the most unusual landmark is the pyramid-shaped Ryugyong Hotel. At 1,083 ft (330 m), it is the tallest structure in the city, but has had serious design flaws that have prevented construction from being completed despite the passing of many years. Kim Il-Sung Stadium is used for football (soccer) matches, while Rungrado May Day Stadium, the world's largest with a seating capacity of 300,000, is used for football too but mostly for *arirang* events (mass games).

Culture and Society. North Korea's society is heavily regimented and is trained to put cooperative action first rather than individual pursuits. This is reflected in mass games, a synchronized gymnastics, dance, and acrobatics sporting event in which 100,000 or more highly practiced participants take part every May Day (May 1) to honor North Korea and its leaders. The athletes on the field are accompanied by 100,000 or more greatly practiced students in the stadium seats who create enormous, fast-changing, highly detailed and colorful pictures by holding up specific-colored placards on cue. The illustrations they create are also political in nature. These participants are sometimes referred to as human pixels.

Statue of Kim Il-Sung in Pyongyang, one of more than 500 statues of the long-time "Great Leader" of the reclusive country. (Linqong/Dreamstime.com)

Like the population of South Korea, the population of North Korea is ethnically very homogeneous—nearly 100 percent Korean. Most people practice no religion, although the country's constitution supposedly guarantees freedom of worship, but Buddhist and Confucian traditions are deeply embedded in aspects of culture. North Koreans are not free to travel abroad, and foreign visitors are encouraged mostly as a source of foreign currency and are greatly restricted as to where they may go and what they can do. Among the many striking scenes that visitors to Pyongyang report on is the presence of uniformed female traffic wardens who direct traffic almost mechanically from perches at the centers of big-street intersections in place of traffic lights. Often, these women go through their signaling motions even though there are few or no vehicles on the streets.

Further Reading

Delisle, Guy. *Pyongyang: A Journey in North Korea.* Montreal: Drawn and Quarterly, 2003.

Demick, Barbara. *Nothing to Envy: Ordinary Lives in North Korea.* New York: Spiegel & Grau, 2010.

Springer, Chris. *Pyongyang: The Hidden History of the North Korean Capital.* Budapest: Entente Bt, 2003.

Q

QUITO

Quito (full name San Francisco de Quito) is the capital of Ecuador, a small country in South America. It is named after a prehistoric tribe that once occupied the site, while Ecuador itself is named for the equator which runs through the country and just to the north of the capital city. The city is the closest to equator compared to any other capital city in the world. The population of Quito is some 2.2 million (2011), making it the second-largest city in the country after the port city of Guayaquil. The city is in the valley of the Guayllabamba River in a highland area in the north-central part of Ecuador. The elevation of the central square in the city is 2,800 m (9,200 ft), making the city the second highest capital city in the world after La Paz in Bolivia. Active volcanic peaks surround Quito, most notably Mt. Pichincha to the west and Cotopaxi in the south.

Historical Overview. Quito was founded in 1534 by Spanish conquistador Diego de Almagro who subjugated the native population and opened the way for Spanish colonial settlement. Roman Catholic churches were built soon thereafter and the natives were forced to convert. The Roman Catholic diocese of Quito was established in 1545. In 1563, the city became a regional administrative center within Spain's Viceroyalty of Peru. Independence from Spain came in 1822 in the wake of the decisive Battle of Pichincha won by South American liberationists under the command of Antonio José de Sucre. The history of Quito and all of Ecuador has long been unsettled, with periods of war, revolution, civil unrest, political assassinations, and military coups, mixed with brief times of peace and democracy. The most recent attempted coup d'état took place in 2010. Major earthquakes devastated the city in 1797, 1859, and 1949.

Major Landmarks. The historic center of Quito is the largest and best-preserved historic district in urban Latin America, and is a rich depository of original colonial churches, convents, government buildings, and residential streets. It consists of some 5,000 individual historic structures. The area was designated as a UNESCO World Heritage Site in 1978, and was one of the first two such designations in the world. Among the most notable structures are the Metropolitan Cathedral, the Church of La Compaña de Jesus, and the Church of San Francisco. The 19th-century Basilica del Voto Nacional was for a time the largest church in the Western Hemisphere. The Virgin of El Panecillo is a tall aluminum monument to the Virgin Mary that was built in 1976 on a hill in the western part of Quito. It can be seen from many parts of the city. Outside Quito are many rugged parks and biodiversity preserves on high-elevation volcanic slopes.

Culture and Society. Most residents of Quito are Roman Catholic and speak Spanish as their first language. About two-thirds of the residents are Mestizos (mixed Amerindian and descendants of Spanish colonialists), while about one-quarter are Amerindians. Less than 10 percent are Criollos (unmixed descendants of European settlers). The northern part of the city is the more prosperous and has a modern downtown with skyscrapers and fashionable shopping malls, while the south is more industrial and working-class in character.

Further Reading

Capello, Ernesto. "The Postcolonial City as Universal Nostalgia," *City* 10, no. 2 (2006): 125–47.

Capello, Ernesto. *City at the Center of the World: Space, History, and Modernity in Quito.* Pittsburgh: University of Pittsburgh Press, 2011.

Karanikolas, Mike. "Journey to the Center of the World," *Hispanic* 21, no. 3 (2008): 16–18.

Ryder, Roy H. "Land Use Diversification in the Elite Residential Sector of Quito, Ecuador," *The Professional Geographer* 56, no. 4 (2004): 488–502.

RABAT

Rabat is the capital city of the Kingdom of Morocco, a country in the Maghreb region of northwest Africa that faces both the western end of the Mediterranean Sea and the North Atlantic Ocean. The city is on the Atlantic coast in the northern part of the country at the mouth of the river Bou Regreg. Across the river is the city of Salé, also an important and historic Moroccan city, but now mostly a commuter suburb for Rabat. The population of Rabat is about 650,000, the seventh-largest total in Morocco. Salé, with about 815,000 inhabitants, ranks third. The population of the Rabat-Salé metropolitan area, which also encompasses the cities of Zemmour and Zaer, is about 2.4 million, second in Morocco after the Casablanca metropolitan area (3.6 million).

Historical Overview. The history of Rabat begins in the third-century BC with a river-bank settlement that was called Chellah. From 40 AD until about 250 AD, the town was part of the Roman Empire and was known as Sala Colonia. In 1146, Almohad ruler of North Africa Abd al-Mu'min fortified the city and used it as a base from where he launched Moorish attacks on Spain. In this context, the city was given the name from which the current word "Rabat" derives: *Ribatu I-Fath*, meaning "stronghold of victory." Near the end of the 12th century, another Almohad ruler, Abu Yusuf Ya'qub al-Mansur, made Rabat his capital and initiated large construction projects that included walls around the city, the *kasbah* (fortress), and what would have been the world's largest mosque. However, he died in 1199 and construction was halted. The ruins of the unfinished mosque and its minaret, Hassan Tower, still stand in Rabat.

Rabat declined in the 13th century when the Almohad dynasty lost Spain and much of its African territory, and Fez, another Moroccan city, increased in influence in its place. In 1627, Rabat was joined with Salé to form the Republic of Bou Regreg where piracy was a principal occupation. As late as 1828, Rabat was shelled by Austrian ships in retaliation for a sinking of an Austrian vessel by Rabat pirates. France invaded Morocco in 1912 and made it a protectorate, moving the capital from Fez back to Rabat. The sultan of Morocco, Yusef ben Hassan, moved his residence to Rabat from Fez as well. The French maintained strict separation between their own parts of the city and the districts of Rabat where the local population was allowed to reside. When Morocco gained its independence from France in 1956, Rabat remained the capital. From the end of World War II until 1963, the United States operated a military base near Rabat in Salé. That site is now a base of the Royal Moroccan Air Force.

Major Landmarks. The main landmarks of Rabat include Ya'qub al-Mansur's Kasbah of the Udayas and the unfinished mosque with its soaring Hassan Tower, the Royal Mausoleum of Mohammed V, the Palace Museum and Andalusian Gardens, and the Royal Palace. There is also the site of ancient Chellah, the National Archeological Museum, and the narrow streets and markets of the old Arab quarter. St. Pierre Cathedral is a remnant of the French era but still active as a place of Catholic worship in an Islamic city.

Culture and Society. The main ethnic group of Rabat is Arab-Berbers and the official language of the city and its country is Arabic. The vast majority of the population is Muslim. As capital city, Rabat attracts many migrants from the countryside. The outskirts of Rabat have squatter settlements and many substandard dwellings. There are still close cultural ties with France, as nowadays many Moroccans from Rabat and other cities live and work in France.

Further Reading

Abu-Lughod, Janet. *Rabat: Urban Apartheid in Morocco.* Princeton, NJ: Princeton University Press, 1981.

Findlay, Anne M. "Rabat-Salé," *Cities: The International Journal of Urban Policy and Planning* 1, no. 4 (1984): 322–27.

Peennell, C. R. *Morocco: From Empire to Independence.* London: Oneworld Publications, 2009.

RANGOON. *See* Yangon.

REYKJAVIK

Reykjavik is the capital and largest city of Iceland, an island country in the North Atlantic Ocean. Located at latitude 64°08′N, it is the northern-most national capital in the world. Warm ocean waters from the Gulf Stream moderate temperatures, so the city is not as cold as it might seem from latitude alone, although there is a pronounced winter season indeed and snow cover. The city is located in the southwestern part of Iceland, on the Seltjarnarnes Peninsula, although many of the newer residential areas have spread off the peninsula to the south and east. The city's international airport is Keflavik, located some 31 miles (50 km) away to the southwest on the Reykjanes Peninsula. The city's harbor is an important fishing and whaling port and one of the mainstays of the economy of Iceland.

The population of Reykjavik proper is about 120,000 (2011). An additional 80,000 people live in the suburban areas, for a total population for the Greater Reykjavik Area of about 200,000, or nearly two-thirds of the population of the entire country. The name Reykjavik means "Smoke Cove" and comes from the steam that was visible from the harbor from the area's natural hot spring.

Historical Overview. The first permanent settlement near what is now Reykjavik is attributed to the Norse chieftain Ingólfur Arnarson in 874 AD, but it is not until nearly 900 years later when Reykjavik emerged as a city. The officially recognized year of the city's founding is 1786. It was mostly a trading and fishing port. Iceland was a part of Denmark and Norway for much of its history until 1918 when the sovereign Kingdom of Iceland was established under the Danish Crown. In 1944, by a vote of more than 95 percent of the adult population, Iceland became an independent republic and Reykjavik a national capital. During World War II, British and American forces had bases in Reykjavik despite Iceland's official policy of neutrality. The British built the airport that today is Reykjavik Airport for domestic flights, while what the Americans built became Keflavik International Airport. In 1972, Reykjavik hosted the world chess championship, and in 1986 it was the site of an important summit meeting between U.S. president Ronald Reagan and the last prime minister of the Soviet Union, Mikhail Gorbachev.

Major Landmarks. Notable landmarks in Reykjavik include the Icelandic Parliament Building, the Reykjavik Cathedral (Lutheran), and Hallgrimskirkja a beautiful modern Lutheran church with a high tower that provides a panoramic view of Reykjavik and surroundings. The historic core of the city is charming and colorful, as individual buildings are painted in various colors. The roofs are colorful too. Perlan is a building on a hilltop that is built around five storage tanks for geothermal hot water. It also offers a panoramic view of the city. There are several museums in the city, including the National Gallery of Iceland and the Reykjavik Art Museum. The main shopping and restaurants street in Reykjavik is Laugavegur.

Culture and Society. Iceland's population is mostly of Scandinavian origin. The official language is Icelandic, although English is very widely understood as well, and the main religion is Church of Iceland Lutheran. The country is generally prosperous and has a very high standard of living, although recent economic crisis has brought new hardship and bank failure. Fishing and whaling remain important to the economy, but Iceland also has fast-growing economies related to services, tourism, technology, and the creative arts. Many Icelanders are outdoors oriented and enjoy the country's national parks and rugged landscapes. Bathing in geothermal hot water swimming pools is popular, even in the coldest weather. Reykjavik has an active nightlife with many bars that become active only after midnight and stay open until dawn. One of Reykjavik's nicknames is "nightlife capital of the north."

Further Reading

Acker, Paul. "Reykjavik Revisited," *Scandinavian Review* 98, no. 2 (2011): 74–77.

Lacy, Terry G. *Ring of Seasons: Iceland—Its Culture and History*. Ann Arbor: University of Michigan Press, 1998.

Magnússon, Sigurdur. *Wasteland with Words: A Social History of Iceland*. London: Reaktion Books, 2010.

Reynarsson, Bjami. "The Planning of Reykjavik, Iceland: Three Ideological Waves—A Historical Overview," *Planning Perspectives* 14, no. 1 (1999): 46-67.

RIGA

Riga is the capital and largest city in Latvia, a country on the Baltic Sea in northern Europe that borders the two other so-called Baltic countries, Estonia and

Lithuania, as well as Belarus and Russia. The city is in the north center of the country on the Gulf of Riga of the Baltic Sea at the mouth of the Daugava River, the current preferred Latvian word for the river that is also known as the Western Dvina from the Russian. In addition to being the administrative center for the Republic of Latvia, Riga is also an important port and industrial center, as well as a leading commercial, cultural, and financial center in its region. The population of Riga is 657,424 (2011), nearly one-third of the national total, while that of the metropolitan area is slightly more than 1 million, about one-half the national total.

Historical Overview. The origins of Riga date to an ancient trade route along the Dvina-Dnipro (Dnieper) rivers between Scandinavia to the north and Byzantium to the south, and the city's excellent natural harbor. In 1201, the German Bishop Albert von Buxhoeven arrived in Riga on a crusade to convert the pagan Livonian population of the area to Christianity, and make the town the seat of his bishopric. Trade with German cities increased, and in 1282 Riga became a member of the Hanseatic League, an economic alliance of trading cities over a wider area in northern Europe. Later, Riga was a Free Imperial City within the Holy Roman Empire. It then came under the rule of the Polish-Lithuanian Commonwealth until the Polish Swedish War of 1621–1625 when it became part of the Kingdom of Sweden until 1710. It was Sweden's biggest city at that time. In 1710, Riga was annexed to the Russian Empire by Czar Peter the Great as a result of the Great Northern War. Under Russia, Riga was made into an industrial city, as well as one of great beauty with many parks and boulevards and expansive new suburbs. The Russians made it a provincial capital. Riga remained Russian until World War I and 1918 when Latvia declared independence, Riga became a national capital. The country looked more to alliances in the West than to Russia, until 1940 and the outbreak of World War II when Latvia and neighboring Estonia and Lithuania were forcibly taken by the Red Army of Russia's successor state, the Soviet Union. Nazi Germany occupied Latvia from 1941 to 1944. The city's Jews were taken to a concentration camp at Kaiserwald in the outskirts of Riga and put to forced labor. Many were later executed. When Soviet troops recaptured Riga on October 13, 1944, Latvia became part of the USSR. Many ethnic Russians moved to the city and a process of russification was started. By 1989, the percentage of Latvians in Riga's population had fallen to 36.5 percent. Latvia gained its independence again in 1991 with the collapse of the Soviet empire. The country has become an open, democratic society with Latvian language and culture on the rise. In 2004, Latvia joined both NATO and the European Union.

Major Landmarks. The center Riga is Old Town, one of the most attractive urban districts of Europe. It is a UNESCO World Heritage Site with many historic buildings and public squares, and the largest collection of *Jugendstil* architecture in the world (German Art Nouveaux). Perhaps the most spectacular façade is that of the House of Blackheads, an old merchants' guild building that now houses a tourist information center and a museum. The same district has Town Hall, St. Peter's Church, which dates to 1209, St. John's Church, and a mix of museums that includes the Museum of the Occupation of Latvia about Soviet-era atrocities in the country. Riga Cathedral, a symbol of the city since 1207, is on Cathedral Square. There are many other beautiful historic churches in the city as well. Another prominent landmark is Freedom Monument, a 138-ft-high (42 m) memorial to freedom fighters for Latvian independence in 1918–1920. Its square has been the main venue for public gathering and political demonstrations in Riga, particularly during the time of Soviet collapse and the winning of Latvian independence. Nearby are many historic houses and other buildings with beautiful art nouveaux flourishes, especially along Alberta and Elizabetes Streets. The Latvian National Opera is another impressive structure. A landmark of a different sort is the Latvian Academy of Sciences Building. Built in 1953–1956, it is a 354-ft-high (108 m) Stalinist-

The beautiful rebuilt House of the Blackheads, an old merchants' guild building in the historic Old Town district of Riga. (Lagartija/Dreamstime.com)

style skyscraper decorated with Communist hammers and sickles and Latvian ethnic motifs that was a "gift" to Latvia from the citizens of the other republics of the Soviet Union. Another Soviet-era monument is Victory Monument dedicated to the victors in the Great Patriotic War (World War II) against the Nazis.

Culture and Society. Riga's population was at a high of just under 1 million in 1991 and has declined dramatically since independence. This is because of birthrates in Latvia that are among the lowest in the world and emigration for Latvia now that there is no longer an Iron Curtain. Many Latvians have gone to live in stronger economies in the West, while other emigrants have been Russians returning to Russia once their privileged position in Latvia came to an end. By 2011, the percentage of Latvians in Riga's population had risen to 42.4 percent of the total, while that of Russians has fallen between 1989 and 2011 from 47.3 percent to 40.7 percent. It was in 2006 that Latvians overtook Russians as the most numerous ethnic group in the Latvian capital. The language on the streets in Riga is now Latvian, the nation's official language. The majority of Latvians are Christians, although only a small percent of citizens attend church services regularly. The main denominations are the Evangelical Lutheran Church of Latvia and Roman Catholicism. The most popular sport is ice hockey.

Further Reading

Angrick, Andrey and Peter Klein. *The "Final Solution" in Riga: Exploitation and Annihilation, 1941–1944.* New York: Berghahn Books, 2009.

Krisjane, Zaiga and Maris Berzins. "Post-Socialist Urban Trends: New Patterns and Motivation for Migration in the Suburban Areas of Riga, Latvia," *Urban Studies* 49, no. 2 (2012): 289–306.

Taylor, Neil. "Riga: Hansa City at the Baltic Crossroads," *History Today* 58, no. 10 (2008): 14–15.

RIYADH

Riyadh is the capital and largest city in the Kingdom of Saudi Arabia, the largest country on the Arabian Peninsula in the Middle East. The city is in the Najd, the highlands area in the center of the country, in Riyadh Province, of which it is also the capital. The population of Riyadh is 5,254,560 (2010), while that of the wider metropolitan area is about 7 million, about one-quarter that of the country. The name Riyadh is Arabic and means "the gardens."

Historical Overview. The first settlement at the site of Riyadh was founded by the Banu Hanifa tribe as early as the third-century AD and was named Hajr. It was the capital of a province called Al Yamamah but waned in importance in the late ninth century. In 1737, Deham ibn Dewwas gained control of the settlement and built a wall around it. The First Saudi State was formed in 1774 with Diriyah as capital. After it was defeated in 1818 and Diriyah destroyed, the Second Saudi State was created in 1823 by Turki ibn Abdallah who made Riyadh the capital. From 1865 Riyadh was ruled by the Al Rashid family whose capital was in the city of Ha'il. King Abdulaziz ibn Saud recaptured Riyadh from the Al Rashid family in a decisive battle in 1902, and in 1932 established the modern Kingdom of Saudi Arabia with Riyadh as capital.

The kingdom was at first a poor desert state, but after vast oil reserves were discovered near the Persian Gulf in 1938, Saudi Arabia stated to become extremely wealthy. The royal family of Saudi Arabia has enjoyed seemingly unlimited resources. By 1976, the country had become the largest oil exporter in the world. It had succumbed to enormous influence from the United States and its major oil companies, and is a major purchaser of American and British weapons. The government's friendship with the West has angered many of the country's citizens, as well as anti-American and anti-Israel radical Islamists in Saudi Arabia and other Muslim countries. As a result, Saudi Arabia has come to be in the center of Islamic terrorism, both as a victim and as a source of recruits for both domestic and global terrorism, including most of those who were responsible for the September 11, 2001, attacks on targets in the United States. The Saudi government has tried to balance its close political and economic ties with the United States against ever louder domestic and regional calls to turn against the United States. In addition, in 2011 the government was facing growing protests by many of its own citizens on demanding a turn to democracy and an end to excessive social and economic inequalities.

Major Landmarks. Riyadh has a small old city district called Al-Bathaa in which the main landmarks are the Fortress, site of the 1902 battle in which Riyadh was recaptured by King Abdul Aziz, and Murabba Palace that the king built afterwards.

The National Museum, opened in 1999, features high technology presentations about Saudi Arabia's history and culture, and about Islam. Next to the Great Mosque is As-Sufaat, or Deira Square, a large expanse where on many Fridays public beheadings of condemned criminals take place. For this reason, the square is sometimes referred to in English as Chop-Chop Square. Souk-al Thimairi is the city's most famous old bazaar. The northern part of Riyadh is more modern, and has many high-rise buildings, upscale shopping malls, and wide highways. The major landmark is Kingdom Center, also called the Burj al Mamlakah, a 99-floor (1,000 ft; 300 m) high-rise office tower with a distinctive opening at the top that is spanned by a top-floor

Kingdom Center, also called the Burj al Mamlakah, a 99-floor (1000 feet; 300 m) high-rise office tower and shopping mall in Riyadh. (Fedor Selivanov/Shutterstock.com)

pedestrian bridge from which visitors are afforded a spectacular view. At the base is a large three-level shopping mall in which one floor is reserved for women shoppers only. Riyadh's second-tallest building is Burj al Faisaliya. It stands out because of a unique ball near the tapered top in which there is a restaurant. The Royal Saudi Air Force Museum is at a freeway exit in the city's outskirts.

Culture and Society. Riyadh is a sprawling, fast-growing, and ever more modern city. At the same time, it is religiously conservative, and is trying to resolve conflicting opinions about issues such as whether or not Islamic law allows women to drive automobiles. Many museums and other attractions in the city have separate visiting hours for men and women. There are more than 4,300 mosques in this heavily Muslim city. The language of Riyadh is a recognizable dialect of Arabic called Nadji Arabic. The mode of dress follows the principles of *hijab*, that is, strict modesty is required.

Further Reading

Al-Hammad. "Riyadh: City of the Future," *Cities: The International Journal of Urban Policy and Planning* 10, no. 1 (1993): 16–24.

Menoret, Pascal. "Development, Planning and Urban Unrest in Saudi Arabia," *Muslim World* 101, no. 2 (2011): 269–85.

Struyk, Raymond J. "Housing Policy issues in a Rich Country with High Population Growth," *Review of Urban & Regional Development Studies* 17, no. 2 (2005): 140–62.

ROME

Rome is the capital and largest city in Italy, a member country of the European Union located in southern Europe. It is among the oldest and most historic cities in the Western world, and has had special roles as center of the large Roman Empire that flourished from the first-century BC into the fifth-century AD, and as seat of the Roman Catholic Church in Vatican City, a sovereign state within the city's borders. Rome is located in the Lazio region of central Italy at an historic ford of the Tiber River about 15 miles (24 km) east of the Tyrrhenian Sea, a part of the Mediterranean Sea. The area is hilly and Rome is said to have been built on seven hills. The city's nickname is the "Eternal City." The population of Rome is about 2.8 million. The Rome metropolitan area has an additional 1 million or so residents.

Historical Overview. Rome's history goes back some two and one-half millennia. According to a famous myth, the city was founded in 753 BC by the twins Romulus and Remus who were suckled by a she-wolf. The actual origins of the city are obscure, but the city may indeed have been founded in the middle of the eighth century. It developed into the capital of the Roman Kingdom, then the Roman Republic, and then the Roman Empire. The empire stretched from the northern part of the island of Great Britain across most of Europe, much of the Middle East, and across northern Africa, with a center on the Mediterranean world. Especially under the Emperor Augustus who ruled for 63 BC to 14 AD the city gained great wealth and was dotted with many palaces, temples, market places, aqueducts, paved roads, and other magnificent construction. It was the largest city in the world during the rule of Augustus. Early Rome was also a great center of the arts. Rome was sacked in 410 by Alaric I, and then the Roman Empire fell in 476 to be replaced by rule from Byzantium. In 756, the Papal States were created with Rome as capital and the Pope as ruler. In 846, the city was invaded by Muslim Arabs and St. Peter's Basilica was looted. During the Renaissance, Rome was graced with beautiful works of art and architecture, most notably by works of geniuses such as Michelangelo, Botticelli, and Raphael. The Papal States were succeeded by the Kingdom of Italy in 1870 at which time Rome became capital of Italy. After World War I, Italy was governed by the Italian Fascist dictator Benito Mussolini who allied with Nazi Germany. During

---

The Roman Forum

The Roman Forum was a rectangular public plaza that was surrounded by the government buildings of the ancient capital of the Roman Empire, Rome. It was the center of Roman public life, including important speeches, elections, processions, executions, and gladiatorial matches, as well as the center for some of the Roman Empire's oldest religious temples. Originally, the space was a marketplace. The area is now a cluster of ruins and fragments of buildings that serves as a striking reminder that even the greatest of empires can pass. It is located in a small valley between the Palatine and Capitoline Hills, and is an important archaeological site within the city of Rome and a crowded tourist attraction.

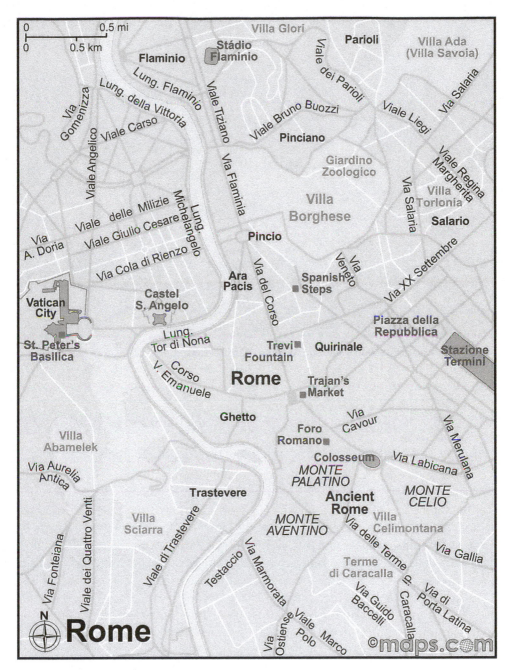

Rome, Italy (Maps.com)

World War II, the Germans occupied the city until liberation by Allied forces. The city grew very quickly in the postwar period and its economy expanded greatly. Rome has once again become one of the great cities of the world, with a highly diversified economy, a pleasing mix of historic architecture and modern life, a

A view across historic Rome. (Matthew Trommer/Dreamstime.com)

cosmopolitan population, and a great many foreign tourists from around the globe. In 1960, the city hosted the summer Olympics Games and in 1990 it hosted the FIFA World Cup.

Major Landmarks. In addition to the superb landmarks of Vatican City, Rome is noted for historic ruins from the time of the Roman Empire, including the spectacular Roman Coliseum, the Arch of Constantine, and the Roman Forum. The historic center of Rome is a UNESCO World Heritage Site. Other prominent landmarks include the famous Catacombs of Rome, the Pantheon, the 131-ft-high (40 m) column of Marcus Aurelius, the Bridge of Angels, the Piazza del Popolo, the Piazza della Repubblica, the monument to Vittorio Emanuele II, the Spanish Steps and the Fontana della Barcaccia below, the Castel Sant' Angelo, the Basilica de Santa Maria Maggiore, the Quirinal Palace (which is the official residence of the president of Italy), and the beautiful Baroque Trevi Fountain, one of the most famous fountains in the world. Important museums include the Villa Borghese, the Museo Nazionale di Villa Giulia, the Galleria Nazionale d'Arte Moderna, the Galleria d'Arte Antica, and the Capitoline Museums. The Parco della Musica is a beautiful modern concert hall designed by Renzo Piano. The Stadio Olimpico is one of Europe's largest with seating for 70,000 for football (soccer).

Culture and Society. More than 90 percent of Rome's population is Italian, and nearly 10 percent is of foreign origin, including Albanian, North African, and Eastern European. The Esquilino neighborhood near the city's busy Termini Railway Station is a highly diverse immigrant district. There are many Roma (gypsies) in the city who live in trailer camps on the city's outskirts. The predominant population is, of course, Roman Catholic, as Rome has long been the seat of the Roman Catholic hierarchy. The Muslim population of the city is increasing with immigration from North Africa and the Muslim regions of Balkan Europe. Rome is the center of a great many arts, music, and fashion festivals, has a lively nightlife, and almost nonstop day and night activity at sidewalk cafés and in public squares.

Further Reading

Agnew, John A. *Rome*. New York: Wiley, 1995.
Bosworth, R.J.B. *Whispering City: Rome and Its Histories*. New Haven: Yale University Press, 2011.
Hibbert, Christopher. *Rome: The Biography of a City*. London: Penguin, 1995.

ROSEAU

Roseau is the capital and largest settlement in Dominica, officially the Commonwealth of Dominca, a small island nation in the Lesser Antilles in the Caribbean Sea. The population of Dominica as a whole is only 72,500, and that of Roseau is about 15,000 (2001). Hence, the capital is not truly a city but more of a small town, although it does embrace more than one-fifth of the country's inhabitants. In addition to being the island's administrative center, the town is also Dominica's principal port. Main exports include bananas, other fruits, and bay oil. Roseau is located on the leeward (west) coast of Dominica at the mouth of the small Roseau River.

Historical Overview. Roseau is located on the site of a Kalinago Indian village that was named Sairi. The first Europeans to sight Dominica were Spaniards sailing with Christopher Columbus. Because they first sighted the island on a Sunday, they named it Dominica, the Latin word for Sunday. France claimed Dominica in 1635, but did not begin settlement until 1715 because for a time they and the English had agreed to leave the island as neutral territory to be inhabited by indigenous Carib peoples. Roseau was founded soon after the first French settlers arrived, and was named for the reeds that grew there. The early town was built around the port and central market, from which roads radiated to various parts of the settlement. French landowners established a plantation economy based on African slave labor. In 1763, as a result of the Treaty of Paris that ended the Seven Years War, Dominica became a British possession. Subsequent attempts by France to regain the island failed. The British also operated plantations with slave labor until slavery was outlawed in 1831. As Roseau grew under British influence, new districts were laid out with a grid street plan. In 1871, Dominica became part of the Leeward Island Federation, and in 1898 it became a British crown colony. In 1958, the island became one of the members of the West Indies Federation until 1962. Independence from Great Britain was achieved on November 3, 1978. Three years later, there was an attempt by United States and Canadian white supremacist mercenaries to overthrow the government of Dominica, but the plan, known by the code name Operation Red Dog, was thwarted by U.S. federal agents and failed.

Major Landmarks. The center of Roseau has a number of historical houses that were built by early French settlers. The Fort Young Hotel was once a British military installation that was built in 1770 by the governor of the time, William Young. Other landmarks are the Roman Catholic Cathedral of Roseau, the Botanical Gardens, the Government House, and Windsor Park, a sports stadium for football (soccer) that was built in 2007 as a result of a donation from the People's Republic of China.

Culture and Society. Most Dominicans are of African descent, but there are also descendants of European settlers (mostly French and English), East Indians, and small numbers of Lebanese, Syrians, and Chinese. There are also small numbers of Caribs, most of whom live in a designated tract on the east coast of Dominica. The population of both Roseau and Dominica as a whole has grown only slowly because many of the island's younger citizens have migrated abroad, mostly to the United States, Canada, the United Kingdom, and France. English is Dominica's official language, but Dominican Creole and French are spoken as well. The economy of Dominica is based heavily on agricultural exports such as bananas and is vulnerable to global price fluctuations and damage from tropical storms.

Further Reading

Atwood, Thomas and J. Johnson. *The History of the Island of Dominica.* Charleston, SC: BiblioLife, 2010.

Honychurch, Lennox. *The Dominica Story: A History of the Island.* Oxford: Macmillan Education, 1995.

S

SAN JOSÉ

San José is the capital and largest city in Costa Rica, a small country in Central America between the Pacific Ocean and the Caribbean Sea. The country also borders Nicaragua and Panama. The city is centrally located within Costa Rica in the fertile Central Valley and is the hub of the nation's economy and transportation networks, as well as seat of government. According to data for 2011, the population of San José is about 288,000 while that of the metropolitan area totals more than 2.3 million. Residents of San Jose are called Josefinos.

Historical Overview. San José was founded in 1737 by Spanish settlers who had come from nearby towns, including Cartago, which had been founded in 1583 and was the area's leading population center. Its original name was Villanueva de la Boca del Monte del Valle de Abra, but San José was adopted later in honor of the settlement's patron saint and the name of its church. In 1821, after Costa Rica achieved independence, there was great difference of opinion within the country as to whether it should join a federation of Central American states led by Mexico or be totally independent. Leading politicians from Cartago favored the former, while San José argued for the latter. The conflict led to a short civil war which was decided on April 5, 1823, with San José's victory in the Battle of Ochomongo Hills. As a result, Costa Rica remained independent and San José replaced Cartago as capital. Internal strife continued for a time afterwards, but in 1837 Josefinos settled the matter with victory against an attempted coup called *La Guerra de la Liga* (the War of the League).

The University of Santo Tomas, the country's first university was opened in 1843 and eventually evolved into the University of Costa Rica. Prosperity came to Costa Rica because of its rich agricultural lands, especially from coffee production and tobacco crops. San José grew as a result, and was able to develop into an impressive, modern city. It became a magnet for migrants from the rural areas, as well as for the country's poorer neighbor Nicaragua. "Nicas," as they are called, constitute a significant part of the population in San José's outlying squatter settlements.

Major Landmarks. San José has many beautiful and historic churches, including the National Cathedral in the city's downtown. Other landmarks are the Teatro Variedades, the National Theater of Costa Rica, and the National Museum. Other museums include the Museum of Pre-Columbian Gold, the Jade Museum, and the Insect Museum at the University of Costa Rica. The Plaza de la Justicia in the center of the city is as a good example of stern modernist architectural form. Escazu City on the outskirts of San José is an upscale residential area with popular shopping centers such as Multiplaza, as well as places for enjoyment of nightlife.

Culture and Society. The population of Costa Rica is overwhelmingly white or *mestizo*, with blacks and Native Central Americans constituting less than 3 percent of the total. Most residents are of Spanish origin, but there are also Costa Ricans who are descendants from other European ethnic groups, including Italian, German, and Dutch. Retirees from the United States are also a significant minority, as are poor migrants from Nicaragua and Colombia. Spanish is the common language. The official national religion of Costa Rica is Roman Catholicism. About 70 percent of the population is Catholic, with most of the rest being Protestant. Costa Rica is generally a prosperous and stable country. Costa Ricans are proud of the natural beauty of the mountains, jungles, and beaches of the country, its productive farmlands, and historic townscapes. Foreign tourism is an important part of the economy.

Further Reading

Low, Setha M. "Urban Public Spaces as Representation of Culture: The Plaza in Costa Rica," *Environment and Behavior* 29, no. 1 (1997): 3–34.

Palmer, Steven and Iván Molina, eds. *The Costa Rica Reader: History, Culture, Politics.* Durham, NC: Duke University Press, 2004.

Warf, Barney. "Do You Know the Way to San José? Medical Tourism in Costa Rica," *Journal of Latin American Geography* 9, no. 1 (2010): 51–66.

SAN MARINO

San Marino is a small town that is the capital of the Republic of San Marino, a tiny enclave country within Italy near the Adriatic Sea that, at 24 square miles (62 sq km), is one of the smallest countries of Europe and indeed the world. The country is also known as the Most Serene Republic of San Marino. The country has no level ground and essentially covers the slopes and summit of Mount Titano, a modest peak in the Appennine mountain range that rises to an elevation 2,457 ft (749 m). There is a total national population of 31,887 residents, of whom approximately 4,493 live in the capital city, the third-ranking town in the country (data for 2003). The largest settlement in San Marino is Dogana with 7,000 residents, while the second most populous is Borgo Maggiore (approximately 5,992), the country's main commercial center. The city of San Marino (officially *Città di San Marino*) is located in the approximate center of the country San Marino.

Historical Overview. The city was founded in 301 by Saint Marinus (the stonecutter Marinus from the island of Rab in present-day Croatia) and other Christian refugees from the persecutions of Rome. They built a small stone church that was the foundation of the state of San Marino. Three great towers were built to protect the city, Guaita in the 11th century, Cesta during the Crusades in the 13th century, and the Montale in the 14th century. In 1600, San Marino enacted a constitution which is still in effect, making the country the world's oldest constitutional republic. In 1631, San Marino was formally recognized as an independent state by the Pope. During the 19th-century unification of Italy, San Marino provided refuge to Italians who were persecuted for their support of unification. As thanks, Giuseppe Garibaldi, a central figure in the making of the

Guaita, the oldest of the famous Three Towers of San Marino on the heights of Mount Titano. (Vladimir Sazonov/Dreamstime.com)

modern Italian state, granted the Sammarinese their wish not to be incorporated into Italy. San Marino remained neutral in World War I and World War II, although was in the control of the Sammarinese Fascist Party for 20 years between the wars, and took in refugees from elsewhere in Italy during World War II. Near the end of that war, San Marino was the scene of a battle between German and Allied forces that resulted in Allied occupation of the country as the war ended. Between 1945 and 1957, San Marino had the world's first democratically elected communist government.

Major Landmarks. The town of San Marino sits prominently on Mount Titano and is accessible from Borgo Maggiore via cable car. It is a walled city with a main gate. The three towers still stand and look over different parts of the country below. Guaita and Cesta are open to visitors, while Montale is privately owned. The main church is the Roman Catholic Basilica of San Marino built in 1836. Other landmarks are the Museum of the Republic of San Marino and the University of the Republic of San Marino. The historic center of San Marino has been recognized as UNESCO World Heritage Site.

Culture and Society. The predominant population group of San Marino consists of ethnic-Italian citizens of the country plus about 4,000 foreigner residents most of whom are also Italians. Italian is the main language of the country. About 97 percent of the population is Roman Catholic. Sammarinese are very aware of their country's history and distinctive characteristics, and have turned both into commodities for tourism, a mainstay of the local economy. For example, the country's Crossbow Corps, once an integral part of the San Marino's military in the Middle Ages, still exists. It is a unit of volunteers who demonstrate crossbow shooting and participate in crossbow competitions. Similarly, although San Marino has open borders, there are various uniformed military units such as the colorful Guard of the Rock whose functions are purely ceremonial and for show.

Further Reading

Eccardt, Thomas M. *Secrets of the Seven Smallest States of Europe: Andorra, Liechtenstein, Luxembourg, Malta, Monaco, San Marino, and Vatican City.* New York: Hippocrene Books, 2005.

Hart, Albert Bushnell. "The Ancient Commonwealth of San Marino," *Nation* 58, no. 1492 (1894): 81–82.

SAN SALVADOR

San Salvador is the capital and largest city in the Republic of El Salvador, the smallest and most densely populated country in Central America. The country borders the Pacific Ocean and both Guatemala and Nicaragua. San Salvador is inland near the center of the country in a highlands valley called Valley of the Hammocks (Valle de las Hamacas) surrounded by volcanic mountain peaks. The local river is the Acelhaute River and to the east is Lake Illopango, a large volcanic lake. The population of San Salvador is about 568,000, making it the third-largest city in Central America, while that of the metropolitan area is more than 2.4 million, second-largest in Latin America (estimates for 2011). The name San Salvador means "Holy Savior" in Spanish. Locals refer to the city as San Sivar, an abbreviation for San Salvador.

Historical Overview. Before the arrival of the Spanish in the early 16th century, there was a Pipil Amerindian capital city named Cuscatlán near the site of San Salvador. It may have been abandoned to avoid conquest. San Salvador itself was founded in 1525 by Pedro de Alvarado, a conquistador, and moved about 20 miles (32 km) to its present site in 1528. It gained formal designation as a city in 1546. It was capital of the province of Cuscatlán during the colonial period, and then from 1834 to 1839 capital of the United Provinces of Central America. It became capital of El Salvador in 1839. In the late 19th and early 20th centuries, the city was endowed with many beautiful buildings and modern infrastructure thanks to profits from El Salvador's coffee production. Earthquakes damaged the city in 1854, 1873, 1917, and 1986. Another wave of urban modernization took place in the 1960s–1980s, although the country had been rocked by many years of civil war, contested elections, and mass protests. Some 75,000 people were killed in the civil war before peace in 1992. A landmark event was the assassination of popular Roman Catholic Archbishop Óscar Romero as he was saying Mass in support of human rights in the country.

Major Landmarks. The National Palace of El Salvador is located in the center of San Salvador's historic core. It was once the seat of government but is now a museum. The new Roman Catholic National Cathedral, where Romero is entombed, is also in the center. There are many monuments in the city, the most famous of which is Monumento al Divino Salvador del Mundo, the "Monument to the Savior of the World." It has a figure of Jesus Christ atop a large globe which is, in turn, perched atop a high pedestal embellished with a cross. Other landmarks are the National Theater, the David J. Guzman Museum of Anthropology, and the Art Museum MARTE. The National University of El Salvador was founded in San Salvador in 1841. San Salvador has quite a few modern high-rise office towers, the tallest of which is Tower 3 at the San Salvador World Trade Center (325 ft; 99 m).

Culture and Society. San Salvador is a diverse city with residents of European, mostly Spanish, heritage, descendant of Amerindians, and people of mixed ancestry. In addition to El Salvadorans, there are migrants from neighboring countries. The major religion is Roman Catholicism. There are also many Protestants and a Mormons. The city has an impressive new Mormon temple. The language of the country is Spanish. El Salvador is a bustling business and financial center in addition to being the national capital and has economic influence on Central America. Crime rates are high, however, and there has been a problem with street gangs called *maras* that were formed in the jails of California among illegal immigrants from El Salvador who where then deported back to their home country.

Further Reading

Didion, Joan. *Salvador*. New York: Vintage, 1984.

Moodie, Ellen. *El Salvador in the Aftermath of Peace: Crime, Uncertainty, and the Transition to Democracy*. Philadelphia: University of Pennsylvania Press, 2010.

Tardanico, Richard. "Post-Civil War San Salvador: Social Inequalities of Household and Basic Infrastructure in a Central American City," *Journal of Development Studies* 44, no. 1 (2008): 127–52.

SANA'A

Sana'a is the capital and largest city of the Republic of Yemen, a country in the Middle East on the south-southwestern margins of the Arabian Peninsula. The city is in the Yemeni Mountains in the western part of the country, and has an elevation of 7,500 ft (2,300 m), making it one of the highest capital cities in the world. Because of the high elevation, there is more rainfall than in other parts of the desert country, which in turn helps to account for the higher population concentration. The population is about 1.9 million in the city itself and 2.2 million in the metropolitan area (2011). The city is growing very rapidly because of migration from the Yemen countryside, and is expected to double within a decade. Such a rate of growth makes Sana'a one of the fastest growing cities in the world. The historic core of Sana'a has many distinctive, old multistory buildings with facades that are beautifully decorated with carved geometric patterns and stained glass, and is a UNESCO World Heritage Site. The name Sana'a (also spelled Sanaa) means well fortified and refers to the ancient walls that surrounded the city.

Historical Overview. Sana'a has been inhabited for more than 2,500 years, making it one of the oldest continuously inhabited cities in the world. According to legend, the city was founded by Shem, the son of Noah. The city was built at a natural crossroad for trade between ports on the Red Sea and the interior. It was walled in the middle of the third century by Sabean ruler Sha'r Awtar.

After the spread of Islam, the city continued its administrative role as the seat of a caliphate. In 1538–1635 and again in 1872–1918, Sana'a was a provincial capital under Ottoman rule. From 1918 to 1962, Yemen was a monarchy ruled by the Hamidaddin family. From 1904 to 1948, Sana'a was the capital of the Imam of Yemen, Imam Yahya. Sana'a became capital of the Yemen Arab Republic, also referred to as North Yemen, following a revolution in 1962. Union with South Yemen (the People's Democratic Republic of Yemen; capital city Aden, a port city on the Gulf of Aden) came on May 22, 1990, to form the current Republic of Yemen, of which Sana'a is capital. In 2011, a period of civil unrest has begun against the government of Yemen. There have been large protest demonstrations and eruptions of violence.

Major Landmarks. There are many landmarks in the Old City of Sana'a, beginning with the Bab al-Yemen, an ancient gate leading into the old walled city. The Great Mosque of Sana'a is one of the oldest mosques in the world, while the Al Saleh Mosque near the Presidential Palace is enormous, with space for 40,000 worshippers. The commercial center of Sana'a is Al Medina. Outside Sana'a in Wadi Dhahr, a beautiful irrigated valley with green fields and traditional villages, is the former palace of Imam Yahya perched high on a rock.

Culture and Society. Yemen's population is of Arabic origin. Arabic is the official language. About 98 percent of the population is Muslim. Some 53 percent of the Muslims are Sunnis and 43 percent are Shiite. Yemen had a large Jewish population since Biblical times, but almost all Jews migrated to Israel in an airlift operation after the formation the Jewish state. Birth rates are very high in this

Old houses in the historic center of Sana'a, a UNESCO World Heritage Site. (Corel)

traditional society, accounting for rapid population growth in Yemen and resultant urbanization. Unemployment is high in both the countryside and in Sana'a, and there is considerable civil unrest.

Further Reading

Dresch, Paul. *A History of Modern Yemen.* Cambridge, UK: Cambridge University Press, 2000.
Lewcock, Ronald. *The Old Walled City of Sana'a.* Paris: UNESCO, 1986.
Macintosh-Smith, Tim. *Yemen: The Unknown Arabia.* Woodstock, NY: Overlook Press, 2001.

SANTIAGO

Santiago is the capital and largest city of Chile, a long, ribbon-shaped country that runs north-south along much of the South Pacific Ocean Coast of South America. The city is also known as Santiago de Chile, and is located in the Santiago Basin, the country's central valley. It is in the approximate middle of Chile's latitudinal extent, about midway between the Pacific coast and the high peaks of the Andes Mountains that define Chile's border with neighboring Argentina. The city is at a latitude of 33°27'S and has a climate that is referred to as a Mediterranean-type climate. The Mapocho River flows from the Andes peaks and divides the city into two.

The population of Santiago is about 5.4 million (2002), while that of the wider metropolitan area is about 7.2 million. In addition to being the national capital, Santiago is the leading industrial and financial center of Chile. The country is generally prosperous and fast developing, and Santiago reflects that progress with its dramatic city skyline, many new buildings, and continual growth.

Although Santiago is the official national capital, since 1990 the National Congress of Chile has met in the city of Valparaíso, a Pacific Ocean port city located about 80 miles (130 km) to the west. With a population of about 276,000, Valparaíso is much smaller and is seen as quieter alternative to the bustle of busy Santiago. It is a very beautiful city.

Historical Overview. Santiago was founded in 1541 by Spanish conquistador Pedro de Valdivia. He was attracted to the location because of climate and vegetation patterns, and because an island in the Mapocho River would be easy to defend against attacks from indigenous tribes. He named the settlement Santiago de Nueva Extremadura, Santiago in honor of St. James, and the rest after his birth city in Spain. As was typical for Spanish cities in the New World, Santiago was laid out in a rigid grid plan with square blocks and a prominent central square. The first years of the city were difficult for the settlers because of conflicts with the indigenous population and the persistent food shortages that resulted. The city's first cathedral was built in 1561 after the threats from attacks subsided.

Chilean independence from Spain was proclaimed on February 12, 1817, after a combined force of Chilean and Argentinean troops led by José de San Martin and Bernardo O'Higgins defeated Spanish royalists at the Battle of Chacabuco north of Santiago. The final break from Spain came on April 5, 1818, when Chilean patriots defeated the last Spanish army at Battle of Maipú, also near Santiago. From 1879 to

1883, Chile was involved in the War of the Pacific against neighboring Bolivia and Peru for control over mineral-rich territory to Chile's north. Chile won that conflict and annexed the disputed land. Santiago grew and prospered largely because of the productive agricultural lands of Chile and its mineral resources. Many of the wealthiest Chileans were large landowners or owners of mines and railroads. The city expanded and was modernized in the 19th century, with railway and telegraph lines, wide paved roads, beautiful parks and gardens, museums, concert halls, and at least two fine universities, the Universidad de Chile founded in 1843 and the Universidad Pontificia Católica founded in 1888. Chilean social and economic progress suffered a setback in 1973 when the elected government of President Salvador Allende was overthrown by a military coup and a military junta led by General Augusto Pinochet Ugarte took over the leadership of the country. Pinochet's repressive regime ruled Chile until 1990. His supporters point to economic growth in the country during his rule, but critics cite increased economic inequality, repression of basic rights, and tens of thousands of instances of torture of political opponents and thousands of "disappearances." A strong earthquake caused considerable damage in Santiago on March 3, 1985.

Major Landmarks. The main square of Santiago is Plaza de Armas. The National Cathedral faces the square, as does the main post office. The Central Market is nearby. Other landmarks are the Palacio de La Moneda, the seat of the president of the Republic of Chile and once a colonial mint; the Municipal Theater of Santiago; the Fine Arts Museum; the Contemporary Art Museum of Santiago; the National Library of Chile; the former congress building; and the two aforementioned universities. San Cristóbal Hill has a beautiful park that includes the Santiago Metropolitan Zoo. O'Higgins Park dates to 1873 and is in the center of the city. Its Movistar Arena is a popular venue for concerts. The city's financial district El Golf has a number of impressive high-rise office towers, most notably Torre Titanium La Portada and the newer Torre Gran Costanera.

Culture and Society. Chile's population consists of many ethnic groups, most notably those who trace ancestry to Spain (including Basques), Germans, Italians, Greeks, Irish, and British among European groups, a small indigenous population, and more recent migrants from nearby Peru and Bolivia. About 70 percent of the population is Roman Catholic, with about one-half of the rest being Protestant. Spanish is the official language of Chile. However, the Chilean form of Spanish is distinctive for its accent and the dropping of final "s" sounds and final consonants. English is widely understood in Santiago and other cities, and is a compulsive subject in public schools. Chile has a rich heritage of music and literature. Chileans call their country the "country of poets." In 1945, Gabriela Mistral, a feminist Chilean poet and educator, was the first Latin American to win a Nobel Prize for Literature. Chileans are justifiably proud of their prosperous and beautiful country.

Further Reading

Berhardson, Wayne. *Santiago de Chile*. Hawthorn, Australia: Lonely Planet Publications, 2000.

Walter, Richard J. *Politics and Urban Growth in Santiago, Chile, 1891–1941*. Palo Alto, CA: Stanford University Press, 2005.

Wood, James A. *The Society of Equality: Popular Republicanism and Democracy in Santiago de Chile*. 1818–1851. Albuquerque: University of New Mexico Press, 2011.

SANTO DOMINGO

Santo Domingo is the capital and largest city in the Dominican Republic, a small country in the West Indies that shares the island of Hispaniola with Haiti, its neighbor to the west. The city is officially named Santo Domingo de Guzmán. It is located on the south coast of the Dominican Republic on the Caribbean Sea, at the mouth of the Ozama River. The river bisects the city. From 1930 to 1961, Santo Domingo was named Ciudad Trujillo after strongman dictator Rafael Trujillo who named the city after himself. The population of Santo Domingo is about 2.5 million, with about 3.7 million (more than one-third of the total population of the Dominican Republic) living in the metropolitan area as a whole (2010).

Historical Overview. The indigenous inhabitants of the site of Santo Domingo were Taino Indians, but their population was decimated after the Spanish began arriving on Hispaniola beginning with the first trans-Atlantic voyage of Christopher Columbus in 1492. The natives died from European diseases to which they had no immunity, as well as from forced labor, torture, and other abuses. The city was founded in 1496 by Christopher Columbus's brother Bartholomew Columbus, and was at first named La Neuva Isabela after the Spanish queen. It was the first European city in the New World. For the Spanish, the city was a base the plantations that they established on the island, and for further exploration and conquest in the New World. However, they failed to hold the settlement. In 1586, it was captured by the English pirate Francis Drake who held it for ransom, the first of many pirate incursions. Starting in 1795, there were many other changes of ownership, too, with either the entire island or its eastern half containing Santo Domino going back and forth at various times between France, rebellious Haitian slaves, Haiti, brief periods of independence, and back to Spain, until finally a lasting Dominican independence was achieved in 1865. However, that landmark ushered in more than a century of internal power struggles, revolutions, political assassinations, and occupations by the United States (1916–1924 and 1965). Hurricane Zenón devastated the city in 1930. The country is now more stable and more democratic, and has developed a stronger economy based on manufacturing, tourism, financial services, and export of agricultural products.

Major Landmarks. The Zona Colonial (Colonial Zone) of Santo Domingo faces the Caribbean Sea on the western banks of the Ozama River, and it is a UNESCO World Heritage Site. It has many beautiful historic churches, including Santo Domingo's historic cathedral, Iglesia Regina Angelorum, and Convento de los Dominicos, as well as Ozama Fort, Alcazár de Colón, and Panteon Nacional, the burial place of national heroes. Plaza de Cultura has five museums and the National Theater. Malecon (George Washington Avenue) is a wide boulevard along the waterfront with casinos, nightclubs, and restaurants. In 1992, on the 500th anniversary of Christopher Columbus's landing in the Dominican Republic, the monumental Faro a Colón, the Columbus Lighthouse, was built in Santo Domingo. It has a powerful cross-shaped light beam and is a mausoleum as well, said to contain the remains of Columbus.

Culture and Society. Nearly two-thirds of the Dominican population is biracial, being a mix of European and African ancestry. There are also Taino elements in the population, as well as a large minority of Haitians. Many American citizens live in the Dominican Republic as well. About 70 percent of the population is Roman Catholic and nearly 20 percent Evangelical Christian. Spanish is the official language, although English is also widely used. Many Dominicans live in the United States, most prominently in New York City, and travel back and forth to their home island. Remittance income from Dominicans abroad is an important element of the national economy.

Santo Domingo is divided between rich and poor. The latter tend to live at the outskirts of the city, where there are extreme conditions of poverty and shantytown slums, while the center has fine residential areas, vibrant nightlife, and a prosperous, modern downtown commercial area called the Poligono Central. The Dominican Republic is in love with baseball and Santo Domingo has many stadiums and baseball leagues. The Dominican Republic is home for quite a few top players in the U.S. Major Leagues, including Albert Pujols, Aramis Ramirez, David Ortiz, and Melky Cabrera.

Further Reading

Fe, Liza Bencosma and Clark Norton. *Santo Domingo, Dominican Republic.* Winston-Salem, NC: Hunter Publishing Company, 2010.

Hoffnung-Garskof, Jesse. *A Tale of Two Cities: Santo Domingo and New York after 1950.* Princeton, NJ: Princeton University Press, 2008.

SÃO TOMÉ

São Tomé is the capital and largest city of the Democratic Republic of São Tomé and Principe, an island nation off the west coast of Africa just north of the equator in the Gulf of Guinea. The country consists of two main islands, one named São Tomé and the other Principe, that are about 87 miles (140 km) apart and 155 and 140 miles (250 and 225 km), respectively, off the coast of Gabon, plus some smaller islands. The city of São Tomé is located in the northeastern part of São Tomé Island, the larger of the two islands, on Ana Chaves Bay. The words São Tomé are Portuguese for St. Thomas, the name chosen for the island by Portuguese explorers who first arrived there on the feast day of the apostle saint. The population of the São Tomé, the city, is about 56,000 (2005), approximately one-third of that of São Tomé and Principe as a whole.

Historical Overview. The first Portuguese arrived on the island of São Tomé in 1470 and reportedly found an uninhabited island. The town of São Tomé was founded by Portuguese settlers in 1485 and became a base for sugar production. Among the Portuguese settlers were Jews who were driven from Portugal. Slave labor was imported from the African continent. In 1522, the island of São Tomé became a colony of Portugal. Principe became a colony in 1573. The sugar industry failed to compete with sugar production in the western hemisphere, so in the mid-17th century the basis of the economy shifted to the slave trade between Africa and the Americas. In the 19th century, coffee and cocoas were introduced as cash crops. Although Portugal had banned slavery in 1876, it continued to be practiced on both main islands as well as into the early 20th century. On February 3, 1953, approximately 1,000 people were killed in São Tomé during a clash between local workers and Portuguese landowners about working conditions. Known as the Batepá Massacre, the incident energized Santomean yearnings for independence and sparked the establishment of the Movement for the Liberation of São Tomé and Principe. The islands achieved independence from Portugal on July 12, 1975. Since independence, São Tomé and Principe has been one of the more stable and more democratic of African nations, although there was an attempted coup in 2009. The plotters of the coup were pardoned by the president they attempted to overthrow. Agriculture is the mainstay of the economy, especially the production of cocoa that accounts for 95 percent of exports. There are hopes that current oil exploration in the waters off São Tomé and Principe will soon be successful and that oil exports could fuel the economy in the future.

Major Landmarks. The main sights of São Tomé are the Cathedral of São Tomé, the 1575 Fort São Sebastião which now houses the São Tomé National Museum, and the São Tomé Palace. Outside the city in the southern part of the island is Pico Cão Grande (Great Dog Peak), a spectacular volcanic plug that rises dramatically about 1,000 ft (300 m) above the surroundings.

Culture and Society. The population of São Tomé and Principe is a mixture of ethnic groups and ethnic mixes, most notably the descendants of Portuguese settlers and African slaves who were brought to the islands, the descendants of African slaves, the descendants of contract laborers who were brought from Angola, the descendants of Chinese from Portuguese Macau, and a small number of Europeans, mostly Portuguese. Portuguese is the most widely spoken language. The main religion is Roman Catholicism. The culture of São Tomé is a blend of Portuguese and African influences as seen in the music and dance of the island, and in a signature theatrical performance called Tchiloli.

Further Reading

Seibert, Gerhard. *Comrades, Clients and Cousins: Colonialism, Socialism and Democratization in São Tomé and Principe.* Leiden, The Netherlands: Brill Academic Publishers, 2006.
Weszkalnys, Gisa. "The Curse of Oil in the Gulf of Guinea: A View from São Tomé and Principe," *African Affairs* 108, no. 433 (October 2009): 679–89.

SARAJEVO

Sarajevo is the capital and largest city of Bosnia and Herzegovina, sometimes written Bosnia-Herzegovina, and commonly referred to as simply Bosnia, a country on the Balkan Peninsula in southeastern Europe. The city is located more or less in the center of the country in a narrow valley on the Miljacka River at the foot of Mount Trebević, one of many peaks of the Dinaric Alps near the city. The population of the city is about 311,000, while that of the metropolitan area is about 439,000.

Historical Overview. The area of Sarajevo has been continuously inhabited since the Neolithic Age. Urban history begins in about the 13th century with a city called Vrhbosna that existed at or near the site of Sarajevo, and that had a cathedral to St. Paul that was built in 1238 and a citadel. The Ottoman Empire gained control of the area in 1429, and Saravejo was founded by the Ottomans in the 1450s or in 1461. The first Ottoman governor of Bosnia was Isa-Bed Ishaković, who had a mosque, marketplace, and public bath erected in the city, as well as a palace, a *saray* in Turkish, the word from which "Sarajevo" originates. The city grew quickly and became a major regional trade and religious center. Most of the local population converted to Islam. In 1697, the city was burned to the ground in the Great Turkish War between the Ottomans and the Austrian Empire. It was rebuilt, but did not immediately regain the importance that it once had. The Austro-Hungarian Empire occupied Sarajevo from 1878 and invested considerably in modernization and urban development. The city became an outstanding blend of Western and Eastern cultures and architectural traditions. On June 28, 1914, Austrian Archduke Franz Ferdinand was assassinated in Sarajevo by a Bosnian Serb, precipitating the onset of World War I. The city was greatly damaged in that war. After the war, Sarajevo became part of the Kingdom of Yugoslavia. The city was captured by Germans in World War II and was liberated by the Yugoslav partisan movement in 1945. Afterward, the city became Socialist Federal Republic of Yugoslavia and was capital of the Socialist Republic of Bosnia and Herzegovina. In 1984, the city gained global fame for hosting the Winter Olympics. The dissolution of Yugoslavia in 1992 was accompanied by several years of warfare and ethnic violence between Bosnians and Serbs that took many innocent lives and destroyed many parts of the city. The city is now at peace and it is developing rapidly as an important regional center and European capital.

Major Landmarks. Old Town Sarajevo has narrow cobbled streets, old shops, and markets, mosques, and in keeping with the religious diversity of Sarajevo, Catholic and Orthodox churches, and a synagogue as well. Sebilj Fountain is one of the main landmarks of this district. The Latin Bridge across the Miljacka River is the assassination spot that triggered World War I. Important museums include the Bosnian Historical Museum, the National Museum, and the Sarajevo City Museum. The Bosniak Institute has a library and other materials about the history of Bosnia and Bosniaks. Other important landmarks include the modernist, new government building of Bosnia and Herzegovina, City Hall, and the iconic Avaz Twist Tower, a 564-ft-high (172 m) skyscraper that is the tallest building in the Balkans and headquarters for a major national newspaper.

The Avaz Twist Tower, the headquarters building of Sarajevo's principal newspaper, the *Dnevni Avaz*. (Zurijeta/Shutterstock.com)

Culture and Society. Sarajevo's culture has been diverse throughout history, and has included Muslims, Orthodox, Catholics, and Jews. One of the city's nicknames was "Europe's Jerusalem" because of this mix. The main ethnic groups are Bosniaks, Serbs, Croats, and other minorities, including small numbers of Jews and Roma. The years of war in the 1990s changed the population mix, as many Serbs left Bosnia. As a result, the proportion of Orthodox declined, and the proportion of Moslems increased, as Islam is the main faith of the Bosniak population.

Further Reading

Demick, Barbara. *Logavina Street: Life and Death in a Sarajevo Neighborhood.* New York: Spiegel & Grau, 2012.

Donia, Robert J. *Sarajevo: A Biography.* Ann Arbor: University of Michigan Press, 2006.

Maček, Ivana. *Sarajevo Under Siege: Anthropology in Wartime.* Philadelphia: University of Pennsylvania Press, 2009.

Makaš, Emily Gunzburger. "Sarajevo," in Emily Gunzburger Makaš and Tanja Damljanović Conley, eds., *Capital Cities in the Aftermath of Empires: Planning in Central and Southeastern Europe*, 241–57. London: Routledge, 2010.

SEOUL

Seoul is the capital and largest city of the Republic of Korea, also known as South Korea, an independent country that occupies the southern half of the Korean Peninsula. The city is in the northwest of the country and is bisected by the lower reaches of the Han River. With a population of more than 10 million and a metropolitan population of approximately 23.6 million (2011), Seoul and its metropolitan area rank among the largest urban concentrations in the world. The entire metropolis is very crowded, as it is pinched for space between the boundary of the Democratic People's Republic of Korea to the north, the coast of the Yellow Sea to the west, and many steeply sloped mountains. As a result, it increasingly consists of densely built up multistory neighborhoods and residential districts. The mountains, however, provide considerable greenery and open space.

Historical Overview. The banks of the lower Han River have been settled since ancient times. In 18 BC, the Baekje kingdom established its capital city Wiryeseong on a site that is within today's city. Baekje became one of the three kingdoms of Korea, along with Goguryeo and Silla, and Wiryeseong was capital until 475. Remains of old walls can still be found in Seoul. In 1104, Goguryeo King Sukjong built a palace in Seoul. The city was named Namgyeong (Southern Capital) at the time. After the Goreyo Dynasty was overthrown, the successor Joseon Dynasty made Seoul its capital until its rule ended in 1897. As the Joseon capital, the city was named Hanseong, "Fortress City on the Han River." The city was surrounded by a 20-ft-high (6.1 m) stone wall, but urban growth soon spilled beyond the wall's confines. From 1910 to 1945, the Korean Peninsula was ruled by Japan, and Seoul was the colonial administrative center. Its name was Gyeongseong (Capital City) at the time (Keijō in Japanese). Under Japanese rule, Seoul was greatly industrialized and many prominent buildings were built.

After World War II, Korea gained its independence from Japan, and the city was re-named Seoul, which also means Capital City. During the Korean War of 1950–1953, Seoul was the scene of fierce fighting between the forces of the North and those of the South, and suffered enormous damage and loss of life. It became the capital of the Republic of Korea (South Korea) after the war, although the city remains heavily militarized as the border with the Democratic People's Republic of Korea (communist North Korea) begins just beyond the city's northern outskirts. In addition to Korean forces, many American troops are stationed in the city and along the boundary zone between the two Koreas.

Seoul grew very quickly in the decades after the Korean War and is now one of the world's largest and most dynamic metropolitan areas. Although there have been long periods of repressive and corrupt government, the economy of Seoul and South Korea has grown impressively, and the city has established itself as one of the main business centers of Asia. Its port city Incheon, located just to the west of Seoul, is a major exporter of Korean manufactured products around the world. In 1988, Seoul hosted the summer Olympics, and in 2002 it cohosted the FIFA World Cup along with Tokyo and other cities in Japan and Korea.

Major Landmarks. Two prominent historic gates to Seoul still stand from the walls of the Joseon city, Namdaemun or South Gate, and Dongdaemun or East Gate, although Namdaemun was destroyed by an arsonist's fire in 2008 and is now under reconstruction. The city also has five "Grand Palaces" that were built during the Joseon Dynasty, one of which, Changdeokgung, has been recognized as a UNESCO World Heritage Site. Another iconic landmark is Seoul Tower atop Namsan Mountain in Central Seoul, a communication tower and visitor attraction that was erected in 1969. Major museums include the National Museum of Korea, the National Folk Museum, the War Memorial, and the Seoul Museum of Art. The Kimchi Field Museum is devoted to displays about *kimchi*, an important element of Korean cuisine. The downtown of Seoul has many landmark buildings, including the Japanese-built Old Seoul Station and Seoul City Hall, as well as spectacular new office and hotel towers, and the headquarters of many of Korea's largest corporations.

Culture and Society. The Korean Peninsula is one of the most ethnically homogenous places in the world, with more than 99 percent of the population being Korean and speaking the Korean language. Koreans refer to their country as the "single race society." Seoul, however, is a bit more diverse than Korea as a whole, because it is the base for foreign diplomats to Korea as well as expatriates in international business with Korea. In terms of religion, Korea is mixed. More than half of the population expresses no religious preference, while the rest are mostly Buddhist or Christian. Buddhism arrived in Korea in 372 AD, and has a strong role in Korean culture, even if people are not practicing Buddhists. Increasingly, South Korea is a prosperous society, and Seoul is a vibrant city with many cultural and recreational opportunities, as well as many shopping malls and lively street markets.

Further Reading

Cho, Mihye. "Envisioning Seoul as a World City: The Cultural Politics of the Hong-dae Cultural District," *Asian Studies Review* 34, no. 3 (2010): 329–47.

Ha, Seong-Kyu, "Housing, Social Capital and Community Development in Seoul," *Cities: The International Journal of Urban Policy and Planning* 27, Supplement 1 (2010): S35–42.

Kim, Hyung Min and Sun Sheng Han. "Seoul," *Cities: The International Journal of Urban Policy and Planning* 29, no. 2 (2012): 142–54.

Kim, Joochul and Sang-Cheul Choe. *Seoul: The Making of a Metropolis.* Chichester, UK: Wiley, 1997.

SINGAPORE

Singapore is a small city-state located at the southern tip of the Malay Peninsula in South East Asia, at a latitude of 1°17′N, that is, about 85 miles (137 km) north of the equator. It consists of one main island, known as Singapore Island, and 62 much smaller ones, and is separated from nearby Malaysia on the peninsula by the Straits of Johor. Indonesia's Riau Islands are across Singapore Strait to the south. The total area of Singapore is 268 square miles (694 sq km), making the country one of the smallest in the world. Singapore keeps growing, however, as land reclamation projects add to ground to the national territory.

As a country that is also a city, Singapore is in essence its own capital. The population is 5,183,700 as of a count in 2011, of whom 3,257,000 are citizens of Singapore. The difference includes many guest workers from nearby countries who are engaged in service work, construction, and other occupations, as well as business executives, foreign diplomats, and their families from around the world.

The word Singapore comes from the Malay word meaning "Lion City." A popular symbol of Singapore is the so-called merlion, a mythical creature with the head of a lion and body of a fish. It is said to reflect the meaning of the word Singapore with the city's distant origins as a fishing port.

Historical Overview. Singapore originated as a fishing port and center of local trade, and was once named Temasek, "sea town." From the 16th to 19th centuries it was part of the Sultanate of Johore centered across the Straits of Johore, but it was not a very important settlement. The modern history of Singapore dates to 1819 when the Englishman Thomas Stamford Raffles arranged a treaty with the sultanate that the southern part of Singapore Island (also called Pulau Ujong) would become a trading post of the British East India Company. Five years later, the entire island became a British colonial possession, and in 1836, Singapore was designated as capital of the British Straits Settlements. A British Crown Colony was established on April 1, 1867. The colony prospered, as did British Malaya, a forerunner of Malaysia, and Singapore grew in population with inmigration of Malays, Chinese, and Indians. After the Battlke of Singapore on February 8–15, 1942, Singapore was ruthlessly occupied by Japanese troops. They renamed the city *Shōnan-tō*, meaning "Light of the South Island" in Japanese. The defeat of Japan in 1945 returned Singapore to British hands. Self-government came in 1959, with Lee Kuan Yew (1923–) as the first prime minister. In 1963, Singapore declared its independence from Great Britain. For two years it was part of the new Federation of Malaysia, but internal conflict within the federation caused Singapore to split off, forming the Republic of Singapore in 1965. Lee Kuan Yew was the prime minister and continued to serve until near the end of 1990.

Mr. Lee had enormous influence in shaping the country and its society, and continues to be an advisor to the country and a major voice after his retirement. Under his strict leadership, Singapore came to be one of the most prosperous nations in the world, despite its small size and small population, and lack of natural resources. It enjoys a highly educated society and a disproportionately large role in Southeast Asian and global business, and is also one of the world's cleanest, safest, and mostly highly technology-savvy cities (or countries) in the world. The current (and third) prime minister of Singapore is Lee Hsien Loong, Lee Kuan Yew's eldest son.

Major Landmarks. The historic core of Singapore is along the short Singapore River where there are restaurants and cafes in colonial-era shophouses in the shadows of the city's tallest office towers. Across the river is the district of government office buildings inherited from the British and a statue of Sir Thomas Stamford Raffles, the founder of the British colony. The Orchard Road district is famous for its many shopping malls and international hotels. There are various ethnic districts in the city, most notably Chinatown, Little India, and Bugis (Malay), among others, in which tourist-oriented shopping and restaurants with ethnic fare are as much the scene as ethnic residence, if not more. The Marina Bay District on reclaimed land offers new attractions such as the architecturally stunning Esplanade Theatres on the Bay and the Singapore Flyer, the world's highest Ferris wheel (541 ft; 165 m). The waterfront area also has a large Merlion fountain. There are many parks in Singapore, including Jurong Bird Park, Mount Faber Park, an excellent zoo, and many fine beaches. Sentosa, an offshore island connected to Singapore Island by cable car, offers a wide range of recreation opportunities and museums, as well as another statue of the fabled Merlion.

A view of the skyscrapers of downtown Singapore and the tile-roofed shop-houses of the Chinatown district in the foreground. (Photo courtesy of Roman Cybriwsky)

Culture and Society. Singapore is known for its high standards of cleanliness and public order, and strict rules that govern everything from traffic control to the chewing of gum and flushing of public toilets. Violators are fined, earning the city the tongue-in-cheek nickname "Fine City." Criminal violations are punished by imprisonment or caning, or the death penalty. Possession of narcotics for distribution carries a mandatory death penalty.

About 76 percent of Singapore's population is of Chinese origin, 14 percent Malay, and 8 percent South Asian. The national language is English, which is spoken at a very high standard, but at home many residents speak Chinese, Malay, or Tamil. There are many Buddhist temples in Singapore, as well as mosques, Hindu temples, and Christian churches of various denominations. Singapore celebrates its ethnic mix in various ways including ethnic holidays and festivals and a rich array of ethnic foods. Government policy encourages ethnic mixing in new housing developments, as opposed to homogeneity.

Further Reading

Kong, Lily and Brenda S. A. Yeoh. *The Politics of Landscape in Singapore: Construction of a "Nation,"* Syracuse, NY: Syracuse University Press, 2003.

Lee, Kuan Yew. *From Third World to First: The Singapore Story: 1965–2000.* New York: HarperCollins, 2000.

Yeoh, Brenda S. A. and Lily Kong. *Portraits of Places: History, Community and Identity in Singapore.* Singapore: Times Editions, 1995.

SKOPJE

Skopje is the capital and largest city of the Republic of Macedonia, a small landlocked country on the Balkan Peninsula in southern Europe. The city is in the northern part of the country on the Vardar River. Its population is about 670,000 (2006), approximately one-third that of Macedonia. In addition to its role as capital, Skopje is an important transportation center and a diversified manufacturing city. Historically, the city has bridged the cultures of Asia Minor and the Mediterranean with those of central and northern Europe.

Historical Overview. The site of Skopje has been inhabited since ancient times. An early settlement was made into a fortified outpost of the Roman Empire in the first- century AD and was known as Scupi, the word from which the name Skopje is derived. After 395, the town came under control of the Byzantine Empire based in Constantinople. Later it fell under the Bulgarian Empire and then to an antecedent of Serbia. In 1392, Skopje was conquered by the Ottoman Turks, who ruled until 1912 when the Kingdom Serbia defeated the Ottomans in the Balkan Wars. During the Ottoman period, the city was known as ÜsKükp. After World War I, Macedonia was part of Yugoslavia, first the Kingdom of Yugoslavia and then after World War II the Socialist Federal Republic of Yugoslavia, and Skopje was the Macedonia capital. In 1943, Skopje's entire Jewish population of 3,286 was killed in Nazi gas chambers in Poland. A major earthquake destroyed most of the city on July 26, 1963. Reconstruction was led by noted Japanese architect Kenzo Tange. Macedonian independence came in 1991 as part of the dissolution of the Yugoslavian federation.

Major Landmarks. The most significant landmark in Skopje is Kale Fortress that was built in the sixth century by the Byzantine emperor Justinian. It stands on the city's highest hill and is a symbol of the city. The Stone Bridge across the Vardar River is also an icon of Skopje from the sixth century. Other landmarks include the 16th-century Clock Tower, the Feudal Tower, the Aqueduct, the Old Bazaar, and the Double Hamam, an old public bath that is now an art gallery. Macedonia Square in the center of the city has a large statue of Alexander the Great. The Old Railway Station is left partly ruined as a monument to the 1963 earthquake and has a museum about the disaster. The city has many historic mosques and Orthodox churches. Mother Theresa House, small chapel and museum, stands on the site where the famous missionary to India lived as a child. The Millennium Cross was built in 2002 to celebrate 2,000 years of Christianity. It stands atop Vodno Mountain in the city and can be reached by cable car.

Culture and Society. Macedonians make up about two-thirds of Skopje's population. The main ethnic minorities are Albanians (about 20%) and Roma (about 5%). The Orthodox faith is the main religion. Skopje's historic landscape reflects the ethnic diversity of the city and its long history of occupation by regional powers.

Further Reading

Berg, Glen Virgil. *The Skopje, Yugoslavia Earthquake, July 26, 1963.* New York: American Iron and Steel Institute, 1964.

Bouzarovski, Stefan. "Skopje," *Cities: The International Journal of Urban Policy and Planning* 28 (2011): 265–77.

SOFIA

Sofia is the capital and largest city of Bulgaria, a country on the Balkan Peninsula in southern Europe. It is located in the western part of the country in a valley between mountains, near Mount Vitosha (7,513 ft; 2,290 m) a prominent local peak. There are many mineral and thermal springs in and near the city. The port city that serves Sofia is Varna, 240 miles (380 km) to the east on the Black Sea coast of Bulgaria.

Sofia's population is 1,232,008 according to an official estimate for 2011, while that of the wider metropolitan area is about 1,370,000.

The name Sofia was first applied to the city in the 14th century. It is taken from Hagia Sophia Church in the city which dates to the sixth century, and whose name means Holy Wisdom. A larger church of the same name existed earlier in Constantinople (now Istanbul, Turkey), and was at the time a guiding center of Orthodox Christianity. The correct pronunciation of Sofia places the accent on the first syllable. Previous names for the city were Serdica, for a local Celtic tribe, and Sredets, a Bulgarian word referring to the concept of "middle."

Historical Overview. As evidenced by archaeological digs, the site of Sofia has been settled since prehistoric times. The walls of a settlement by Thracians from the seventh-century BC have been unearthed at a local mineral spring that is still in use. The city was conquered by the Romans in about 29 BC and became a regional administrative center with protective walls and high turrets. In the third-century AD, the city was Christianized and became the capital of the Roman province Dacia Mediterranea. In 447, the city was destroyed in an invasion by the Huns. It was then rebuilt and fortified by the Emperor Justinian of Byzantium. Ruins of Justinian's walls are still in place. In the Middle Ages, Sofia was alternately part of the Kingdom of Bulgaria and the Byzantine Empire until it was captured after a long siege by Ottoman forces in 1382. The Ottomans consolidated their hold on the city after a battle in 1442, and made Sofia the capital their province called Rumelia. Many mosques were built in the city during Ottoman rule. Russians captured Sofia in 1878 as a result of the Russo-Turkish War, which led to reestablishment of the Kingdom of Bulgaria in 1908. From 1918 to 1943, the kingdom was ruled by Boris III, who is credited with saving Sofia's Jews from the Nazis despite the fact that Bulgaria was allied with Germany. Sofia was bombed by American and British planes in 1943 and 1944, and was captured by Russian forces at the end of the war in 1945, at which time the Kingdom of Bulgaria was overthrown. In 1946, Bulgaria became a satellite state of the Soviet Union, the Bulgarian People's Republic. The city then began to develop along socialist lines of land use, architectural style, and economic organization. Prefabricated apartment blocks called *panelki* (panel buildings) were a main feature of Soviet-influenced construction. One-party Communist rule ended in 1989. In 2004, Bulgaria became a part of NATO, and in 2007, it joined the European Union.

Major Landmarks. The enormous and very beautiful Alexander Nevsky Cathedral, built in the late 19th century, is perhaps Sofia's best-known landmark and most visited tourist destination. Nearby is the historic St. Sophia Church from the sixth century. The Boyana Church, constructed in the 19th or early 11th century is a UNESCO World Heritage Site. Other attractions are the National Archaeological Museum, the SS Cyril and Methodius National Library, the National Historical Museum, the Ivan Vazov National Theater, and the National Opera and Ballet. Vitosha Boulevard, popularly called Vitoshka, is a popular shopping street.

Culture and Society. Sofia's population is about 96 percent Bulgarian. The main minority groups are Roma, Turks, Russians, and Greeks, although all in small numbers. Many Turks were repatriated after the fall of the Ottoman Empire. The majority religion is Orthodox Christian and the official language is Bulgarian. The city is rapidly developing into a leading economic center of the European Union.

Further Reading

Gigova, Irina. "The City and Nation: Sofia's Trajectory from Glory to Rubble in WWII," *Journal of Urban History* 37, no. 2 (2011): 155–75.

Hirt, Sonia, "Suburbanizing Sofia: Characteristics of Post-Socialist Peri-Urban Change," *Urban Geography* 28, no. 8 (2007): 755–80.

Hirt, Sonia. "Landscapes of Postmodernity: Changes in the Built Fabric of Belgrade and Sofia since the End of Socialism," *Urban Geography* 29, no. 8 (2008): 785–810.

Staddon, Caedmon and Bellin Molov. "Sofia, Bulgaria," *Cities: The International Journal of Urban Policy and Planning* 17, no. 5 (2000): 379–87.

SOUTH TARAWA

South Tarawa is the official capital of the Republic of Kiribati, a nation comprised of 32 atolls (coral reef rings) and one additional island that straddle the equator in the mid-Pacific Ocean and border the International Date Line. It is a populated district along the southern rim of Tarawa Atoll, one the 32 atolls that make up most of the country, in the north center of the expansive island group just a little more than 1° north of the equator. South Tarawa is made up of small islets that stretch for about 12 miles (20 km) between Bairiki in the west and Temaiku-Bonriki in the east. The islands are linked by causeway. It is the largest population center in Kiribati. In 2005, its population was 40,311, about 40 percent of that of the country as a whole. The formal name for South Tarawa is Teinainano Urban Council, or TUC.

Historical Overview. Micronesians settled the islands many centuries ago, and with time became intermixed with Melanesians and Polynesians who had invaded the islands and also settled. British and American ships came in the late 18th and early 19th centuries, and by 1820 the main group of islands came to be called the Gilbert Islands after an English ship captain. The first English settlers arrived in 1837. In 1892, the Gilbert Islands were joined with the nearby Ellice Islands chain into a British protectorate administered from Fiji, and in 1916 they became a British Crown Colony. Japanese forces occupied the islands in World War II. On November 20–23, 1943, U.S. forces landed on the Japanese in the Battle of Tarawa, which the Americans won after heavy casualties on both sides. The battle was one of the turning points of the War in the Pacific in favor of Allied forces. The Gilbert and Ellice Islands were granted self-rule in 1971. In 1975, they were separated, and in 1978 the Ellice Islands became the independent nation of Tuvalu. The Gilbert Islands became independent as Kiribati on July 12, 1979, "Kiribati" being a Gilbertese pronunciation of "Gilberts."

Because of low elevation, the islands of Kiribati are in special danger from rising sea levels that result from global warming. In 2008, the government of Kiribati entered into discussions with Australia and New Zealand about accepting refugees from Kiribati should the nation become uninhabitable. Kiribati officials have also negotiated with the government of Fiji about purchasing higher grounds to resettle its population.

Major Landmarks. The major landmarks of Tawara are the Kiribati Parliament House in Bairiki and the Presidential Residence. The Umanibong or the Kiribati Cultural Museum is a museum in Bikenibeu in South Tarawa that is devoted to the culture and history of Kiribati. The National Stadium is in Bairiki. It seats 2,500 for football (soccer). Another landmark is the Kiribati campus of the University of the South Pacific. The shores of Tawara also have some old Japanese gun positions still in place from World War II.

Culture and Society. The people of Kiribati are called I-Kiribati and are Micronesians. Their language is called Gilbertese. English is the official language but it is not widely used. Most I-Kiribati are Christians, with Roman Catholics being the predominant denomination. There are also Congregationalist Protestants and Jehovah's Witnesses, as well as Mormons. Most people of the islands are villagers.

The country is poor, with copra and fishing being the main industries. Much of the economy is supported by assistance from donor counties such as Australia and New Zealand, and by remittance income from Kiribati citizens working abroad. Kiribati has a rich traditional musical and dance culture.

Further Reading

Jones, Paul and John P. Lea. "What Had Happened to Urban Reform in the Island Pacific? Some Lessons for Kiribati and Samoa," *Pacific Affairs* 80, no. 3 (Fall 2007): 473–91.
Storey, Donovan and Shawn Hunter. "Kiribati: An Environmental 'Perfect Storm,'" *Australian Geographer* 41, no. 2 (June 2010): 167–81.

ST. GEORGE'S

St. George's is the capital and largest city of Grenada, an island country consisting of the island of Grenada and six smaller islands in the southeastern Caribbean Sea off the northern coast of Venezuela, South America. The country was once a colony of Great Britain and is a member of the British Commonwealth of Nations. St. George's is on the southwest coast of Grenada Island, the largest, most populous, and southern-most of the seven islands that make up the country. It is on a horseshoe-shaped bay called The Carenage that provides sheltered mooring from all but the strongest of tropical storms. The population of St. George's is about 89,000, more than 80 percent of the total national population of 110,000.

Historical Overview. Grenada was inhabited originally by indigenous Carib people, but their population was decimated by European contact that began on this island in 1498 during the third trans-Atlantic voyage of Christopher Columbus. In 1609, there was a short-lived English settlement on the island, and then permanent settlement by the French beginning 1649. The French subjugated the remaining islanders and pushed survivors to other islands. They named their colony La Grenade and raised sugar and indigo as export crops, using slaves imported from Africa as labor. They established Fort Royal as their capital, the town that would later become St. George's. The English captured the island in 1762 during Seven Years War and then held it as their colony until 1950 except for a brief return of French rule from 1779 to 1783 following their naval victory over the British in the July 1779 Battle of Grenada. In 1877, Grenada was made a British Crown Colony.

Grenadian independence movements intensified with open rebellions after 1950 and succeeded in 1974. There was considerable political instability afterwards, however, with rival political parties locked in struggle for control. The New Jewel Movement led by Maurice Bishop took control of the government in an overthrow in 1979 and moved Grenada toward socialism. Bishop was assassinated in 1983, and a more overtly pro-Communist government took root, which in turn caused the United States to invade the island and force political change. The U.S. invasion was condemned by the United Nations General Assembly as a "flagrant violation of international law." On September 7, 2004, Hurricane Ivan landed ashore in Grenada and destroyed some 90 percent of the buildings, while nearly a year later on July 14, 2005, Hurricane Emily did great damage to the northern part of the island. The People's Republic of China built a $40 million new stadium in

St. George's for the 2007 World Cricket Cup as a foreign aid project after the 2004 hurricane, but was then greatly offended when the national anthem of the Republic of China (Taiwan) was played by mistake at the opening ceremony.

Major Landmarks. In addition to the aforementioned National Stadium (also known as Queen's Park), the main landmarks of the city are Fort George built in 1705 by the French, St. George's Roman Catholic Cathedral (1818), St. George's Anglican Cathedral (1825), and the Grenada National Museum housed in a former French army barracks and prison that were built in 1704. Other attractions are Grand Anse Beach, the Parliament Building, and St. George's University School of Medicine.

Culture and Society. The majority of Grenada's population is descendant from African slaves. There are also descendants from indentured laborers who were brought to the island from India in the mid-19th century, and small numbers of English and French. English is the official language of the country, although most people speak an English-based Grenadian Creole in daily life. There is a French-based Creole that is spoken on the island as well. About one-half of the population is Roman Catholic and one-half is Protestant. The Anglican Church is the largest Protestant denomination. Because of outmigration, more Grenadians live abroad on other islands of the Caribbean and in the United Kingdom and the United States than on Grenada. In the second of week of August every year, St. George's is venue for Grenada's Carnival, a raucous street party that celebrates emancipation of slaves.

Further Reading

Sinclair, Norma. *Granada: Isle of Spice*. Northampton, MA: Interlink Publishing Group, 2003.
Steele, Beverley A. *Grenada: A History of Its People*. Oxford: Macmillan Education, 2003.

ST. JOHN'S

St. John's is the capital and largest city of Antigua and Barbuda, a country comprised mainly of the two islands it is named for plus a number of other much smaller islands in the Lesser Antilles where the Caribbean Sea meets the wider Atlantic Ocean. The city itself is on the northwest coast of Antigua, the larger of the two islands, and is the nation's main commercial center as well as capital, and the island's chief port. Its economy is based on tourism, especially higher-end luxury resorts and cruise ships, offshore investment banking, and an American university, the medical school called the American University of Antigua, with 10,000 students. St. John's also distills rum and hosts Internet gaming. The population of St. John's is about 43,000 (2007).

Historical Overview. Antigua was settled in ancient times by the Ciboney people, and later by Arawaks and, in about 1500, by Caribs. Christopher Columbus landed on the island in 1493 and named it Santa Maria de la Antigua after an icon in the cathedral in Seville, Spain (*antigua* being Spanish for "old"). European colonization began in 1632 with the arrival of English planters who raised tobacco, indigo, ginger, and sugar cane. St. John's was the British administrative center of the island from the start. Sugar became the biggest business and large plantations were established. Slaves imported from Africa were used as the labor force. The economy of the nearby island of Barbuda was similar and the two islands shared a common past. In the 18th century, Antigua was an important base for the British navy. The slaves in Antigua and Barbuda (and other English colonies) were emancipated in 1834, but many remained on the land afterwards in employ of their former masters. On November 1, 1981, Antigua and Barbuda became an independent nation.

Major Landmarks. Among the main landmarks of St. John's are St. John's Anglican Cathedral, the Antigua Recreation Ground (the national cricket stadium), and the St. John's Antigua Lighthouse. The present cathedral was built in 1845, as earthquakes in 1683 and 1745 destroyed earlier cathedral buildings. Fort James stands at the entrance to St. John's harbor. Other landmarks include the Museum of Antigua and Barbuda, the Museum of Marine Art, and the Botanical Garden.

Culture and Society. More than 90 percent of the population of Antigua is of West African descent. There are small minorities of mixed-race persons, whites (mostly of Irish or British descent), and South Asians. In addition, Antigua has a small number of residents from Portuguese background who are descendants of mid-19th-century settlers from Portuguese Madeira. About three-quarters of the population is Christian, with Anglicans being the largest single denomination. English is the official language, but Antiguan Creole is commonly spoken too. Cultural borrowings from English culture are especially strong on Antigua; for example, the most popular sport by far is cricket. Many citizens of Antigua and Barbuda have emigrated to the United Kingdom in search of economic opportunity and send remittance income home. Because of its natural beauty, fine weather, and advantages as a tax haven, Antigua is or has been a second home to many rich and famous people around the world.

Further Reading

Baldwin, J. "Tourism Development, Wetlands degradation, and Beach Erosion in Antigua, West Indies," *Tourism Geographies* 2, no. 2 (2000): 193–218.

Baldwin J. "The Contested Beach: Limits to Resort Development," in C. Cartier and A. A. Lew, eds., *The Seduction of Place: Geographical Perspectives on Globalization and Touristed Landscapes*, 338–60. Malden, MA: Blackwell, 2007.

Dodman, David. "Developers in the Public Interest? The Role of Urban development Corporations in the Anglophone Caribbean," *The Geographical Journal* 174, no. 1 (2008): 30–44. http://www.antigua-barbuda.org/index.htm

SRI JAYAWARDENEPURA KOTTE. *See* Colombo.

STOCKHOLM

Stockholm is the capital and largest city in Sweden, a Scandinavian country in northern Europe, and the largest urban center in Scandinavia as a whole. It is located on the east coast of south-central Sweden between Lake Mälaren and the Baltic Sea. The historic core of the city is series of near shore 14 islands that make up the Stockholm archipelago, while subsequent growth took place inland. Much of the city's territory is water, earning Stockholm the occasional nickname "Venice of the North." In addition to be being the center of government for Sweden, Stockholm is a major cultural, educational, financial, and industrial center. It has a reputation for being clean, green, and beautiful city, as one that is cosmopolitan and sophisticated. The population of Stockholm is about 871,000, while that of the metropolitan area is about 2.1 million (2010).

Historical Overview. The narrative of Stockholm's history begins in the middle of the 13th century with the construction of fortifications by Swedish ruler Birger Jarl to protect other settlements and Sweden's emerging mining trade from foreign invasion. The city prospered from European trade, especially with the German city of Lübeck, and grew accordingly. It became capital of Sweden in 1436. In 1523, Gustav Vasa became King of Sweden. The country became a significant European power in the mid-17th century, during which time the city expanded greatly and many buildings and monuments were built. As fires destroyed neighborhoods of wooden dwellings, much of the city was rebuilt in stone. In the 19th century, Stockholm's growth was fueled by industrialization and rural to urban migration. In 1897, the city hosted the General Art and Industrial Exhibition on the island of Djurgården as a show case of new technology and economic opportunity for the future. Stockholm has since become a major center of information technology and a headquarters city for Swedish industry, banking, insurance, and other companies. It is now an overwhelmingly white-collar city, with most people being employed in the service sector of the economy.

Major Landmarks. Gamla Stan is the city's Old City section. It occupies the small islands where the city first developed and has a medieval street plan and a number of historic palaces and mansions. The Riddarholmen Church dates to the late 13th century and is one of the oldest buildings in the city. It is the burial place for Sweden's monarchs. Storkyrkan, meaning "the great church," is even older and is the seat of the Episcopal bishop of Sweden. The Stockholm Palace is on Stadsholmen Island in Gamla Stan and is the official residence of the King of Sweden. The City Hall of Stockholm is an iconic brick structure that was constructed from 1911 to 1923 on the Old City island of Kungsholmen. It is a popular tourist attraction and the site for Nobel Prize award banquets. The city's main public square is Sergels torg (Sergel's Square), named after an 18th-century sculptor whose workshop was once there. It is surrounded by various prominent shops, office buildings, and theaters, and has at its center a broad circular fountain that is centered by distinctively shaped obelisk. Other important landmarks are the Natural History Museum, the Museum of Modern Arts, and the Vasa Museum, a unique institution that focuses on a bad-luck 17th-century warship that sank in the city's harbor just as it was launched. Main theaters are the Royal Dramatic Theater, and the Royal Swedish Opera. The Ericsson Globe is an iconic hemispherical building that is the city's principal indoor arena.

Culture and Society. Increasing numbers of Stockholmers are of non-Swedish background, now estimated to be about 26 percent of the total. The main foreign-born minorities include Finns, Dutch, Germans, Turks, Kurds, Bosnians, Serbians, Croatians, and Syrians. Some of the city's suburbs have concentrations of poor immigrants from poor countries that had come to Sweden as economic migrants. The city as a whole, however, is quite prosperous and lives very comfortably. The national language is Swedish, but English is very widely understood and spoken by Stockholm's generally very highly educated population. The majority of the population is Christian, with about 70 percent of the population belonging to the Lutheran Church of Sweden. The main spectator sports are football (soccer) and ice hockey. Participation sports are also popular, as Stockholm residents in general are active and outdoors oriented no matter what the season.

Further Reading

Griffiths, Tony. *Stockholm: A Cultural History*. New York: Oxford University Press, 2009.
Hall, Thomas. *Stockholm: The Making of a Metropolis*. London: Routledge, 2009.
Khakee, Abdul. "From Olympic Village to Middle-Class Waterfront Housing Project: Ethics in Stockholm's Development Planning," *Planning Practice & Research* 22, no. 2 (2007): 235–51.

SUCRE

Sucre is the official capital of Bolivia, a landlocked country in central South America. It shares the duties of capital city with La Paz, the national's second largest city located some 250 miles (402 km) to the northwest, which is referred to as the "de facto" capital because so many of the nation's government offices are actually located there. Sucre is more properly the historical capital and the seat of Bolivia's Supreme Court. The city is located in the south-central part of Bolivia at an elevation of about 9,000 ft (2,750 m), and has a population of approximately 250,000.

Historical Overview. Sucre was founded on November 30, 1538, by Pedro Anzures, an official of Spain's early colonization efforts in the center of South America. Its first name was Ciudad de la Plata de la Nueva Toledo. It was a true Spanish colonial city with a grid plan of streets, a central square, and around the square the main church and administration buildings. In 1601, the Recoleta Monastery was founded in the city by Franciscans, and in 1609 the city was given its first archbishop. St. Francis Xavier University of Chuquisaca was founded in 1624 and is one of the oldest universities in the Americas. The fight for Bolivian independence from Spain began in Sucre and succeeded in 1825. Indeed, it is said that the movement for independence of all South and Central America from Spain originated in Sucre. The city became Bolivia's capital in 1826. In 1839, President José Miguel de Velasco renamed the city in honor of revolutionary leader Antonio José de Sucre. When the silver mining industry in nearby Potosí declined, Sucre lost strategic importance and many of the functions of government were switched to La Paz in 1898.

Major Landmarks. The Metropolitan Cathedral built between 1559 and 1712, the Archbishop's Palace built in 1609, the House of Freedom (Casa de la Libertad), and the National Library are among the many historic buildings in the center of Sucre. The central plaza of the city is Plaza 25 de Mayo. The main government building in the city is the Bolivian Supreme Court of Justice. There are many other historic churches, cemeteries, and residences, as much of Sucre retains a colonial look and feel and is unspoiled by modern intrusions. St. Francis Xavier University is still operating is another major historical landmark. In 1991, the city was designated as a UNESCO World Heritage Site.

Culture and Society. Sucre's population reflects that of Bolivia. Quechua and Aymara Indians are most numerous, but there are also many other native ethnic populations, as well as a sizable minority of mestizos (mixed Amerindians and whites) and whites, most of who descend from the early Spanish colonists. Whites are at the top of the economy and political power, while the poorest residents are native Amerindians. The main languages as Quechua, Aymara, and Spanish, but other native languages are spoken as well, and the main religion is Roman Catholicism. Sucre is an important repository of living indigenous culture in South America, including the rich weaving and other handicrafts of the local population and a distinctive musical culture.

Further Reading

Klein, Herbert S. *A Concise History of Bolivia.* New York: Cambridge University Press, 2011.
Zulawski Ann. *They Eat from Their Labor: Work and Social Change in Colonial Bolivia.* Pittsburgh: University of Pittsburgh Press, 1994.

SUVA

Suva is the capital and largest city in the Republic of Fiji, a small country consisting of 332, mostly of volcanic origin, islands (110 inhabited) and 500 or so islets in Melanesia in the South Pacific Ocean. Suva is located on a peninsula on the southeast coast of an island named Viti Levu, the country's largest island. This location is more or less central with respect to the country as a whole. The population of Suva is about 86,000 and that of the urban area about 172,000. In addition to being the national capital, Suva is the main regional urban center of the South Pacific between Hawaii and New Zealand, and an emerging center of export-oriented manufacturing.

Historical Overview. The Fiji Islands became a part of the British Empire in 1874 and Suva, a small village, was designated as the colonial capital in place of Levuka, a town on the island of Ovalau to the east that had been the seat of Fijian power. Levuka was squeezed between the sea and high cliffs, and lacked room for expansion. The transfer of capitals was made official in 1882. The British developed a sugar industry and imported labor from India. Suva grew as a result and in 1910 acquired municipal status. In 1952, after enlargement of the municipal boundaries and incorporation of adjacent towns, Suva became a city. Fiji achieved independence from Great Britain in 1970 and Suva became the national capital. In 1963, 1979, and 2003, Suva hosted the South Pacific Games, the first city to have been host three times.

Major Landmarks. Major landmarks in Suva include Government House, which was once the residence of colonial governors of Fiji but is now the official residence of the nation's president, the Parliament building, the Fiji Museum, Holy Trinity Anglican Cathedral, and Sacred Heart Roman Catholic Cathedral. The city also has a campus of the University of the South Pacific and the Fiji Institute of Technology. The downtown of Suva has many office buildings for banks, airline and travel companies, international shipping and trade, and Pacific regional NGOs. The tallest building in Suva is the 14-story Reserve Bank of Fiji Building. Amidst the government buildings of Suva is a memorial statue of Ratu Sir Lala Sukuna, a noted Fijian chief, soldier, scholar, and statesman from the first half of the 20th century.

Culture and Society. The majority of Suva's population is comprised of indigenous Fijians and Fijians who descended from 19th century labor migrants from India. Relations between the two groups have often been strained. There are also small minorities of Europeans and Chinese. Most people are Christian, but Hinduism predominates among the Indo-Fijians. The languages of the island nation are English, Bau Fijian, and Fijian Hindi. The national sport of Fiji is rugby, but football (soccer) is also popular.

Further Reading

Mamak, Alexander. *Colour, Culture & Conflict: A Study of Pluralism in Fiji*. Rushcutters Bay: Pergamon Press Australia, 1978.

Pabon, Laura, Nithin Umapathi, and Epeli Waqavonovono. "How Geographically Concentrated is Poverty in Fiji?" *Asia Pacific Viewpoint* 53, no. 2 (2012): 205–17.

Trnka, Susanna. *State of Suffering: Political Violence and Community Survival in Fiji*. Ithaca, NY: Cornell University Press, 2008.

TAIPEI

Taipei is the capital city and hub of the largest metropolitan area on Taiwan, a large island in the western Pacific Ocean off the southeastern coast of the mainland China. The island is also known as the Republic of China and functions in many ways as an independent country. However, the political status of Taiwan is disputed, as the People's Republic of China considers the island to be a renegade province instead of an independent country, so there is disagreement as to whether Taipei is truly a national capital or not.

The city's population is about 2.6 million, while the metropolitan area that it centers totals some 6.9 million inhabitants (2010). In addition to Taipei proper, the metropolitan area consists of New Taipei, a fast-growing residential zone that completely surrounds the city of Taipei, and the industrial port city of Keelung. The metropolitan area occupies the Taipei Basin at the northern tip of the island and is surrounded by mountainous terrain, volcanic peaks, and rugged national parks. The Tamsui River flows through the heart of the city.

Historical Overview. The Taipei Basin has been settled for centuries by local tribes. Migration from China began in 1709, after which Taipei began to develop as a city. The port in Tamsui, a part of what is today New Taipei, exported tea. From 1975 until 1895 when Japanese colonial rule commenced, Taipei was a Chinese prefectural capital. The Japanese designated the city as their colonial capital and called it Taihoku. They invested heavily in construction and infrastructure in Taipei, and laid the basis for industrialization. Colonial rule ended in 1945 when Japan was defeated in World War II. On December 7, 1949, Taipei was declared to be the provisional capital of the Republic of China by the Kuomintang (KMT) Party led by General Chiang Kai-shek, who fled mainland China after the civil war won by the Communists. With time, KMT claims on China weakened, and Taiwan established its own economy independent of the mainland. Taipei is now the hub of a prosperous and fast-growing industrial economy on Taiwan that includes the textile and apparel industries, shipbuilding, electronics, and high technology, among other exports.

Major Landmarks. The center of Taipei is the site of the National Chiang-Kai-shek Memorial Hall, an imposing monument honoring the military and political leader who established the rival government to China on Taiwan. Also in the Zhongzheng District of the city are the National Theater and National Concert Hall, 228 Peace Memorial Park (a monument to a massacre on February 28, 1947, by KMT troops by antigovernment protestors), and the beautiful Japanese-built Presidential Office Building. The National Sun Yat-sen Memorial Hall is a large and beautiful building that honors the founder of the Republic of China. The National Palace Museum houses some 600,000 artifacts from the art and culture of ancient China.

Taipei also has many modern-era office towers and shopping malls. The tallest building is Taipei 101, a landmark skyscraper that opened in 2004. Counting the antenna spire at the top, the structure is 1,670.6 ft high (509.2 m), and was for a short time the tallest building in the world. It takes its name because it has 101 stories above ground.

Culture and Society. Taipei is a modern city with an affluent consumer society. It also honors Chinese traditions, and celebrates festivals such as the Lantern Festival, Tomb-Sweeping Day, the Dragon Boat Festival, Mid-Autumn Festival, and the Ghost Festival. There are many beautiful Buddhist temples in the city, including Longshan Temple built in 1738. A night market in the Wanhua District of Taipei is famous for its Snake Alley, a place selling snake blood and meat among other exotic delicacies. The Chinese population of Taipei includes ethnic Hoklos, Mainlanders, and Hakkas. The population of the city also includes a small minority of aboriginal Taiwanese.

Further Reading

Allen, Joseph R. *Taipei: City of Displacements.* Seattle: University of Washington Press, 2011.

Lin, Cheng-Yi and Woan-Chiau Hsing. "Culture-Led Regeneration and Community Mobilization: The Case of the Taipei Bao-an Temple Area," *Urban Studies* 46, no. 7 (2009): 1317–342.

Marsh, Robert. *The Great Transformation: Social Change in Taipei, Taiwan since the 1960s.* Armonk, NY: M.E. Sharpe, 1996.

Wang, Jenn-hwan and Shuwei Hunag. "Contesting Taipei as a World City," *City: Analysis of Urban Trends, Culture, Theory, Policy, Action* 13, no. 1 (March 2009): 103–109.

TALLINN

Tallinn is the capital and largest city in Estonia, the smallest and northernmost of the three so-called Baltic countries of northern Europe that face the Baltic Sea to the west. The city is in the north of the country about midway along its shore with the Gulf of Finland. Helsinki, the capital of Finland, is located across the gulf about 50 miles (80 km) to the north. The population of Tallinn is 415,416 (2011), approximately one-third that of Estonia as a whole. Before 1918, the city was known as Reval.

Historical Overview. Archeological evidence indicates that the site of Tallinn has been inhabited for at least 5,000 years. The first castle at the site was erected in 1050 on a hill called Toompea that is now the center of the city. The city was connected to early trade routes between Russia and the Scandinavian countries of Sweden and Denmark, and then fell under Danish rule from 1219 during the Livonian Crusades in which Scandinavians tried to impose Christianity on the pagan regions of northern Europe. In 1285, Tallinn (Reval) became one of the cities of the Hanseatic League, an association of northern European trading ports, and prospered because of midway position between Russia and ports in Western Europe. In 1346, the Danes sold the city and northern Estonia to the Teutonic Knights and the region became part of the State of the Teutonic Order, the Ordensstaad. Because of German influence, Tallinn became a Lutheran city. In 1561, Tallinn and northern Estonia came under Swedish control. As a result of the Great Northern War, the city became part of the Russian Empire in 1710. In 1816, serfdom was abolished. An Estonian nationalist movement developed in the 19th century despite Russian attempts to enforce russification. In 1918, after the Russian Revolution in 1917, Estonia declared its independence and Tallinn became a national capital with its new, Estonian name.

The independent Estonian Republic existed until World War II when the country was occupied by Soviet troops 1940, then by Germany from 1941 to 1944, and then again by the Soviets from 1944 until the collapse of the Soviet Union in 1991. Tallinn became capital of the Estonian Soviet Socialist Republic, and the city's name was seen more commonly on English as Tallin, after the single-n spelling of the city's name in the Russian language. World War II had taken a great many lives in Estonia, but after the war, Estonia continued to suffer from Soviet repression of both Estonian nationalism and religion. Many Estonians were murdered in a Soviet campaign to subjugate the country, and tens of thousands were sent to forced labor camps in Soviet Siberia. In their place, many Russians were resettled into Estonia, particularly Tallinn and other cities, and took command of influential positions. Estonia achieved independence again in 1991, with Tallinn as capital. Estonian replaced Russian as the main language and many Russians repatriated. Estonia is now a member of NATO and the European Union.

Tallinn has prospered since independence and has emerged as a leading center of high-technology industry, as well as center of research and development. Estonia has been referred to as a "Baltic Tiger" and a "Silicon valley on the Baltic Sea." Tallinn is also growing as a popular destination for international tourism. The city compares favorably with other cities in the European Union in income levels and other economic indicators.

Major Landmarks. Most of Tallinn's most famous landmarks are in its beautiful and well-preserved medieval Old City. The district centers on the hill Toompea and is on the UNESCO World Heritage Sites lists. Highlights include remains of the old city walls and turrets; Raekoja Plats, the central square in the heart of the Old City; Reakoja, the old town hall built in 1371 that now houses the Tallinn City Museum; the onion-domed Russian Orthodox Alexander Nevsky Cathedral built in the 19th century; and Toomkirk, St. Mary's Cathedral, built as a Roman Catholic Church in 1229 and converted into a Lutheran Church in 1561. The spire of the old town hall has an historic weather vane of a warrior named Old Thomas that is a symbol of Tallinn. St. Olaf's Church with a tall Gothic spire was the tallest church building in the world from when it was constructed in 1549 until 1625. It reaches to 522 ft (159 m). Estonia's pretty pink Parliament Building, the Riigikogu, is also in Old City. Other landmarks are the Museum of Occupations, which focuses on the horrors of Nazi and Soviet rule in Estonia, and the former headquarters of the Soviet KGB (secret police), now the Estonian Interior Ministry building. There are various places on Toompea from where to enjoy a panoramic view of Tallinn. Viru Gate connects the Old City with Viru Street in All-Linn, the Lower Town, the

A view of Tallinn's well-preserved Old City district, a UNESCO World Heritage Site. (Maigi/ Dreamstime.com)

city's main shopping street. The Tornimäe business district is in another part of the center of Tallinn and is noted for new high-rise office buildings, modern shopping center, and international hotels.

Culture and Society. Tallinn's population is comprised of about 52 percent Estonian people and 38 percent Russians, the latter a product of the Soviet occupation of the country from World War II to 1991. The percent of Estonians has increased somewhat since Estonia's independence in 1991, as Estonians have become free to resettle in the capital from small towns and rural areas, and many Russians have emigrated. The official language is Estonian, a Finnic language, although Russian is widely understood and is used by many Russian residents. The population of Tallinn is generally well educated and increasingly prosperous as the economy of Estonia continues to strengthen.

Further Reading

Hackmann, Jörg. "Mapping Tallinn after Communism: Modernist Architecture and Representation of a Small Nation," in John Czaplicka, Nida Gelazis and Blair A. Ruble, eds., *Cities after the Fall of Communism: Reshaping Cultural Landscapes and European Identity*, 105–36. Washington: Woodrow Wilson Center Press, 2009.

Hannula, Helena, Slavo Radoševic, and Nick von Tunzelmann, eds. *Estonia, the New EU Economy: Building a Baltic Miracle?* Burlington, VT: Ashgate, 2006.

Miljan, Toivo. *Historical Dictionary of Estonia.* Lanham, MD: Scarecrow Press, 2004. http:// www.tallinn.ee/eng

TASHKENT

Tashkent is the capital and largest city of Uzbekistan, a doubly landlocked country (the country itself plus all countries that border it are without direct access to a sea) in Central Asia that was part of the Russian-Soviet Empire until 1991. The city is in the east of Uzbekistan near part of the country's long border with Kazakhstan, and is on the Chirchik River and some of its tributaries. To the east are the Altai Mountains. The population of Tashkent is officially about 2.2 million, although estimates say that the true population might be twice that. The city is the main cultural and business center of not just Uzbekistan, but of Central Asia more generally.

Historical Overview. Tashkent began as an oasis settlement at the foot of the Altai Mountains and was once known as Chach. In the middle of the seventh century, the city was converted to Islam and was renamed Binkath. Later, during the rule of the Turkic Kara-Kanide Khanate (999–1211), the city was named Tashkent, meaning "city of stone." In the Middle Ages, the city was an important stopover on caravan routes on the Great Silk Road between Europe and East Asia. It was conquered by Genghis Khan's Mongols in 1219 and destroyed. The city was rebuilt under the rule of the Timurid and then the Shaybānid dynasties. In 1809, the city was annexed by the Khanate of Kokand, and in 1865 it was conquered by the Russian Empire in a daring and unauthorized attack led by General Mikhail Chernyayev. The Russians designated the city as the capital of Turkistan and appointed General Konstantin von Kaufman as the first Governor-General. Russians began to settle the city and built European-style districts beside those that were indigenous. In 1918, Tashkent became the capital of the Turkestan Autonomous Soviet Socialist Republic. When that territory was split in 1924, Tashkent fell within the newly formed Uzbek Soviet Socialist Republic and temporarily ceased being a capital, as that designation went to Samarkand. In 1930, however, the capital of the Uzbek Republic was moved to Tashkent. During the Soviet period, many Russians and other Slavs moved to the city, and Tashkent was made into an important industrial center. It became the fourth-largest city of the USSR. On April 26, 1966, the city suffered a massive earthquake that took many lives and destroyed much of the city. The earthquake and subsequent reconstruction destroyed much of the pre-Russian fabric of the city, making the city even more Soviet in appearance. In 1991, the Soviet Union fell apart and Tashkent became capital of an independent Uzbekistan. The city has since been undergoing reconstruction and modernization to undo its Soviet look.

Major Landmarks. One of the two main squares in Tashkent is Mustakillik (Independence) Square. In the Soviet period, a 30-m statue of Vladimir Lenin stood here but it was joyously pulled down with the coming of independence for Uzbekistan and a new monument is in its place, a large world globe on which the sovereign territory is prominently engraved. The monument symbolizes Uzbekistan as a distinct country in the world, as well as Uzbekistan's membership in a world of nations. A second monument on the square is a mother and child, also a symbol of nationhood as well as a symbol of the future. Entrance to the square is through the Ezgulik, a symbolic archway decorated with storks taking

Samarkand and the Silk Road

Samarkand is Uzbekistan's second-largest city. In history it occupied a central position on the Silk Road, the storied trade route between China and the West, and in 14th century it became the capital of the powerful Central Asian ruler Timur (Tamerlane). The city is also known as an historic center of Islamic learning and culture. The Registan is the beautiful ancient center of Samarkand, and is a registered as a UNESCO World Heritage Site. The Bibi-Khanym Mosque, completed in 1404, is one of the main landmarks of the city, as is the mausoleum of Tamerlane, the spectacular azure-domed Gūr-e Amīr. In the 15th century, the city had an advanced astronomical observatory with a sextant that was 36 ft (11 m) long which was calibrated by the scholar Ulugh Beg.

flight. The second main square is Amur Timur Square. Its center is an equestrian statue of the great conqueror Timur (also known as Tamerlane; 1336–1405). A Russian Orthodox Church and a Soviet memorial honoring heroes of World War II were pulled down from this square after independence as part of a broader campaign to replace Russian cultural influences with those that are Uzbek or Islamic. However, the Soviet-built Monument of Courage which honors the bravery and perseverance of Tashkent residents after the 1966 earthquake still stands. Among the many other landmarks in the city are the Amur Timur Museum, the Tashkent History Museum in the building that was once the Lenin Museum, the Alisher Navoi Opera and Ballet Theater, the Prince Romanov Palace, and Tashkent Tower, a tall communications tower. The city also has a great many mosques, many of them newly built, Islamic schools, and traditional marketplaces, as well as many modern skyscrapers. Outside the city is Tashkentland, a popular amusement park.

Culture and Society. More than 80 percent of the population of Tashkent is ethnically Uzbek. Russians were once in the majority, but their numbers have declined with outmigration after Uzbekistan achieved independence. Nevertheless, Russians are still a significant minority, as are Tajiks other Central Asians. The official language of the country is Uzbek. Russian is still widely understood and is the main language of interethnic communication. More than 90 percent of the city's residents are Muslims and about 5 percent practice the Russian Orthodox faith. There are various forms of Islam in practice in Tashkent, including fundamentalist strains that have been linked to incidents of terrorism in the city. Islam is an ever more important aspect of life in Tashkent, and the city is coming to be one of the world's main centers of Islamic learning and thought. The city is growing rapidly with migration from the countryside of Uzbekistan.

Further Reading

Alexander, Catherine, Victor Buchli, and Caroline Humphrey, eds. *Urban Life in Post-Soviet Asia.* London: University College London Press, 2007.

Stronski, Paul. *Tashkent: Forging a Soviet City, 1930–1966.* Pittsburgh: University of Pittsburgh Press, 2010.

Tokhtakhodzhaeva, Marfua. *The Re-Islamization of Society and the Position of Women in Post-Soviet Uzbekistan.* Boston: BRILL/Global Oriental, 2008.

TBILISI

Tbilisi is the capital and largest city in Georgia, a small country on the eastern shores of the Black Sea in the Caucuses Mountains region of Eurasia. The city lies in the east-central part of the country on both banks of the Mt'k'vari River, in a valley with mountains on three sides. It was known as Tiflis until 1936. The population of both the city and the metropolitan area is about 1.5 million (2012).

Historical Overview. Tbilisi was founded in the fifth century by Vakhtang Gorgasali, the Georgian king of Causasian Iberia, a region that corresponds to a part of what is Georgia today. His eldest son Dachi I, who reigned from 522 to 534 moved the capital to Tbilisi from Mtskheta. From the second half of the sixth century into the 10th century, a succession of foreign occupiers ruled the city: Persians, Byzantines, Arabs, and Seljuk Turks. In 1122, Georgian troops under King David the Builder recaptured Tbilisi from the Turks and the city became the capital of a unified Georgian state. The ensuing almost 100 years were a golden age for the flowering of Georgian arts and culture. The great Georgian writer Shota Rustaveli (1172–1216) wrote his landmark epic poem "The Knight in the Panther's Skin" during this time while living in Tbilisi.

From the 13th century there were more conquests of Georgia, most notably by the Mongols in 1236, and then after a brief return to self-rule from 1320 to 1386, there were defeats of Georgia at the hands of other invaders and more centuries of foreign rule. In 1795, the city was burned to the ground by Persians. In 1801, Georgia joined in an alliance with the Russian Empire for protection against Persia and fell under Russian rule that lasted until the Russian Revolution in 1917. Georgia gained its independence at that time, and the Democratic Republic of Georgia was created in 1918 with Tbilisi as capital. In 1921, however, Communist Russian forces took command of Georgia and the country was annexed to the Soviet Union as part of the Transcaucasian SFSR and then as the Georgian Soviet Socialist Republic. The demise of the Soviet Union in 1989 led to Georgian independence again, but only after a period of turmoil and a brief civil war. There has been more stability, economic growth, and optimism since the Rose Revolution of 2003, a time of mass protests by citizens for honest elections and democracy. There was a setback in 2008 when Russia attacked Georgia in the South Ossetia War and instigated secessionist politics in one of the country's border regions.

Major Landmarks. The symbolic center of Tbilisi is Freedom Square located at the eastern end of Rustaveli Avenue. In the times of Imperial Russia, the space was known as Erivan Square after a conquering Russian general who was made count of Erivan (Yerivan) in neighboring Armenia, and during Soviet times it was first named Beria Square after the head of the Soviet Union's secret police, and then Lenin Square after the Russian revolutionary leader and founder of the Soviet Union. The name Freedom Square came after independence in 1989. A large statue of Lenin was torn town by happy crowds in 1991, and in 1996 a high Liberty Monument depicting St. George slaying a dragon was erected. Freedom Square was the site of a bank robbery in 1907 in which Lenin, Stalin, and other Bolshevik revolutionary robbed a bank in order to finance their movement, as well as the location of many Soviet Communist rallies and celebrations during the Soviet years, and of the Rose Revolution.

The city also has many beautiful churches, including Metekhi Church with a striking location at the edge of a gorge that the Mt'k'vari River has cut through central Tbilisi, the Holy Trinity Cathedral of Tbilisi (Sameba Cathedral), built in 1995–2004 and the third-tallest Orthodox cathedral in the world, and historic Sioni Cathedral. The oldest surviving church in Tbilisi is the sixth-century Anchiskhati Basilica of St. Mary. Overlooking the city from the top of a high hill is historic Narikala Fortress. Just below are the distinctive domes of Tbilisi's old sulfur baths. Other landmarks in Tbilisi include the Parliament of Georgia, the Supreme Court, the Tbilisi Opera and Ballet Theater, the Tbilisi State Conservatory, and the Shota Rustaveli State Academic Theater. The former headquarters of the Georgain Ministry of Highways (now owned by the Bank of Georgia) is an eye-catching unique building form from 1965, while in the center of the Tbilisi there is a graceful modern Bridge of Peace across the Mt'k'vari River.

Culture and Society. Nearly 90 percent of the population is ethnically Georgian, but Tbilisi also has a great variety of other ethnic groups, the most numerous being Russians, Armenians, and Azeris. More than 95 percent of the population is Christian, with the Georgia Orthodox Church being the main denomination. Georgians are proud of a history of religious tolerance, and point to a location in the center of Tbilisi where there is a rare instance in the world of a mosque and a synagogue standing next to one another. Because of the combination of a beautiful natural landscape and beautiful historic architecture, plus Tbilisi's role through history as a center of learning and intellectual life, the city has been the subject of great landscape painting more than most other cities that are not Venice or Paris. Prominent works from the Russian czarist period are "Tiflis" by Mikhail Lermontov (1837), a view of Metekhi cliff and surroundings by N. G. Chernetsov

The Presidential Palace in Tbilisi, Georgia. (Photo courtesy of Roman Cybriwsky)

(1839), and another view of Tiflis by Ivan Aivazovsky. There is considerable art that is sold in the city, some of it junk for tourists, while other pieces are exquisite.

Further Reading

Shavishvili, Nick. "An Exemplary Scheme to Breathe New Life into a Crumbling City," *Architectural Review* 228, no. 1364 (2010): 27–28.

Van Assche, Kristof and Joseph Salukvadze. "Tbilisi Reinvented: Planning, Development, and the Unfinished Project of Democracy in Georgia," *Planning Perspectives* 27, no. 1 (2012): 1–24.

Van Assche, Kristof, Joseph Salukvadze, and Nick Shavishvili, eds. *City Culture and City Planning in Tbilisi: Where Europe and Asia Meet.* Lewiston, NY: The Edwin Mellen Press, 2009.

TEGUCIGALPA

Tegucigalpa is the capital and largest city of Honduras, a country in Central America with shorelines on both the Pacific Ocean (Gulf of Fonseca) and the Caribbean Sea. The city is commonly called Tegus for short, and has in inland location on the Choluteca River in the south center of the country. Across the Choluteca is Comayagüela, the so-called "twin sister" city of Tegucigalpa, with whom some of the functions of national capital are shared. The population of Tegucigalpa is about 1.1 million, while that of the metropolitan area totals about 1.3 million (2010).

Historical Overview. Tegucigalpa was founded in 1578 by Spanish settlers as *Real de Minas de San Miguel de Tegucigalpa* on a site that had been a settlement of indigenous Amerindians. The settlers were drawn by the silver that could be mined in the area, hence the original name *Teguz-galpa* which means "silver mountain" in the local Nahuatl language. In 1762, the mining settlement was raised in status to *Real Villa de San Miguel de Tegucigalpa y Heredia*, and in 1821, at the time of Honduran independence from Spain it was designated as a city. Tegucigalpa and Comayagüela alternated duties as capital until 1839 when Comayagüela was designated sole capital until 1880 when the functions of capital were assigned to Tegucigalpa. The 1936 Honduran constitution united both cities as one political entity called the Municipality of the Central District (DC for short), making both cities the capital. Chapter 1, Article 8 of the current Constitution of Honduras, states that "the cities of Tegucigalpa and Comayagüela, jointly, constitute the Capital of the Republic." For much of the 20th century, the politics and economy of Honduras were heavily influenced by U.S. business interests and the country was called a "banana republic."

From October 29 to November 3, 1998, Hurricane Mitch devastated both Tegucigalpa and Comayagüela, along with the rest of Honduras, causing considerable damage from landslides and flooding and taking many lives. Neighborhoods of poor people on unstable hill slopes and in the flood zones of rivers and streams were most vulnerable. Infrastructure in the Capital District was severely disrupted for several years afterwards. The destruction was so widespread that even now, almost 15 years after the disaster, Honduras continues recovery and rebuilding efforts.

Major Landmarks. The main landmarks of Tegucigalpa include old churches from the colonial period such as San Miguel Cathedral (1765) and Los Dolores Church (1735), as well as the former presidential palace, which is now a museum, and the Manuel Bonilla Theater. Central Park, located in the heart of the historic center of the city, has an equestrian statue of General Francisco Morazán, an important political figure after Central America's break from Spain. Paseo Liquidambar is a popular pedestrian street in the historic center. The Mall-Multiplaza is a modern upscale shopping mall that is popular with the wealthier residents of Tegucigalpa. It stands in contrast to the slum neighborhoods that characterize much of the Central District.

Culture and Society. Honduras is characterized by extreme polarization of incomes, with a small very wealthy minority and a large population that is impoverished. Tegucigalpa and Comayagüela is mostly a poor urban area that is not only still recovering from the 1998 hurricane, but that never had proper housing and urban infrastructure in the first place. The metropolis is characterized by unplanned sprawl of squatter settlements, most of which lack basic services. The Choluteca River is contaminated by raw sewage. Industrial plants pay low wages to workers who manufacture clothing and other consumer items for export, mostly to the United States. Many Hondurans have emigrated to the United States and support the economy by sending remittances home.

Further Reading

Angel, Shlomo et al. "Rapid Urbanization in Tegucigalpa, Honduras," Woodrow Wilson School of Public and International Affairs, Princeton University, February 2004, http://wws.princeton.edu/research/final_reports/wws591g_f03.pdf

Leonard, Thomas. *The History of Honduras.* Santa Barbara: Greenwood/ABC-CLIO, 2011.

Meyer, Harvey K. and Jessie H. Meyer. *Historical Dictionary of Honduras.* Lanham, MD: Scarecrow Press, 1994.

THE HAGUE. *See* Amsterdam.

TEHRAN

Tehran (sometimes spelled Teheran) is the capital and largest city of the Islamic Republic of Iran, a country in western Asia between the Caspian Sea and the Persian Gulf. The city is in the northern part of Iran near the Elburz (also spelled Albroz) Mountains. The population of Tehran is about 12.2 million and the metropolitan area is about 13.8 million (2012). The city is growing rapidly with migration from the Iranian countryside. In addition to being national capital, Tehran is a major commercial and industrial center, a center of Islamic learning and political influence, and the largest city of western Asia.

Iran has had many capitals in its history and Tehran is the country's 32nd. Because Tehran is acknowledged to be too big and crowded, as well as "for security and administrative reasons," the Iranian government announced a desire in 2010 to move the capital to another city. Sharoud, Esfahan, and Semnan are candidates to be the next capital of Iran.

Historical Overview. The origins of Tehran are linked to the 11th-century city of Ray (also spelled Rayy), for a time the capital city of the Seljuk Empire. Ray is now a suburb of Tehran and is known for historic buildings and religious shrines. In the 13th century, Tehran was known as a regional market town, known especially for pomegranates. In 1796, the city became the seat of Āghā Mohammad Khān, the founder of the Qājār dynasty. In the 19th century there was growing Russian and British influence in Iran and much greater social stratification and inequality. Reza Shah founded the Pahlavi dynasty (1925–1979) after a coup d'état in 1921, and quickly consolidated power by reorganizing bureaucracy and establishing a new armed forces. The Anglo-Persian Oil Company (the predecessor to British Petroleum) produced great wealth for the Iranian ruling family and for British investors. In the 1930s, new palaces were built in the city and large parts of Tehran were modernized, most notably by addition of wide new roads. Reza Shah was forced to abdicate in 1941 because of his support of Nazi Germany in World War II, and his son Mohammed Reza Shah ruled the country until the dynasty was overthrown in 1979. He had considerable support from the United States. Tehran's modernization continued during his rule, but the social divide worsened, leading to the revolution that toppled the monarchy and propelled Ayatollah Ruhollah Khomeini into power on April 1, 1979. Iran was proclaimed an Islamic republic. For 444 days from November 4, 1979, to January 20, 1981, 66 Americans were held hostage in the U.S. Embassy in Tehran by Islamist students and militants who had taken control of the embassy in support of the Iranian Revolution. In the mid-1980s, the city was under frequent terrorist attack and suffered air assaults from Iraq.

Major Landmarks. Azadi (Freedom) Tower is the most iconic landmark of Tehran. It was built in 1971 and celebrates 2,500 years of the Persian Empire. Milad Tower (1,427 ft; 435 m) opened in 2008 and is a communications tower that ranks as the world's fourth-tallest concrete tower. Golestan Palace is one of the oldest historical monuments in Tehran, and consists of 17 palaces, museums, and halls. Niavaran Palace is another significant landmark. Important museums include the National Museum of Iran, the Tehran Museum of Contemporary Art, Iran's National Rug Gallery, and the Reza Abbasi Museum. The gigantic Mausoleum of Ayatollah Khomeini is in the south of the city. The Crown Jewels of Iran are on display in the Central Bank of Iran. Azadi Stadium which seats about 100,000 is the fourth-largest football (soccer) stadium in the world.

Culture and Society. Tehran's population reflects the great ethnic population diversity of Iran. Persians are the majority ethnic group and Azeri, Kurds, Arabs, Baluchis, Armenians, Bakhtiatis, and Assyrians are among the most populous minority groups. The mixed population is the result of rural to urban migration from all regions of the country. The main language in the city is the Tehrani accent of Persian. The religion of most people is Twelver Shia Islam. Since the Iranian Revolution of 1979, Iranians have lived under Islamic religious law. Women lack many basic

The iconic Azadi (Freedom) Monument in Tehran and Milad Tower. (Borna Mirahmadian/ Dreamstime.com)

rights. Many Iranians fled the country after the revolution and settled in the United Kingdom, the United States, and Canada, among other countries. The population enjoys skiing in the nearby mountains, football (soccer) as a spectator and participant sport, watching movies, and many other diversions.

Further Reading

Madanipour, Ali. *Tehran: The Making of a Metropolis*. New York: Wiley, 1998.
Madanipour, Ali. "Tehran," *Cities: The International Journal of Urban Policy and Planning* 16, no. 1 (1999): 57–65.
Nafizi, Azar. *Reading Lolita in Tehran: A Memoir in Books*. New York: Random House, 2008.

THIMPHU

Thimphu is the capital and largest city of Bhutan, a small, landlocked, Himalayan Mountain kingdom that borders the Republic of India and the Peoples Republic of China in South Asia. The city is located in the west center of the country along the west bank of the Thimphu Chuu River (also known as the Wang Chuu), and is elongated north south along the contours of the narrow valley. The population of Thimphu is 79,185 (2005), making the city one of the smallest capital cities of Asia. The city's name can also be spelled Thimpu.

Historical Overview. Thimphu became capital in of Bhutan in 1961, after a decision to move the seat of government from the ancient capital of Punakha. It is only then that Thimphu began to develop as a city. The site of Thimphu is where Ugyen Wangchuk, the first King of Bhutan, won a decisive battle that allowed him to unify the country under his control. The Changlimithang Sports Stadium built in 1974 marks the spot of the battle. In 2001, a National Constitution Committee began meeting in Thimphu to draft a constitution for Bhutan, and in 2008, under the fifth king of the Wangchuk dynasty, Jigme Khesar Namgye Wangchuk, made a peaceful change from being an absolute monarchy to a becoming a parliamentary democratic constitutional monarchy. A national objective is to elevate gross national happiness (GNH), identified in Bhutan as a better indicator of true quality of life than economic indicators such as gross national product (GNP) that are used around the world.

Major Landmarks. The main landmark in Thimphu is Tashiccho Dzong, a Buddhist monastery and fortress in the northern part of the city that is seat of Bhutan's government, and houses the monarchy's throne room and the offices of the king. There are numerous other monasteries in Thimphu and it is in the hills overlooking the city, including Simtokha Dzong, Dechen Phodrang Monastery, the Tango Monastery, and the Cheri Monastery. The Memorial Chorten, also known as the Thimphu Chorten (a chorten is a stupa), is the focus of an important traffic circle near the center of Thimphu. It was built in 1974 to honor the Third King of Bhutan. A large bronze statue of the Buddha called the Buddha Dordenma is under construction near the city. At 169 ft (51.5 m), it will be one of the world's largest statues of the Buddha. The statue will be gilded in gold and will contain 100,000 8-in. tall (20 cm) and 25,000 12-in. tall (30 cm) gilded Buddhas, respectively. Clock Tower Square is surrounded by shops and restaurants. The Motithang Takin Preserve in Bhutan is a wildlife preserve for the takin, the national animal of Bhutan.

Tashiccho Dzong, a 17th-century Buddhist monastery and fortress that is the seat of Bhutan's government. (Attila Jandi/Dreamstime.com)

Culture and Society. Bhutan is a very traditional society in which great efforts are made to protect Bhutanese culture from the onslaught of foreign influences and globalization. Tourism from abroad was first permitted only in 1974 and still is kept at fairly small numbers by limited access to the country, few hotels and other facilities, and high costs. Bhutanese people are mostly Ngalops or Sharchops, the two main ethnic groups of the country also referred to as Western Bhutanese and Eastern Bhutanese, respectively. The national language is Dzonkha, a language in the Tibetan language family, but there are also 24 other indigenous languages spoken in the country. Most Bhutanese follow Buddhism, the state religion, but there are also many Hindus. Most people in Bhutan dress in traditional clothes, the a knee-length robe for men that is called a *gho*, and the *kira* for women, an ankle-length dress that is clipped at one shoulder and tied at the waist.

Further Reading

Berthold, John. *Bhutan: Land of the Thunder Dragon.* Somerville, MA: Wisdom Publications, 2005
Pommaret, Françoise. *Bhutan: Himalayan Mountains Kingdom.* Hong Kong: Odyssey Books and Guides, 2009.
Wolcott, Susan M. "One of a Kind: Bhutan and the Modernity Challenge," *National Identities* 13, no. 3 (2011): 253–65.

TIRANA

Tirana is the capital and largest city of Albania, a small country on the Balkan Peninsula in southeastern Europe. The country borders the Adriatic Sea. Tirana is located about 30 miles (48 km) inland just to the north of the very center of Albania. It is in an area of hills at an elevation of about 360 ft (110 m). The main local rivers are the Ishëm and the Tiranë. The population of Tirana is about 620,000, while that of the metropolitan area is a little more than 1 million, about one-third of the national total (2011).

Historical Overview. The basin in which Tirana is situated has been occupied for many centuries. The Roman Emperor Justinian built a castle in the area in 520 AD. However, the city itself is said to have been founded as recently as 1614 by Sulejman Bargjini as an Ottoman trading town with a mosque and a Turkish bath. The first Christian settlers arrived in 1800. There was a revival of Albanian national consciousness in the late 19th century, and under the leadership of Ismail Qemali, Albania achieved independence from Ottoman rule in 1912. Tirana was designated to be the national capital in 1920. Between 1917 and 1923, architects from Austria developed a street and land use plan for the city. In 1929, Zog of Albania was crowned King Zog I. In World War II, Tiranba was the scene of battle between Albania Communists led by Enver Hoxha against Italian and German fascists. After the war, from 1946, Albania became a communist country, the People's Socialist Republic of Albania, with Hoxha as ruler until his death in 1985. Communism collapsed in 1991 and the Republic of Albania was established.

Major Landmarks. The central square in Tirana is named Skanderbeg Square (Sheshi Skënderbej), named after 15th-century Albanian leader against Ottoman expansion, George Kastrioti Skanderbeg. His statue stands in the square, as does an iconic clock tower and the National Historic Museum. The Museum of Albanian Clock Towers is nearby the E'them Bey Mosque, completed in 1821, is also located in Skandenberg Square. One more landmark from Skanderbeg Square is Piramida, a pyramid-shaped structure that was built at great cost in 1987 as a museum honoring Enver Hoxha. It has since been converted into a conventions and exhibitions space. The Resurrection of Christ Orthodox Cathedral is a newly built place of worship in the center of the city. Mother Theresa Square honors the well-known Albanian missionary nun who worked among the poor in India. The University of Tirana faces that square. The ruins of Justinian's castle, now referred to as Tirana Castle, are still another attraction.

Culture and Society. The majority religion of Albania is Islam, accounting for perhaps 70 percent of the national population. About 20 percent of Albanians are Orthodox Christians. The official language is Albanian, and Albanians are the dominant ethnic group, forming more than 95 percent of the population. Since the 1990s, ethnic Albanian refugees from Kosovo have settled Albania. Minority groups include Greeks, Bulgarians, and Macedonians.

Tirana is undergoing transition from an autocratic communist rule to democracy and a capitalist economy. There is considerable new construction and modernization, but critics point to lack of careful planning and illegal building in parklands and destruction of traditional neighborhoods to make way for new tall buildings by developers with political connections.

Further Reading

Abitz, Julie. *Post-Socialist Development in Tirana*. Roskilde: Roskilde University, 2006.

Gentiana, Kera, "Tirana," in Emily Gunzburger Makaš and Tanja Damljanović; Conley, eds., *Capital Cities in the Aftermath of Empires: Planning in Central and Southeastern Europe*, 108–22. London: Routledge, 2010.

Pojani, Dorina. "Tirana," *Cities: The International Journal of Urban Policy and Planning* 27 (2010): 483–95.

Pojani, Dorina. "Urbanization of Post-Communist Albania: Economic, Social, and Environmental Challenges," *Journal of Contemporary Eastern Europe* 17, no. 1 (2009): 85–97.

TOKYO

Tokyo is the capital and largest city in Japan, a nation of multiple islands located in the Pacific Ocean off the east coast of the Asian continent. The city is located on the eastern coast of the island of Honshū, Japan's largest island, at the head of Tokyo Bay, and is the focus of an enormous metropolitan region that spans all of Tokyo Prefecture (approximately 13 million in population in 2011) and the neighboring prefectures of Kanagawa, Chiba, and Saitama. With some 32.4 million inhabitants, the Tokyo metropolitan zone is far and away the world's largest. The region is also called Kantō, after Japan's largest plain. In addition to its role as the government center of Japan, Tokyo is a major business and financial center of global significance, and headquarters city for many Japan's largest corporations, including manufacturing companies, investment banks, advertising agencies, the Tokyo Stock

Exchange, insurance companies, and newspapers, television, and other media. It is also an important center of higher education, a major port, a tourism center, and center for world-class arts and entertainment. Manufacturing is also important, although it has been declining as much factory production has relocated to sites where land or labor costs are cheaper.

Historical Overview. The origins of human settlement in the Tokyo region are lost in prehistory, but the rise of Tokyo as an urban center is actually quite recent: the year 1457 when a minor war lord named Ōta Dōkan chose the head of Tokyo Bay as the site for his castle. The town that grew around the castle was named Edo, but remained small and insignificant until 1600 when the powerful warlord Tokugawa Ieyasu unified Japan and decided to make Edo his capital. An enormous castle was built in place of the original one. The city grew very rapidly because of the construction project and other population policies of the shoguns, and soon became one of the world's largest. The shoguns ruled Japan from Edo from 1600 to 1868. During most of this time, Japan was closed to the world and Japanese could not travel abroad. The shogunate fell in 1868 and imperial rule was restored, with the emperor moving his capital from ancient Kyoto to Edo, which was renamed Tokyo, meaning "Eastern Capital." A drive for modernization and Western learning commenced thereafter. It was strategically centered in Tokyo and resulted in Japan's rise as a global military and industrial power. On September 1, 1923, an enormous earthquake leveled Tokyo and took more than 100,000 lives. Japan's

The Shogun's Castle

The construction of Edo Castle by Tokugawa Ieyasu, the all-powerful first shogun of Japan, instantly propelled the small town of Edo into one of the largest cities in the world. The castle was enormous, and required workers and materials from all of Japan's provinces, plus the workers and materials to feed and house the construction workers. Construction began in 1603 and was finished in the 1640s. In 1657, however, a huge fire swept across Edo. Hot embers were carried across the moats and over the walls, and Edo castle was destroyed. So were all accurate drawings. The power of the Tokugawa shogun's was already assured, so the castle was no longer needed and was not rebuilt. Stones were hauled off for other construction, and only outer walls and bits and pieces of the shogun's fortress remain. The residence of the Emperor of Japan is located on the grounds of the old castle. Edo later became the great city of Tokyo. Throughout history construction has been a driving force in the urban economy and the politics of Japan, so much so that Japan is often referred to as a "construction state."

imperial ambitions ended with defeat in World War II, but Japan and Tokyo specifically emerged from the ashes and a period of occupation by U.S. troops to become one of the world's leading economic powers. In 1964, Tokyo hosted the summer Olympics. Much of the country's business is based on exports of automobiles, electronics goods, and other high-quality and newest-technology products. Since the 1990s, however, Japan has experienced a significant economic slowdown. This fact, coupled with the human and economic costs from the March 11, 2011, earthquake and tsunami off the coast of Honshu; north of the city, have cut deeply into Tokyo's once undisputed primacy as the leading business center of Asia.

Major Landmarks. The very center of Tokyo is the site of the Imperial Palace where the emperor of Japan lives. A part of the grounds is open as a park with lovely gardens and some ruins of the enormous castle that Tokugawa Ieyasu had built. The castle was destroyed not long after it was completed in a fire that swept the city in 1657, and most of the stones were carted off afterwards for other construction around the city. Nearby is the landmark Tokyo Station, an enormously busy and architecturally fascinating central passenger rail facility. Tokyo Tower is a major landmark from the period of reconstruction after World War II, as are some sports facilities that remain in the center of the city from the 1964 Olympics. Tokyo Sky Tree is a newly opened communication and observation tower that opened recently in another part of Tokyo. At 2,080 ft (634 m), it is the tallest structure in Japan and the second tallest in the world, and is meant to signify a resurgence of Japanese spirit after hard times. Important districts of Tokyo include Ginza known for its shops and nightlife, Shinjuku known as an area of office skyscrapers, shopping, and nightlife, Shibuya known as a popular district for young people, Akihabara known for its emporia selling the newest electronics goods and stores that appeal to anime fans, and Roppongi, a once-glamorous international nightclub district that is being redeveloped into a zone of fancy shopping malls, office towers, and museums. Odaiba, reached by via the landmark "Rainbow Bridge," is a new island reclaimed from Tokyo Bay that features a new beach, new shopping centers, amusement parks, and a replica of the Statue of Liberty.

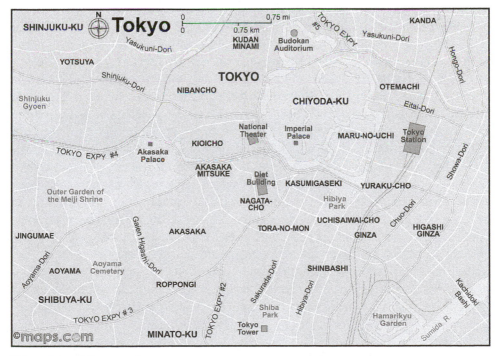

Tokyo, Japan (Maps.com)

A view across the Sumida River and the crowded city of Tokyo taken from atop Tokyo Sky Tree, a high new landmark in Japan's capital. Tokyo Tower is in the distance on the left side. (Photo courtesy of Roman Cybriwsky)

Culture and Society. Tokyo is the most international city in Japan and has foreign residents from around the world. Koreans and Chinese are the largest minority groups. There are also many Westerners who work for foreign and Japanese companies, or who are engaged in education, arts, entertainment, modeling, and other industries. Except for some small districts where there are more foreigners than usual, it is the Japanese who predominate in the population, as they do in the country at large. Along with nearby Korea, Japan is one of the most ethnically homogenous countries in the world. For many people in Tokyo, life is one of long commutes by train and subway to work or school in the center of the city from distant bedroom communities. The Tokyo public transportation system is fast and efficient, but many people have long and taxing commutes nonetheless. Like the rest of Japan, there is a sizable proportion of the population that is elderly, as Japanese tend to live longer than others while birth rates have been low. Because of a shortage of labor, as well as because many older people are short of money after retirement, quite a few elderly people have taken up postretirement part-time employment. Restaurants, nightlife, and cultural events in Tokyo are among the best in the world. The city is one of the world's safest cities from crime and quite clean and orderly.

Further Reading

Cybriwsky, Roman Adrian. *Roppongi Crossing: The Demise of a Tokyo Nightclub District and the Reshaping of a Global City*. Athens: University of Georgia Press, 2011.

Cybriwsky, Roman. *Tokyo: The Shogun's City at the Twenty-First Century*. Chichester, UK: Wiley, 1989.

Seidensticker, Edward, Richie, Donald, and Paul Waley. *Tokyo from Edo to Showa, 1867–1989: The Emergence of the World's Greatest City*. Tokyo: Tuttle Publishing, 2010.

TRIPOLI

Tripoli is the capital and largest city in Libya, a country in the Maghreb region of North Africa on the Mediterranean Sea. The city is located on a rocky site along the Mediterranean coast of the country in the far northwest, and is bordered by desert to the south. In addition to being the national capital, Tripoli is the largest manufacturing and business center of Libya and a busy port. The population of Tripoli is about 2.2 million (2011). The word "Tripoli" comes from the Greek and means "Three Cities."

Historical Overview. Tripoli was founded in about 500 BC by the Phoenicians and was known as Oea. It then passed to the rulers of Cyrenaica, a Greek colony on the south coast of the Mediterranean, and later to the Romans by the second-century BC. In the third-century AD, Oea was one of "three cities" in addition to Sabratha and Leptis Magna that comprised the Roman province of Regio Tripolitana (Region of the Three Cities). It supplied Rome with grain and slaves. In the fifth century, the region was sacked by Vandals, an East Germanic tribe, and Tripoli was heavily damaged. With the first Arab invasion in the seventh century, Tripoli became Muslim and was subsequently rebuilt. It was ruled for a time from Cairo by the Fatimid and Mamluk dynasties. After the second Arab invasion in 1046, the city's old walls were rebuilt using the Roman ruins as foundations. Between the 16th and 19th centuries, Tripoli was part of the Ottoman Empire. During that time, a number of the city's most prominent mosques and market places were constructed. In the late 18th and early 19th centuries, the city was a base of pirate activity in the Mediterranean Sea and scene of the Barbary Wars between the United States and the Barbary States of North Africa. The frigate USS *Philadelphia* was burned off the shore of Tripoli during the 1804 Second Battle of Tripoli Harbor.

Tripoli came under Italian control in 1911 as a result of a naval battle on October 1 of that year between Italian ships and the ships of the Ottoman Empire, and remained under Italian rule until 1943 when Italy's North African provinces of Tripolitania and Cyrenaica where captured by Allied forces in World War II. After the war, Libya was under the trusteeship of the United Kingdom and France until gaining its independence on December 24, 1951. On April 15, 1986, the United States conducted a bombing raid against Tripoli and Benghazi, another Libyan city, as retaliation for Libyan involvement in a terrorist bombing of a nightclub in West Berlin that was frequented by American servicemen and women. In 2011, Tripoli was the scene of fierce battles during a Libyan civil war that eventually resulted in the killing of long-time strong-armed dictator of Libya Muammar Gaddafi on October 20 of that year and the establishment of a new government. Gaddafi had ruled Libya from a military barracks compound known as Bab al-Azizia in the southern suburbs of Tripoli. As Gaddafi vainly struggled to regain power, he announced in a radio address on September 1, 2011, that the capital of Libya was being moved from Tripoli, which had been captured by rebels on August 23, to Sirte, but that plan ended with the rebel victory and the killing of Gaddafi a short time later.

Major Landmarks. The Red Castle Museum, or Assaraya Alhamra Museum, is located on Tripoli's promontory to the Mediterranean Sea, and was established in 1919 in the Red Castle fortress of the city. It displays more than 5,000 years of Libyan history. Nearby is Martyr's Square, formerly known as Green Square under the regime of Muammar Gaddafi and Piazza Italia (Italy Square) during the Italian colonial period. The city's old town, the Medina, is a landmark district of the city with narrow streets and old markets. The Ottoman Clock Tower is a well-known landmark in the Medina. The Maidan al Jazair Square Mosque was built as the Roman Catholic Cathedral of Tripoli in 1928 and was converted in 1970 into a mosque. The Gurgi and Karamanli mosques stand out for their beautiful decorations and tile work. Near the Gurgi mosque is the Arch of Marcus Aurelius, an architectural remnant from Roman times.

Culture and Society. Native Libyans are mostly Arabs or a mixture of Arab and Berber ancestries. The country also has many foreign residents, including as many as 1 million illegal migrants from neighboring Egypt and various countries of sub-Saharan Africa. There are also many Chinese, Filipinos, and Bangladeshis in Libya. The vast majority of the population is concentrated along a narrow Mediterranean coastal strip, and near the country's oil fields. The official language of Libya is Arabic (Libyan dialect), while Berber is spoken as well. About 97 percent of the population is Moslem. At present, the country is recovering from the 2011 civil war, which took at least 30,000 lives, and building a new government to replace that of Muammar Gaddafi who ruled from 1977 until he was killed.

Further Reading

St. John, Ronald Bruce. *Historical Dictionary of Libya.* Lanham, MD: Scarecrow Press, 2006.
St. John, Ronal Bruce. *Libya: From Colony to Independence.* Oxford: Oneworld Publications, 2008.
Vandewalle, Dirk. *A History of Modern Libya.* Cambridge: Cambridge University Press, 2006.

TUNIS

Tunis is the capital and largest city in Tunisia, formally the Tunisian Republic, a country on the Mediterranean Sea in the Maghreb region of North Africa that also borders Algeria and Libya. The city is in the north of the country near the Mediterranean's Gulf of Tunis, on a large (14 square miles; 37 sq km) and shallow lake named the Lake of Tunis (El Buhayra in Arabic) that was once the natural harbor of the city. The city is connected to canal and rail to its Mediterranean port city La Goulette (Hal al Wadi) approximately 6 miles (10 km) to the east. The historic city of Carthage (population 21,000) is about 12 miles (20 km) to the east and north of La Goulette, and is a now considered a suburb of Tunis. The population of Tunis is about 728,000, while that of the metropolitan area exceeds 2.4 million (2011).

Historical Overview. Tunis was founded soon after the founding of nearby Carthage in 814 BC. Carthage grew into a prosperous Phoenician trading city and the capital of the an expanding empire along the shores of the Mediterranean until the Third Punic War (149–146 BC) when it was conquered and destroyed by the Roman Empire. The small settlement at Tunis was destroyed in that war as well, and was then reestablished as a Roman city. It became a Muslim city during the seventh century spread of Islam, and then flourished from about 800 to 909 as a capital city of the Aghlabid Dynasty. From 1236 to 1574 it was the capital city of the Hafsid Dynasty, during which time the city was greatly built up and population levels increased significantly. In the 16th-century control of Tunis passed between the Spanish control and Turkish. The Turks gained long-term control and made the city a North African administrative center of the Ottoman Empire from the mid-16th to the mid-19th centuries. After the Ottomans lost the city, it became part of a French protectorate in 1881 that lasted until Tunisian independence in 1956. Under French administration, there was considerable industrial expansion and construction of urban infrastructure. From 1979 to 1990, the Arab League, an organization of 22 Arab nations, was headquartered in Tunis, and from the 1970s to 2003, the city was headquarters of the Palestinian Liberation Organization. In 2010–2011, Tunis was the focus of the Tunisian revolution that ousted longtime President Zine El Abidine Ben Ali and turned the country to greater democracy.

Carthage: Hannibal's City

Carthage was referred to as the "shining city" and was the powerful capital of more than 300 other cities in the Phoenician (Punic) world of the western Mediterranean Sea region. Hannibal, one of its most famous military commanders, launched an invasion of Europe in 218–201 BC that famously crossed the Alps into northern Italy with the help of war elephants and threatened the Roman Empire. Afterward, the city was completely destroyed by the Romans in the Third Punic War of 149–146 BC. Carthage is now a complex of historic ruins near Tunis, the capital of Tunisia. It is a UNESCO World Heritage Site, a popular visitor attraction, and an important archaeological preserve.

Major Landmarks. In addition to the ancient ruins of Carthage, which are designated as a UNESCO World Heritage Site and are reached easily from the center of Tunis by local train, the main landmarks of Tunis include Zitouna Mosque, which is Tunisia's largest mosque and dates back to the eighth century, the Roman Catholic Cathedral of St. Vincent de Paul, built in 1882 during the beginning of the French colonial period, and Bardo Museum, a museum about many centuries of Tunisian history that is housed in a former palace of an Ottoman ruler. Other landmarks are two famous gates to the city, Bab Saadoun and Bab Bahr, also known as the French Gate, and a statue of Tunis-born scholar Ibn Khaldun that stands in Independence Square. The Tunis Municipal Theater built in 1902 is a beautiful and historic venue for opera, ballet, and various concerts. The main street of Tunis is Avenue Habib Bourguiba, the city's main economic, cultural, and political center. It is often compared with the Champs-Élysées of Paris. To the southeast of Tunis, along the valley of the Wadi Milyān, are the remains of an ancient Roman aqueduct.

Culture and Society. The vast majority of Tunis residents are of Arab-Berber descent. Their language is Arabic, the national language of the country, but spoken with a Tunisian-Maghrebi dialect called Derja. French and English are understood by many Tunis residents. The official religion of the country is Islam, although the culture with respect to modes of dress and grooming is generally secular. Faiths other than Islam are permitted.

Tunis is an important industrial city. Major products include textiles, carpets, and olive oil. Tourism is also an important part of the economy, with many foreign visitors going to the ruins of Carthage, Mediterranean beaches, and shops in the old marketplaces of the city.

Further Reading

Keating, Michael. "Traditional Tunis and the City of the Century," *Washington Report on Middle East Affairs* 26, no. 9 (2007): 39–40.

Larbi, Hedi and Josef Leitmann. "Tunis," *Cities: The International Journal of Urban Policy and Planning* 11, no. 5 (1994): 292–96.

Woodford, Jerome S. *The City of Tunis: Evolution of an Urban System.* Wisbech, UK: Middle East and North African Studies Press, 1990.

ULAANBAATAR

Ulaanbaatar is the capital and largest city of Mongolia, a large landlocked country in the heart of Asia that is bordered by China to the south and Russian Siberia to the north. The city's name is also spelled as Ulan Batar. The city is located in the north-central part of Mongolia on the Tuul River at an elevation of 4,430 ft (1,350 m). In addition to being capital, Ulaanbaatar is the main industrial and financial center of Mongolia and the hub of its road, rail, and airline networks. The population of the city is about 1.2 million (2012), more than 40 percent of that of all Mongolia which itself is a lightly settled country outside the capital.

Historical Overview. Mongolia's capital was once migratory along with the population and Mongolian princes, and the city was founded in 1639 as a movable monastery. It was first named Da Kure and then Örgöö. After 28 changes in location, including one to what is now Inner Mongolia in China, it settled at its present site in 1778. In the 19th century, Russians called the city Urga. For them it was a base for trade between Russia and China. In 1911, Mongolia achieved independence from Qing China and the city was renamed Niislel Khureheh (Capital of Mongolia). In 1921, it was occupied by troops of Mongolia's revolutionary leader, Damdiny Sükhbaatar, and the Red Army of the Soviet Union. The name Ulaanbaatar, which means "Red Hero" in honor of Damdiny Sükhbaatar, was given to the city in 1924 when Mongolia was declared a people's republic. The city was subsequently built up along Soviet lines with a planned central square and monumental government buildings, large industrial districts and tracts of apartment-block housing for workers, and a wide range of cultural and educational institutions. All in all, the Soviet influence produced a drab city. A peaceful democratic revolution took place in Mongolia in the early 1990s and a new constitution for the country was put into effect in 1992, dropping "People's Republic" from the name of the country. The economic development of the country relies heavily on Mongolia's rich mineral resources such as the enormous deposits of gold and copper at the Oyu Togoi mine in the Gobi Desert in the south.

Major Landmarks. The central square of the city is the Sükhbaatar Square, a vast open space in front of the Parliament Building of Mongolia. There is an equestrian statue of the hero Sükhbaatar in the center, and a large seated statue of Genghis Khan, the great 13th century leader of the Mongol Empire, at the top of the stairs to the Parliament Building. Outside the city on a bank of the Tuul River is a 131-ft-high (40 m) statue of an equestrian Genghis Khan that is covered with 250 tons of stainless steel. The Zaisan Memorial is a Soviet-period monument that honors Russians and Mongolians who fought together in World War II. It is on a hill at the city's edge and offers a fine view of Ulaanbaatar. Other landmarks are the Gandan Monastery, which is the city's main monastery, the Choijin Lama Monastery, the Natural History Museum,

The central square of the city is the Sükhbaatar Square, a vast open space in front of the Parliament Building of Mongolia. (Simone Matteo Giuseppe Manzoni/Dreamstime.com)

the Bogd Khan Museum, the Zanabazr Museum of Fine Arts, and the Russian-built Opera House. The center of the city is now sprouting modern high-rise buildings.

Culture and Society. Mongolian society has traditionally depended on herding and agriculture for a living, but that is changing with urbanization, modernization, globalization, and internationalization. Mining is becoming a mainstay of the economy, and foreign mining firms are now intertwined with the economy of Mongolia. Many Mongolians work abroad, particularly in South Korea, and send remittances home. About 95 percent of the population of Mongolia is ethnic Mongol. Mongolian is the principal language. Because of Russian influence it is written in the Cyrillic script as opposed to the historic indigenous script of Mongolia. However, there are now efforts underway to gradually replace Cyrillic with Mongolian script. The main foreign language that is spoken in Mongolia is Russian, although there is considerable attention nowadays to learning English and Chinese, and Russian is slipping. The main religion is Buddhist, especially Mongolian Tibetan Buddhist. The Soviet influence repressed religious practice, with the result that only about one-half of the population identifies with any religion.

Further Reading

Kamata, Takuya, James Reichert, Tumentsogt Tsevegmid, and Yoonhee Kim. *Managing Urban Expansion in Mongolia: Best Practices in Scenario-Based Planning.* Washington: The International Bank for Reconstruction and Redevelopment/World Bank, 2010.

Rossabi, Morris. *Modern Mongolia: From Khans to Commissars to Capitalists.* Berkeley: University of California Press, 2005.

Sabloff, Paula L. W. *Modern Mongolia: Reclaiming Genghis Khan.* Philadelphia: University of Pennsylvania Museum of Archaeology and Anthropology, 2001.

V

VADUZ

Vaduz is the capital and second-largest urban settlement, in the Principality of Liechtenstein, a tiny landlocked country in Central Europe that is bordered by Austria and Switzerland. It is located in the west of the country on the right (east) bank of the Rhine River, just south of Schaan, the largest settlement in the country (population 5,800). With only about 5,300 inhabitants (2009), Vaduz is, like Schaan, more of a town than a city. Liechtenstein itself has only about 36,000 inhabitants.

Historical Overview. Vaduz was probably founded in the 13th century by the counts of Werdenberg, at the time a county of the Holy Roman Empire that covered what is today Liechtenstein and adjoining parts of Austria and Switzerland. A castle existed there in 1322, which was sacked by the Swiss in 1499 during the Swabian War. Liechtenstein itself began with the Liechtenstein family of Austria which was able to purchase the land in 1699 and in 1712 in order to be eligible for a seat in the Diet, the *Reichstag*, that is, the general assembly of the Holy Roman Empire. On January 23, 1719, Charles VI, the Holy Roman Emperor, decreed that this purchased land (now 6.7 square miles; 17.3 sq km) shall henceforth be Liechtenstein, a sovereign state within the Empire. It took some 120 years after that date for any one of the Princes of Liechtenstein to set foot in their new principality.

Major Landmarks. The main landmark of Vaduz is the picturesque Vaduz Castle that sits high on a hill within the town limits and can be seen from almost anywhere within the town. Construction began in the 12th century and was incremental afterwards. It was purchased by the Liechtenstein family in 1712 and has been the home of the reigning prince of Liechtenstein and his family since 1938. Other family members have resided in the castle as well, perhaps accounting for the many additions that have been made to the structure over the centuries and renovations. Other landmarks in Vaduz are the Roman Catholic Cathedral of St. Florin constructed in 1874, the Government House, City Hall, The National Art Gallery, and the National Museum.

Culture and Society. Liechtenstein is a unitary parliamentary democracy and a constitutional monarchy ruled by the Liechtenstein family. The national language is German, with most people speaking the Alemanni dialect. More than three-quarters of the population is Roman Catholic. The economy is based on foreign tourism and on financial services. Liechtenstein has very low business taxes that attract foreign companies that wish to have an official registration address in the principality. The country is quite prosperous, even by the high standards of Europe, and ranks first or second in the world GDP per capital. Increasingly, the workforce of Liechtenstein is made up of foreigners, including many workers from Turkey.

Vaduz Castle overlooking the city of Vaduz. (Bekaze/Dreamstime.com)

Further Reading

Beattie, David. *Liechtenstein: A Modern History*. London: I. B. Tauris, 2004.
Eccardt, Thomas M. *Secrets of the Seven Smallest States of Europe: Andorra, Liechtenstein, Luxembourg, Malta, Monaco, San Marino and Vatican City*. New York: Hippocrene Books, 2004.

VALLETTA

Valletta is the capital of the Republic of Malta, a small archipelago country in the mid-Mediterranean Sea between Italy's island of Sicily and the coast of North Africa. The city is located on the central-eastern coast of the main island, also called Malta like the country, and has a population of 6,966 (2011). It consists of a walled historic city, and newer accretions outside the walls.

Historical Overview. Malta's location in the mid-Mediterranean Sea has been strategically important for many centuries and underlies the history of the country's succession of governing powers: Phoenicians, Greeks, the Roman Empire, Arabs, Normans, the Kingdom of Aragon, Hapsburg Spain, the Knights of St. John, Napoleon's France, and the British Empire. The British used Malta as a shipping base midway between Gibraltar and the Suez Canal. The country gained its independence from the United Kingdom in 1964 and became a republic in 1974. In 2004, Malta was admitted into the European Union, and in 2008 it adopted the Euro as its currency.

Valletta itself was founded shortly after 1565 when the island was successfully defended against an attack by Ottoman forces by the ruling Order of St. John of Jerusalem, also known as the Knights Hospitaller, a military and religious organization connected with the Crusades and the Christian defense of Jerusalem. Valletta was founded by the Order in order to better fortify the island, and was named after the Grandmaster of the Order, Jean Parisot de la Valette. Its official name is Humilissima Civitas Valletta, the Most Humble City of Valletta. The first building completed was Our Lady of Victories Church in thanks for intercession in the epic battle against Ottoman invaders. La Valette's body was originally interred in this church. The city was designed by Francesco Laparellli, and was given a rectangular grid street pattern and wider streets than the norm for the times. The Royal Opera House was erected in 1866, but was destroyed in air raids during the German and Italian Siege of Malta in World War II. In 1980, Valletta hosted the world Chess Olympiad.

Major Landmarks. Valletta is a UNESCO World Heritage Site with many historic buildings and a distinctive cityscape. The city stands out for its historic defenses, high walls, and watchtowers, and for its many historic churches and palaces, museums, theaters, and lively streets and piazzas. The main churches are our Lady of Victories Roman Catholic Church, the Co-Cathedral of St. John where La Valette's body is presently entombed, St. Francis of Assisi Church, the Church of the Jesuits, Christ the Redeemer Church, and others. The most famous palaces are the Magisterial Palace of the Grandmaster, which is now the seat to the Maltese House of Representatives and the office of the president of Malta, the Auberege de Castille, Auberge d'Aragaon, Auberge d'Italie, now the office of the Maltese Tourism Authority, and Casa Rocca Grande. Other landmarks are the National Museum of Fine Arts, the Grandmaster's Palace Armory Museum, the National Museum of Archaeology, and Manoel Theater. The Valletta Waterfront is a contemporary commercial center with restaurants and shops in old warehouses at the water's edge. There are two natural harbors in the city, Marsamxett and Grand Harbour.

Culture and Society. More than 95 percent of the population is ethnically Maltese. The official languages are Maltese, a Semitic language related to a strain of Arabic, and English. Some 98 percent of the population is Roman Catholic and Roman Catholicism is the official state religion, although the constitution provides for freedoms of worship by other faiths as well. Christianity has a long history in Malta, beginning most notably, it is said, with the shipwreck in AD 60 of the Apostle Paul who spent three months on the island performing miracle cures of the sick. Malta is a member of the European Union's Schengen Agreement. As a consequence of its midway position between the Mediterranean coasts of Europe and Africa, it has been the destination of many African refugees and prospective immigrants, at least as a stopover before going to the European continent. The economy of Malta is based on tourism, banking and finance, and increasingly, film production.

Further Reading

Azzopardi, A. *A New Geography of the Maltese Islands*. Valletta, Malta: Progress Press, 1995.
Cassar, Godwin. *A Concise History of Malta*. Msida, Malta: Minerva Publications, 2000.

Chapman, David and Godwin Cassar. "Valletta," *Cities: The International Journal of Urban Policy and Planning* 21, no. 5 (2004): 451–63.

VATICAN CITY

Vatican City is the capital of a sovereign state with the exact same borders that is also called Vatican City, or the State of Vatican City, or as it is known officially by its Italian name, Stato della Città del Vaticano. It is an enclave within the city of Rome, the capital of Italy, and with an area of just 0.17 square miles (0.44 sq km), is the world's smallest fully independent nation-state. For most of its circumference, Vatican City is walled. The head of state of Vatican City is the Roman Catholic Pope, presently Benedict XVI. The city is the seat of the Holy See, the headquarters of the global Roman Catholic and Eastern Catholic Churches. The Holy See is headed by the Bishop of Rome, also the Pope. The population of Vatican City is only 832, is overwhelmingly male, and consists mostly of the hierarchy of the Roman Catholic Church, other Church officials, papal guards, and various assistants, and service workers. As an independent country, Vatican City issues passports, but only very few. The Holy See also issues passports, but mainly diplomatic ones, and also very few.

Historical Overview. The Roman Catholic Church has been governed from Rome for most of its nearly 2,000-year history beginning with the tenure of Peter, the first pope, who was in office from 32 to 67 AD. The Vatican also governed the Papal States on the Italian Peninsula during their existence from roughly the sixth century until unification of Italy in 1861. However, Vatican City as a sovereign national entity dates only to 1929. It was established in that year by the Lateran Treaty that was signed by Cardinal Secretary of State Pietro Gasparri on behalf of Pope Pius XI and by Italian Prime Minister and Head of Government Benito Mussolini on behalf of Italy's King Victor Emmanuel III, and ceded what had been Italian sovereign territory to the Roman Catholic Church.

The Pope, who is elected by the College of Cardinals and serves until death, has full and absolute executive, legislative, and judicial power over Vatican City. He is the only absolute monarch in Europe. Legislative authority is given to the Pontifical Commission for Vatican City State, a commission that is made up of cardinals that are appointed for 5-year terms by the Pope. The president of that commission holds executive power in the Vatican. The Vatican's foreign relations are in the hands of the Secretariat of State.

Major Landmarks. The main landmark of Vatican City is the magnificent St. Peter's Basilica, consecrated in 1626 and the center of the Roman Catholic World. It has a dome designed by Michelangelo and the largest interior space of any Christian church in the world. Outside is St. Peter's Piazza (Square) where many faithful gather on special occasions to receive blessings from the Pope. The boundary between Vatican territory and the territory of Italy is marked by a white line painted on the square. The Vatican Museum is one of the great treasures of the world and is highlighted by the Sistine Chapel that was decorated by Michelangelo and other renaissance artists. The Sistine Chapel is part of the Apostolic Palace, the residence of the Pope. Castel Gandolfo is an extraterritorial property of the Vatican in the outskirts of Rome that is best known as the summer residence of the Pope.

Culture and Society. The population of Vatican City is comprised mostly of priests and nuns of the Roman Catholic Church and lay workers who work in Church administration or in various service capacities. Of the 557 citizens of Vatican City who resided in the country in 2005, 74 percent were clergy, 58 of them cardinals. The business of the Holy See is conducted in Latin, while Italian is the language of administration of Vatican City. The population of Vatican City is global in origin (although Italians predominate), so many other languages are spoken at various times. The 101 Swiss Guards who guard the Pope work in German. All of the residents of Vatican City are Roman Catholic.

Further Reading

McDowell, Bart and James L. Stanfield. *Inside the Vatican.* Washington, DC: National Geographic, 2009.

Williams, Paul L. *The Vatican Exposed: Money, Murder, and the Mafia.* Amherst, NY: Prometheus Books, 2003.

VICTORIA

There are many cities in the world named Victoria but this one is the capital and largest city of the Republic of Seychelles, an independent country that is comprised of 115 small islands in the Indian Ocean some 932 miles (1,500 km) east of the African continent. As a country, Seychelles is generally considered to be a part of Africa. With a population of less than 90,000 the country has the smallest population of any African state. The population of Victoria is about 25,000 (2009), the smallest of all African capitals. The city is situated on the northeastern side of the granitic island of Mahé, the main island of the Seychelles archipelago. It is sometimes called Port Victoria.

Historical Overview. The islands of Seychelles are located between Africa and Asia, and have been visited for centuries by navigators and traders between the two continents, as well as by Maldivians and Arabs. Pirates also had bases on the islands. The first Europeans to sight the islands were probably those who accompanied Portuguese Admiral Vasco da Gama during his explorations of 1502. In 1756, the islands became a possession of France. They were named after King Louis XV's Minister of Finance, Jean Moreau de Séchelles. Great Britain gained control of the islands in the early 19th century. French and English land owners exploited African slaves on their plantations. After slavery was outlawed in 1835, the British imported Indians to work as indentured laborers. The main plantation crops were sugar, as well as cinnamon, copra, and vanilla.

Seychelles gained its independence from Great Britain in 1976 and Victoria became a national capital. Its harbor exports vanilla, coconuts, coconut oil, fish, and guano. Tuna canning is an important local industry. The mainstay of the economy, however, is tourism, as the islands are known for spectacular beaches and sparkling ocean waters. Seychelles International Airport was completed in 1971 and supports the tourism economy. The city experienced significant damage from the Indian Ocean tsunami toward the end of 2004.

Major Landmarks. The center of Victoria has a clock tower that is modeled after the Vauxhall Clock Tower in London. Other points of interest are the Victoria

Botanical Gardens, the Victoria National Museum of History, and the Victoria Natural History Museum. Other landmarks are the court house, the state house, a national stadium, and the local market selling fish, fruits, and produce.

Culture and Society. Seychelles is a diverse society comprised of migrants and their descendants. The main ethnic groups are of African origin, Indians, Chinese, and French. The official languages are English and French, and Seychellois Creole, a local derivation based mostly on French. More than 90 percent of the population is Christian, with Roman Catholics comprising the large majority. Despite the historical connection to Great Britain, French culture and language is disproportionately stamped on Seychelles. For example, an estimated 70 percent of the population has a family name of French origin.

Further Reading

Franda, Marcus F. *The Seychelles: Unquiet Islands.* Boulder, CO: Westview Press, 1982.
Scarr, Deryck. *Seychelles since 1770: History of a Slave and Post-Slavery Society.* Trenton, NJ: Africa World Press, 1999.

VIENNA

Vienna (German: Wien) is the capital and largest city of Austria, a small landlocked country in central Europe. It is located on the Danube River in the Vienna Basin in the northeast of Austria. The population of the city is about 1.7 million, while that of the metropolitan area is about 2.4 million (2011). The city is nicknamed the "City of Music" because of its rich legacy of classical music and "The City of Dreams" because it was the home of Sigmund Freud, founder of psychoanalysis. Vienna consistently ranks as one of the world's best cities in terms of overall quality of life. It is a popular tourism destination, and the city center of Vienna is a UNESCO World Heritage Site.

Historical Overview. The site of Vienna was settled as far back as 500 BC by Celts. In 15 BC, the Romans established a fortification at the site on the Danube River as protection for their northern frontier against Germanic tribes. From the mid-15th century until the mid-18th century, Vienna was the seat of the House of Hapsburg, one of the most powerful royal families of Europe. It ruled the Austrian Empire and the Spanish Empire, and was the de facto capital of the Holy Roman Empire. In 1529 and again in 1679, the armies of the Ottoman Empire were stopped in epic battles near the city, critical historical events that arrested the further spread of Islam into the heart of Europe. From 1804, the city became capital of the Austrian Empire and in 1867 until 1918 it was capital of the Austro-Hungarian Empire. During that time, its population exceeded 2 million and Vienna was one of the largest and most powerful cities in the world. From September 1814 to June 1815, the city hosted the Congress of Vienna, a major meeting of European heads of state and their ambassadors that redrew the political map of Europe in the wake of the Napoleonic Wars and other turmoil. Throughout the 19th century and well into the 20th century, Vienna was a leading center of European cultural and intellectual life, and home of many leading composers, writers, scientists, and political thinkers. After the First Austrian Republic was formed in 1919, Vienna's housing conditions and social services for the mass of citizens were greatly improved by the Red Vienna socialist movement. In World War II, Vienna and Austria were occupied by Nazi Germany, and after the war the city was contested between the victorious Allies of the West and the Soviet Union until the Soviets relinquished their claims in 1955. In the 1970s, the Vienna International Center was opened as an area for the offices of international organizations such as the International Atomic Energy Agency, the United Nations Industrial Development Organization, and the Organization of Petroleum Exporting Countries.

Major Landmarks. The center of Vienna is in a historic square called Stephensplatz which is dominated by the majestic 12th century St. Stephen's Cathedral, the seat of the Vienna Diocese of the Roman Catholic Church. The enormous Hofburg Palace was built in stages between the 13th and 20th centuries and was the royal palace of the Hapsburg court. It now houses the official residence of the president of Austria among many other uses such as library, the Vienna Hofburg Orchestra concert hall, museums, and government offices. An equestrian statue of Emperor Joseph II stands outside the Imperial Library. Schönbrunn Palace, with 1,441 rooms, was the imperial Hapsburg summer residence. The structure known as the Gloriette is a major landmark on the palace grounds. Other Vienna landmarks are the Belvedere Palace; the mid-19th-century Vienna State Opera in the center of city; Burgtheater, the Austrian national theater; the Kunsthistorisched Museum, also called Museum of Art History or the Museum of Fine Arts, and the Museum of Natural History of Vienna. A visit to the Spanish Riding School to see the Lipizzan horses in demonstration is a must for all visitors to Vienna. Modern Vienna is represented by Donau City, an enormous new zone of office buildings, hotels, and other prominent structures along the Danube River that was constructed starting 1995.

Culture and Society. Vienna is a German-speaking city. The major religion, as in Austria as a whole, is Roman Catholicism. The city is known for its rich cultural life that centers on classical music performances and opera, and on many museums and galleries. The Vienna Boys' Choir is an especially famous aspect of the city's cultural life. The city is also known for its many cafés and distinctive coffees, and for cuisine that includes Wiener Schnitzel (veal cutlet) and superb cakes and other desserts. The city is green with many parks, walking trails, and bicycle paths, and has an excellent public transportation system. Historically there was a large and influential Jewish population, but it was greatly reduced by emigration and the horrors of the Nazi Holocaust. The population of the city now includes many immigrants from other countries, most notably Serbs and Turks.

Further Reading

Hatz, Gerhard. "Vienna," *Cities: The International Journal of Urban Policy and Planning* 25 (2008): 310–22.

Lichtenberger, Elisabeth. *Vienna: Bridge between Cultures.* New York: Wiley, 1993.

Parsons, Nicholas. *Vienna: A Cultural History.* Oxford: Oxford University Press, 2008.

VIENTIANE

Vientiane is the capital and largest city of the Lao People's Democratic Republic, a landlocked Southeast Asian country that is most commonly called Laos. The city is located in the western part of the country on a bend of the Mekong River which forms a part of the boundary between Laos and Thailand. The First Thai-Lao Friendship Bridge, built in the 1990s, spans the river and connects the two countries 11 miles (18 km) downstream from Vientiane. The population of Vientiane is about 754,000 (2009). In addition to being the capital, Vientiane is the main economic center of Laos.

Historical Overview. According to the Laotian epic *Phra Lam Phra Lam*, Vientiane was founded by Prince Thattaradtha on the west side of the Mekong River at what is today Udon Thani, Thailand, and then after being instructed to do so by a seven-headed *naga*, reestablished the city on the eastern bank of the river at the present site. Historical evidence, however, suggests that the site of Vientiane was once an ancient Khmer settlement centered around a Hindu temple. Afterward, when the Lao people settled the area and the Lan Xang kingdom was founded by Fa Ngum in 1354, Vientiane became an important administrative city. In 1563, King Setthathirath made the city the capital of Lan Xang. In 1707, the city became the capital of a separate Kingdom of Vientiane, and in 1779 it was conquered by the Siamese kingdom and incorporated into neighboring Siam. The Siamese burned the city to the ground in 1827 while putting down a Laotian rebellion. In 1893, the city was annexed into French Indochina. The French rebuilt Vientiane and added considerable colonial architecture. In World War II, Vientiane was occupied by the Japanese. Laos declared independence when the war ended in 1945, but France continued to rule until 1950 when Laos was granted autonomy within the French Union. In 1953, Laos achieved full independence from France. During the years 1964–1975, the Laos Civil War raged between the forces of the Communist Pathet Lao and those of the Royal Lao Government, and the country became a spillover battleground of the Vietnam War. There was covert involvement in the conflict by the U.S. Central Intelligence Agency; hence, the war is known by Americans as the Secret War in Laos. The Lao People's Democratic Republic was proclaimed on December 2, 1975, with the abdication of the last king of Laos.

Major Landmarks. Perhaps the best-known landmark in Vientiane is Patuxai, also known as the Victory Gate, a war memorial in the center of the city that is dedicated to those who fought for Lao independence from France. It was constructed between 1967 and 1968 and resembles the Arc de Triomphe in Paris in general form, but is highlighted with traditional Lao design motifs and Buddhist mythological figures. Other prominent landmarks are Buddhist monuments such as Pha That Lunag, a 148-ft-high (45 m) stupa that was originally built in 1566 by King Setthathirath and restored in 1953, the temple Wat Si Muang built in 1563, and Buddha Park, an outdoor collection of Buddhist and Hindu sculptures developed in 1958 along the river about 17 miles (28 km) south of Vientiane. There are also the large stupa That Dam, the temple Wat Si Saket, the monastery Wat Ong Teu Mahawihan, and the National Assembly

Patuxai Gate in Vientiane, a monument that commemorates the Laotians' victory against French colonialism. (Photo courtesy of Roman Cybriwsky)

Building. Other attractions in the city are the Lao National Museum and the Talat Sao morning market.

Culture and Society. The main ethnic group of Laos comprises the Lao people, accounting for about 60 percent of the national population and a higher fraction in Vientiane. There are many other ethnic groups among the Laotian population, particularly in rural areas and the mountains. Vientiane has a Chinese minority that is disproportionately engaged in commerce. The official language of the country is Lao, although only a little more than half of the population can speak it. The rest speak local ethnic languages. French is still common in Vientiane and use of English is growing rapidly. About two-thirds of the population of Laos identifies with Theravada Buddhism. The Buddhist faith has long had an important impact on the culture of Laos and on daily routines such as morning offerings to Buddhist monks and prayer rituals at Buddhist temples.

Further Reading

Askew, M., W. S. Logan and C. Long. *Vientiane: Transformations of a Lao Landscape.* London: Routledge, 2007.

Raffiqui, Pernilla S. and Michael Gentile. "Vientiane," *Cities: The International Journal of Urban Policy and Planning* 26 (2009): 38–48.

Rigg, Jonathan. *Living with Transition in Laos—Market Integration in Southeast Asia.* London: Routledge, 2005.

Walsh, John and Nittana Southiseng. "Vientiane: A Failure to Exert Power?" *City: Analysis of Urban Trends, Culture, Theory, Policy, Action* 13, no. 1 (March 2009): 95–102.

VILNIUS

Vilnius is the capital and largest city of Lithuania, a country in northern Europe that borders the Baltic Sea. The city is in Vilnius County in the southeast of Lithuania at the confluence of the Vilnia and Neris Rivers, not far from the country's border with Belarus. The location of Vilnius is commonly regarded to be the geographical center of Europe. The population of the city is about 544,000, while that of Vilnius County is about 839,000 (2011).

Historical Overview. Vilnius began to grow into a city in the late 13th and early 14th centuries during the reign of Vytenis, Grand Duke of Lithuania. The first written records of Vilnius date to 1323 and 1324 in letters by Grand Duke Gediminas to German cities inviting settlers, to Pope John XXII, and to the Dominican and Franciscan orders of Catholic priests, among others. These letters described Vilnius as a capital city. A legend says that Gediminas had received a sign to make Vilnius into a great capital city in a dream about a howling iron wolf on a hilltop. During his reign (1316–1341), Gediminas greatly enlarged the Grand Duchy of Lithuania, making it a powerful state. His grandchildren, however, fought one another in the Lithuanian Civil War of 1389–1392, which resulted in the razing of Vilnius. In the early 16th century, the city expanded and was enclosed with walls for protection. There were nine gates and three towers. In 1569, as a result of the Treaty of Lublin, the Polish-Lithuanian Commonwealth was formed. The dualistic state was one of the largest and most populous states of Europe until it was dissolved in 1795, and was ruled by a single monarch. Kraków, Warsaw, and Vilnius were all capitals at one time or another. Vilnius prospered during his period and developed into a significant center of culture and learning.

A series of wars in the 17th century that was collectively referred to as Deluge, devastated Vilnius, as did an outbreak of bubonic plague in 1710 and outbreaks of fire. In 1795, Vilnius was annexed into the Russian Empire. The city was captured by the armies of Napoleon Bonaparte in 1812 as they advanced toward Moscow, but it was then returned to Russia with Napoleon's defeat. After a nationalistic uprising in 1863, the Lithuanian language was banned, and the city became primarily Russian, Belarusian, and Jewish. In the aftermath of World War I, control of Vilnius changed hands frequently between an independent Lithuanian state, Poland, Russia, and its successor the Soviet Union, with the city being annexed in 1922 by Poland and becoming capital of the Wilno Voivodship. In 1939, the city was taken by the Soviet Union and was briefly incorporated into the Byelorussian Soviet Socialist Republic. From the summer of 1941 to the summer of 1944, Vilnius was occupied by the Nazi Germans. During that time, two Jewish ghettos were established in Vilnius and the great majority of Lithuania's 265,000 Jews were murdered. Some 80,000 victims were Jews from Vilnius. After the Soviet Red Army regained control of the city, Lithuania was annexed into the Soviet Union and Vilnius became capital of the Lithuanian Soviet Socialist Republic. Lithuanian independence was declared on March 11, 1990, as the Soviet Union was in the process of collapse. A last-ditch show of force by Soviet troops in January 1991 failed, and Moscow recognized Lithuanian independence in August of the year.

Major Landmarks. The red-brick Gediminas' Tower, is the last remaining part of the old Vilnius castle that was erected by Gediminas, and is one of the city's main historical landmarks. The city's Old Town district, once the hub of Jewish Vilnius, has historic buildings from the 13th to 19th centuries and has been designated as a UNESCO World Heritage Site. Vilnius University in the Old Town district is the largest and oldest university in Lithuania, with 23,000 students. The Three Crosses monument memorializes seven Franciscan monks who were martyred in 1333 at the time of Lithuania's conversion to Christianity. The original monument was built in 1916 and was destroyed by Soviet authorities in 1950. The present structure dates to post-independence Lithuania in 1989. Major churches in Vilnius include Vilnius Cathedral, St. Ann's Church, and St. Peter's and Paul's Church. The major museums include the National Museum of Lithuania and the Museum of Genocide Victims in the former headquarters building of the Soviet KGB (secret police). In 1995, the world's first statue of musician Frank Zappa was erected in Vilnius.

Culture and Society. Vilnius is a multinational city with many Poles, Russians, Belarusians, and Ukrainians in its population, in addition to the Lithuanians who make up nearly 60 percent of the total. The Lithuanian proportion has risen markedly since the country's independence. About 80 percent of the population is Roman Catholic and 4 percent Orthodox. The Jewish population, once dominant in the cultural life of the city, is now very small. The official language of Lithuania is Lithuanian, but Russian is still spoken by some residents. There has been a flowering of Lithuanian arts, literature, and other culture since the end of the Soviet period.

Further Reading

Aleksandravičius, Egidijus. "Post-Communist Transition: The Case of Two Lithuanian Capital Cities," *International Review of Sociology* 16, no. 2 (July, 2006): 347–60.

Briedis, Laimonas. *Vilnius: City of Strangers*. Budapest, Hungary: Central European University Press, 2009.

Vaisviaite, Irena. "The Changing Face of Vilnius: From Capital to Administrative Center and Back," in John Czaplicka, Nida Gelazis, and Blair A. Ruble, eds., *Cities after the Fall of Communism: Reshaping Cultural Landscapes and European Identity*, 17–52. Washington, DC: Woodrow Wilson Center Press, 2009.

Venclova, Tomas. *Vilnius: A Personal History*. Riverdale-on-Hudson, NY: Sheep Meadow Press, 2009.

WARSAW

Warsaw is the capital and largest city of Poland, a country in east-central Europe that faces the Baltic Sea to the north. The city is on the Masovian Plain in the east center of the country, and is divided into two unequal parts by the Vistula River. To the west of the river is the bulk of Warsaw, including its main commercial and historic core, while to the east is a smaller district of residential and industrial land uses called Praga. The population of Warsaw is 1,708,491, while that of the Warsaw metropolitan area is 2,666,278 (2011). In Polish, the name of the city is Warszawa, which is thought to come from the diminutive form of a Polish masculine name. A folk legend says that Warsz was a fisherman who fell in love with a mermaid Sawa who lived in the Vistula River.

Historic Overview. A fortress stood at the site of Warsaw as early as the 9th or 10th century, but the modern city is said to have been founded in about 1300 by Prince Bołeslaw II of Masovia. It became capital of Masovia in 1413 and then was seat of the Polish Sejm (lower house of Parliament) in 1569. From 1596 to 1795 the city was capital of the Polish-Lithuanian Commonwealth. It prospered during this period and was known for beautiful palaces and churches, and as an important center of commerce, education, and cultural and artistic life, particularly during the 1764–1795 reign of Stanisław August Poniatowski, the last king and grand duke of the commonwealth. In 1795, Poland was partitioned and was annexed into the Kingdom of Prussia, with Warsaw becoming the capital of the province of South Prussia. In 1806, the city became capital of the Duchy of Warsaw created by Napoleon, and then in 1815 capital of the Congress Kingdom of Poland that was created by the Congress of Vienna. From 1830 until World War I, Poland was under Russian rule. There were demonstrations and revolts against Russian authority in the 1860s. Warsaw grew during the 19th century as an industrial city and railroad hub, and was greatly modernized during the 1875–1892 administration of Russian-appointed Mayor Sokrates Starynkiewicz. Poland became independent in 1918, with Warsaw as capital. On August 12–25, 1920, Polish troops defeated an advance by Red Army troops in the heroic Battle of Warsaw just east of the city.

World War II devastated Warsaw. The city was bombed by the Nazi forces on September 1, 1939, and fell after hard-fought resistance on September 28. The Nazis occupied the city for more than five years until it was liberated by Soviet forces on January 17, 1945. By that time, most of the city was in ruins and some 800,000 residents, more than one-half of the city's population before the war, had been killed. The Germans had employed special troops known as "Burning and Destruction Detachments" in order to destroy Warsaw before they capitulated to the Red Army. On orders of the German Governor-General Hans Frank on October 16, 1940, the large Jewish population of the city and surrounding towns was rounded up and herded into a small area near the city center that was designated as the Warsaw Ghetto. Some 400,000 Jews were crammed into 1.3 square miles (3.4 square km). From there, most were taken for extermination in Nazi death camps. The Warsaw Ghetto Uprising was launched against German occupation on April 19, 1943, and held strong for nearly one month until it was finally defeated. Almost all survivors were massacred. On August 1, 1944, Polish civilians in Warsaw launched the Warsaw Uprising against approaching Soviet troops, hoping to restore Polish independence. They were defeated as well, with as many as 200,000 civilian casualties in the resistance.

A Communist government was set up in Poland after World War II and the People's Republic of Poland was officially proclaimed in 1952, with Warsaw as capital. The city was rebuilt as a socialist city. Because of a housing shortage, large tracts of prefabricated apartment blocks were constructed around the city. Warsaw's historic Old Town was reconstructed as well, and in 1980 it was entered into the UNESCO World Heritage Sites list. Opposition to Communist rule grew through the Roman Catholic Church, a powerful influence in Polish life, and visits to Poland in 1979 and 1983 by Pope John II, a native Pole, as well as through the trade union movement Solidarity. By 1980, Solidarity had evolved into a potent political force that challenged Communist authority and forced the first free elections in 1989. In 1990, Solidarity leader Lech Wałęsa was elected president. In 2004, Poland became part of the European Union. The economy of Poland has been comparatively strong since the end of the communist period, and Warsaw has enjoyed a construction boom and considerable economic growth.

Major Landmarks. The Old Town (Stare Mesto) of Warsaw centers on Castle Square (Plac Zamkowy). In the center is a tall column atop which is a statue of King Sigismund III Vasa who moved Poland's capital to Warsaw from Krakow. The Royal Castle that was founded in the 14th century faces the square on the east side. It was totally destroyed in World War II under the direct orders of Adolf Hitler, but was painstakingly rebuilt afterward to the glory that it displayed during Warsaw's heyday as capital of the Polish-Lithuanian Commonwealth. Other landmarks in Old Town are St. John's Cathedral, the Historical Museum of Warsaw, the Literature Museum, and the Barbican, an historic red-brick defensive tower. In Warsaw's financial district is the Palace of Culture and Science, a Stalinist-era colossus that was built as a gift of friendship from the Soviet Union to the people of Poland. It is much reviled in Warsaw for its symbolism, as well as for its overly large size and "Russian wedding cake" design. The site of the Warsaw Ghetto has some surviving buildings from the time of World War II, particularly along Prózna Street, the surviving Nozyk Synagogue, and various monuments to the Warsaw Ghetto Uprising, as well as a new museum. In another part of Warsaw, the beautiful baroque Wilanów Palace was built in the last quarter of the 17th century as the principal residence of Polish King John III Sobieski.

Culture and Society. Until the Holocaust of World War II, Warsaw had long been a multicultural city with a mix of ethnic groups and religions. The city was an especially vibrant center of Jewish cultural and intellectual life, with as much as one-third of the population or more being Jewish. However, the decimation of the Jewish population during the war and resettlement of other ethnic groups such as Germans and Ukrainians afterward resulted in an unusually homogeneous population for Poland and its capital city. More than 90 percent of the people are Poles and nearly 90 percent are Roman Catholics, making Poland one of the most heavily Roman Catholic countries in the world. Among the many globally famous individuals who called Warsaw home are Marie Skłdowska-Curie (1867–1934), the scientist noted for her research about radioactivity; the Polish composer and virtuoso pianist

The 1955 Palace of Culture and Science. (Photo courtesy of Roman Cybriwsky)

Frédéric Chopin (1810–1849); and the popular Jewish writer Isaac Bashevis Singer (1902–1991). The city is once again a thriving center of intellectual and artistic life, with many galleries, museums, music festivals, concert halls, theaters, and other venues for creative arts and performance.

Further Reading

Arens, Moshe. *Flags Over the Warsaw Ghetto: The Untold Story of the Warsaw Ghetto Uprising.* Springfield, NJ: Gefen Publishing House, 2011.
Crowley, David. *Warsaw*, London: Reaktion Books, 2003.
Niemczyk, Maria. "Warsaw," *Cities: The International Journal of Urban Policy and Planning* 15, no. 4 (1998): 301–11.

WASHINGTON, DC

Washington, also known as Washington, DC, the District of Columbia, the District, or simply DC, is the capital of the United States of America. It is located in the eastern part of the country on the northern banks of the Potomac River on land that had been ceded in 1790 by the state of Maryland for the purpose of building a national capital. The city is, therefore, not in any of the 50 states, but within a specially designated district for the capital that is called the District of Columbia. Originally, there was a city of Washington within the District of Columbia, but

Pierre Charles L'Enfant: Washington Architect

Pierre Charles L'Enfant (1754–1825) was a French-born American surveyor and civil engineer who was most famous for designing the iconic street layout for Washington, DC. He was appointed for the task in 1791 by President George Washington, and consulted also with Thomas Jefferson, arguing for a grand city with monumental architecture and dramatic vistas. The result was a grid street plan atop which was a series of wide diagonal streets that were named after the states of the union, and prominent traffic circles at their intersections. L'Enfant's concepts were not all adopted directly, but he was responsible for the basic form of the center of the national capital, including an early version of the National Mall which he had envisioned as a wide "grand avenue."

with the passage of the Organic Act of 1871 the city's charter was revoked, along with that of the neighboring city of Georgetown, and "Washington" and the District of Columbia became coterminous.

The population of Washington is 617,996 (2011 estimate), the 24th most populous city in the United States. The Washington DC Metropolitan Area, which includes parts of Maryland and Virginia in addition to the District, totals 5.58 million inhabitants, the seventh-largest metropolitan total in the United States. The city is named after George Washington (1732–1799), the first president of the United States.

Historical Overview. Washington has been the U.S. capital since 1800, with the first session of Congress in the city taking place on November 17 of that year. Before then eight other cities, most notably Philadelphia, Pennsylvania, had served as meeting places for Congress and can be considered to have been previous capitals. The origins of Washington as national capital trace to July 16, 1790, when the U.S. Congress approved a legislation called the Residence Act of 1790 to create a special federal district that would serve as the seat of national government. The original District of Columbia was a square 10 miles by 10 miles in area on either side of the Potomac River on land that had been ceded by both Maryland (the larger portion) and the state of Virginia, and included the preexisting cities of Georgetown in Maryland (founded in 1751) and Alexandria in Virginia (founded in 1749), but the Virginia portion was returned by Congress in 1846. After the capital was moved to Washington in 1800, Congress passed the Organic Act of 1801 which officially recognized the District of Columbia and put the entire district under direct federal control.

Washington was attacked by British forces on August 24–25, 1813, during the War of 1812, and the Capitol building, Treasury, and White House were burned. The buildings were restored afterward, with the Capitol being completed in its present form in 1868. The city grew quickly during the American Civil War (1861–1865) because the functions of government expanded and because many freed slaves moved into the nation's capital. On April 15, 1865, U.S. president Abraham Lincoln was assassinated in Washington while attending a play at Ford's Theater. Washington and the U.S. government grew rapidly again during the New Deal of the 1930s, during which time many government buildings and monuments were constructed. Such constructions were executed during World War II, too. In 1961, the 23rd Amendment to the U.S. Constitution was passed to give residents of the District of Columbia three votes in the Electoral College. On August 28, 1963, Dr. Martin Luther King, Jr. delivered his historic "I have a Dream" speech to a crowd of well over 200,000 on the National Mall during the Great March on Washington for Jobs and Freedom. In 1973, the District of Columbia Home Rule Act provided an elected mayor and city council in Washington, and in 1975, Walter Washington became the first elected and the first Africa-American mayor of the city.

The city of Washington was planned by Pierre Charles L'Enfant, a French-born architect and urban planner who was commissioned for the task in 1791 by President George Washington. He planned broad streets and avenues that radiated from public squares, and a broad "grand avenue" that grew into today's National Mall. Because of a dispute between L'Enfant and the president, the planning of Washington was completed by Andrew Ellicott. The center of the city has the main government buildings of the United States and monuments. The city is divided into four quadrants of unequal size that radiate from the U.S. Capitol Building: Northeast (NE), Southeast (SE), Southwest (SW), and Northwest (NW). These designations are used in the city's address system, with house numbers indicating the approximate number of blocks that the structure is from the Capitol.

Major Landmarks. Washington is replete with landmarks, including the U.S. Capitol Building, the White House (the residence of the president of the United States), the Washington Monument, and the National Mall, along which are the various museums of the Smithsonian Institution and various memorials such as the Lincoln Memorial, the National World War II Memorial, the Korean War Veterans Memorial, and the Vietnam Veterans Memorial. Other landmarks are the National Gallery of Art, the U.S. Supreme Court Building, the Library of Congress, the Martin Luther King, Jr. Memorial, and the National Archives. Massachusetts Avenue has many foreign embassies and is referred to as "Embassy Row." Arlington National Cemetery is across the Potomac River in Arlington, Virginia. The John F. Kennedy Center for the Performing Arts is home to the National Symphony Orchestra, the Washington National Opera, and the Washington Ballet. The U Street Corridor is known for its famous jazz clubs and other night spots. Major universities in Washington include Georgetown University, American University, and George Washington University.

Confederacy Capitals

Eleven slave states broke off from the Union in 1861 and declared themselves to be the Confederate States of America, triggering the Civil War that the rebels ultimately lost to Union forces in 1865. The new country had two capital cities during its short history. The first capital was Montgomery in Alabama until May 29, 1861, and the second was Richmond, Virginia, until April 3, 1865. The choice of Richmond was driven in part by the presence in the city of the Tredegar Iron Works which supplied Confederate forces with iron and artillery. The Confederate Congress met in the Virginia State Capitol Building and the executive mansion of President Jefferson Davis, the "White House of the Confederacy," was two blocks away. The city was greatly damaged in battle on April 2, 1865, as the Civil War drew to a close.

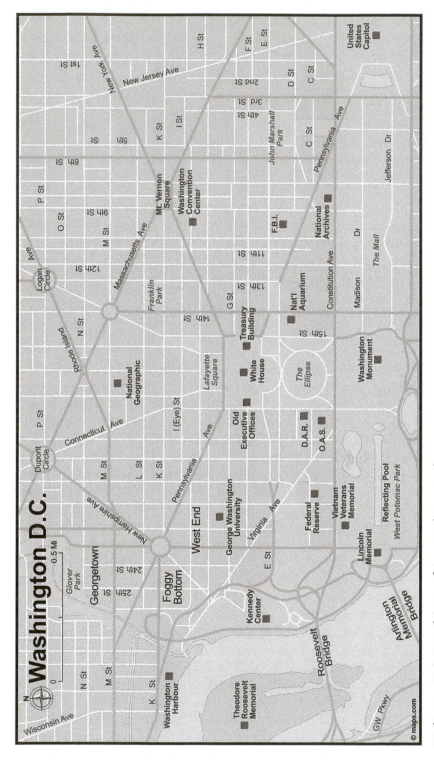

Washington, DC, United States of America (Maps.com)

Culture and Society. Washington reached a peak population in the 1950 census year when it counted 802,178 residents. It declined with every decade since until the 2010 census, which counted a gain of more than 5 percent from 2000. In 1970, about 70 percent of the population was African American. That figure had declined to about 51 percent by 2010, as many African American families have been moving to suburbs outside the District of Columbia, while the city's white population has been increasing with gentrification of many neighborhoods that had been traditionally African American. The city also has many immigrants from abroad, especially from El Salvador, Vietnam, and Ethiopia, and a thriving restaurant culture based on ethnic fare. There is a Chinatown neighborhood in the center of Washington, although it is more a commercial district than residential.

> ### San Juan: Pride of Puerto Rico
>
> San Juan is the capital and largest city (population 395,000) in Puerto Rico, a territory of the United States since it was acquired from Spain in 1898 in the Spanish-American War. The city was founded by Spanish settlers in 1521 and is the second-oldest European-built city in the Western Hemisphere after Santo Domingo, the capital of the Dominican Republic. Its original name was Ciudad de Puerto Rico, "city of the rich coast." The relationship between Puerto Rico and the United States is being reevaluated in 2012 in a plebiscite of voters of Puerto Rico, all of whom are also citizens of the United States. They could vote to continue territorial status, to recommend statehood, or to separate and to form an independent country. There have been separatist feelings in Puerto Rico since the start, but only a minority has voted for Puerto Rican independence in the past.

Residents of the District of Columbia have no voting representation in the U.S. Congress. They are represented in the House of Representatives only by a nonvoting delegate. There is no representation for Washington in the U.S. Senate. Because Washingtonians are subject to U.S. taxes like every other U.S. citizen, they have adopted an unofficial motto "Taxation without Representation" to describe what they consider to be an unfair situation. Their efforts for enactment of a District of Columbia Voting Rights Amendment to the U.S. Constitution have been unsuccessful.

Further Reading

Bordewich, Fergus. *Washington: The Making of the American Capital*. New York: HarperCollins Publishers, 2008.

Knox, Paul L. "The Restless Urban Landscape: Economic and Sociocultural Change and the Transformation of Metropolitan Washington DC," *Annals of the Association of American Geographers* 81, no. 2 (1991): 181–209.

Ruble, Blair A. *Washington's U Street: A Biography*. Baltimore: The Johns Hopkins University Press, 2010.

WELLINGTON

Wellington is the capital and third-largest city in New Zealand. It is located near the geographical center of country on the southern tip of New Zealand's North

Island and faces South Island across Cook Strait. The city is the southernmost capital city in the world (latitude 41°17′S), as well as the capital city that is located furthest from any other national capital. The city was named by English settlers after Arthur Wellesley, the first Duke of Wellington in England. New Zealanders call it "Windy Wellington" because of strong winds that blow in from Cook Strait. The population is about 395,600 (2012). The urban area of Wellington comprises the city of Wellington plus the nearby communities of Porirua, Lower Hutt, and Upper Hutt. The city is known for its beautiful setting between mountains and coastline, and a high standard of life that consistently ranks among the best in the world.

Historical Overview. The original inhabitants of the Wellington area were the Maori people. The first English settlers arrived in 1839, with a second group arriving in January 1840. They called their first settlement Britannia. It was on wet terrain at the mouth of the Hutt River on Wellington Bay, so the community moved to the present site of Wellington soon thereafter. Wellington was made the capital of New Zealand in 1865, replacing Auckland which was thought to be too far from the country's center. Major earthquakes devastated the city in 1848 and 1855. The latter tremor, known as the Wairarapa Earthquake, was so strong that it permanently changed the configuration of Wellington's coast line. City status was conferred on Wellington in 1886.

Major Landmarks. One of the most distinctive buildings in Wellington is the Executive Wing of the New Zealand Parliament Building constructed between 1969 and 1981. It has a distinctive circular-conical shape and is referred to as the "Beehive." The Museum of New Zealand Te Papa Tongarewa is the national museum and art gallery. The Maori words are translated as "the place of treasures of this land." Other landmarks include the Museum of Wellington City & Sea, Old St. Paul's Church, the Wellington Citizens War Memorial, the Wellington Cable Car, and the Botanic Garden. Lambton Quay is the heart of downtown Wellington. The tallest buildings are the Majestic Centre and State Insurance Building (formerly named BNZ Centre), both on Willis Street in a part of the Central Business District known locally as the Golden Mile. Cuba Street, named after an early settler ship, is a popular pedestrian-oriented commercial street in the downtown. Mount Victoria (196 m) offers an excellent lookout. Zealandia is a protected wildlife sanctuary in suburban Wellington.

Culture and Society. Approximately three-quarters of the Wellington population is from European backgrounds, mostly British. Most of the rest of the residents are Maori (7.4% in the city and 12.3% in the metropolitan area as a whole, as Maori comprises a large fraction of the population of Porirua), Pacific islanders, and immigrants and their descendants from Asia, mostly ethnic Chinese and Indians. The city has an unusually large arts and culture scene for a city of its size, with many theaters, museums, music and arts festivals, and a thriving local film industry that is sometimes referred to as "Wellywood." The New Zealand Symphony Orchestra and the New Zealand Ballet add to Wellington's cultural life. The population also enjoys a good selection of restaurants and cafés. Cuisines reflect the ethnic variety of Wellington's population. Much of Wellington's nightlife is concentrated along Courtenay Place, a road that runs from the city's Central

Business District. Wellington is promoted to tourists as the "coolest little capital in the world" and via the slogan "Absolutely Positively Wellington."

Further Reading

Brand, Diane. "Surveys and Sketches: 19th Century Approaches to Colonial Urban Design," *Journal of Urban Design* 9, no. 2 (2004): 153–75.

Pearce, Douglas G. "Capital City Tourism: Perspectives from Wellington," *Journal of Travel and Tourism Marketing* 22, no. 3/4 (2007): 7–21.

Peirce, Sophie and Brent W. Ritchie. "National Capital Branding: A Comparative Case Study of Canberra, Australia and Wellington, New Zealand," *Journal of Travel and Tourism Marketing* 22, no. 3/4(2007): 67–79.

Smith, Philippa Mein. *A Consice History of New Zealand.* Cambridge: Cambridge University Press, 2012.

WINDHOEK

Windhoek is the capital and largest city of the Republic of Namibia, a country in southern Africa that borders the South Atlantic Ocean. The city is located in the very center of the country in the Khomas Highland plateau at an elevation of about 5,600 ft (1,700 m). In addition to being the capital, Windhoek is Namibia's main commercial and cultural center, and its main port of entry by air by way of Windhoek Hosea Kutako International Airport. The population of Windhoek is about 322,500 (2011).

Historical Overview. Windhoek was founded in about 1840 by Jonker Afrikaner, leader of the Orlam subtribe of the Nama people, and approximately 300 fellow settlers who migrated from Cape Colony in what is today the Republic of South Africa. They were attracted by the hot springs that are still found in an otherwise very dry area. The settlers arrived with missionaries and soon built a large stone church. The town flourished until warfare broke out in 1880 between the local Nama and Herero peoples and much of the city was destroyed and abandoned. Germans arrived in 1884 to intervene in the fighting and took Namibia as a colony which they named German South-West Africa. From 1904 to 1907, the African population rebelled against German rule, resulting in the Herero and Namaqua Genocide in which the majority of the Namibia's Africans, totaling tens of thousands of victims, were killed by the Germans. Survivors where then confined into African "homelands." In 1915, during World War I, South Africa took control of German South-West Africa on behalf of the British Empire, and then continued to administer the territory as a League of Nations mandate after Germany lost the war. The mandate territory was called South-West Africa.

Namibian independence from South Africa was hard-won and came in 1990. After the United Nations was formed in 1946, South Africa refused to give up its administration of Namibia despite UN pressure to do so. In 1971, the International Court of Justice declared in an advisory ruling that South Africa's hold on Namibia was illegal. The South-West Africa People's Organization (SWAPO) and its military-guerilla win, the People's Liberation Army of Namibia, began an armed struggle that lasted until 1988 when a truce was reached and South Africa agreed to Namibia's independence. Namibia has since made a transition from white minority rule to being a multiparty parliamentary democracy.

Major Landmarks. Tintenpalats (German for "Ink Palace") is a striking building constructed in 1912–1913 to house the German colonial administration of Namibia. Today it is the seat of Nambia's legislature, including both the National Council and the National Assembly. The Alte Feste (Old Fortress), located in the center of Windhoek's modern downtown, was completed in 1890 during the German colonization of Namibia and is the city's oldest standing structure. It now functions as a museum. The Germans also built three other castles in the city: Heinitzburg, Sanderburg, and Schwerinsburg. Other landmarks are Christ Church, a Lutheran church that dates to 1910, St. Mary's Roman Catholic Cathedral, and the modern, African-style Supreme Court of Namibia. Reiterdenkmal is the German name for the Equestrian Monument, a highly controversial statue that was inaugurated in 1912 to honor German soldiers who died fighting African opponents of colonization. Many Africans resent the monument. The statue was moved from its original site in 2009 to make way for construction of an independence museum, and was later put up again outside Alte Feste, not far from its original site.

Culture and Society. Windhoek is a fast-growing city with migration of Namibia's rural population to the national capital, and the formation of informal settlements at the city's edges. The African population comprises about two-thirds of the population total and includes many ethnic groups, including Ovambo, Hereo, Himba, and Damara peoples. Some 17 percent of the population comprises the Colored and Basters, two categories of mixed-race people, while approximately 16 percent is white, mostly of German, English, or Portuguese origins. English is the official language, but many African languages are widely spoken, including Afrikaans by much of the white population. More than 80 percent of the population is Christian, with Lutherans representing the largest denomination. Namibia has one of the highest incidences of HIV/AIDS infections in the world.

Further Reading

Ankomah, Baffour. "The Trouble with Namibia," *New African* 507 (June 2011): 40–44.

Frayne, Bruce. "Migration and the Changing Social Economy of Windhoek, Namibia," *Development Southern Africa* 24, no. 1 (March 2007): 91–108.

Müeller-Friedman, Fatima. "Beyond the Post-Apartheid City: De/Segregation and Suburbanization in Windhoek, Namibia," *African Geographical Review* 25 (2006): 33–61.

Y

YAMOUSSOUKRO

Yamoussoukro is the capital of the Republic of Côte d'Ivoire, also known as Ivory Coast, a country in West Africa on the shores of the Gulf of Guinea. It has been the capital since 1983, replacing the city of Abidjan, Côte d'Ivoire's largest urban center and hub of the national economy. As opposed to Abidjan, which has a port and is located in the southeast of the country, Yamoussoukro is centrally located. The population of Yamoussoukro is approximately 243,000 (2010). The correct pronunciation of Yamoussoukro is a lot easier than the spelling; Ivoirians pronounce their capital city as "yam-so-kro."

Historical Overview. Côte d'Ivoire was a colonial possession of France from the late 19th century until independence was achieved in 1960. Earlier capitals were Grand-Bassam from 1893, Bingerville from 1900, and then Abidjan from 1933. All three cities are in the same coastal area in the southeast. Félix Houphouët-Boigny was the first prime minister of Côte d'Ivoire (for a little more than 4 months in 1960) and then the first president from November 3, 1960 until his death on December 7, 1993. It was Félix Houphouët-Boigny who made Yamoussoukro the capital of Côte d'Ivoire.

The history of Yamoussoukro begins with a village of the indigenous Akoué people named N'Gokro. In 1909, the local French colonial administration elevated the status of N'Gokro from village to an administrative center, and renamed it Yamoussoukro in thanks to Queen Yamousso, a village leader who along with her uncle, the chief Kouassi N'Go, had convinced local people to not wage war against France. Félix Houphouet-Boigny was born in Yamoussoukro in 1905 when it was still N'Gokro. In 1939, he became the village chief. His career rose from there. Not long after he became president, he began the project of shifting the capital to his home village, and to turn the Yamoussoukro into a source of great Ivoirian pride. The announcement that Yamoussoukro was replacing Abidjan as the official center of government was made in March, 1983. However, while Yamoussoukro is now the official capital and many new buildings have been constructed to support that role, Abidjan continues to be the unofficial capital. Most of the foreign embassies in Côte d'Ivoire are still in Abidjan.

There has been considerable political turmoil in Côte d'Ivoire since the death of President Houphouët-Boigny, including a civil war in 2002 that divided the nation in two, and renewed violence in 2011 following a disputed presidential election that is referred to as the Second Ivoirian Civil War. Death tolls have been high, and there are many internally displaced refugees. Yamoussoukro itself has been the scene of intense fighting.

Major Landmarks. The Roman Catholic Basilica of Our Lady of Peace of Yamoussoukro is the city's most famous landmark. It was constructed between 1985 and 1989 at the insistence of President Houphouët-Boigny at a cost of $300 million. Its design was inspired by that of the Basilica of Peter in the Vatican City. It is believed to be the largest church in the world, with a height of 518 ft (158 m), although St. Peter's can accommodate more worshippers. Another landmark is the House of Deputies, an enormous new government building. Yamoussoukro is still a project in the making, as there are vast empty spaces between completed buildings and wide streets in a grid plan with many vacant blocks and streets that are lightly trafficked.

Culture and Society. The essential fact about Côte d'Ivoire that no amount of oversized architecture can hide is that most citizens are very poor despite the natural wealth that the country has contained, and that decades of war and political strife have impoverished Ivoirians even more. Yet, when Mr. Houphouët-Boigny died, his personal wealth is estimated to have totaled between $7 and $11 billion. As reported in the French-language *Encyclopædia Universalis,* he reportedly said in 1983 that "People are surprised that I like gold. It's just that I was born in it." The population of Côte d'Ivoire is ethnically diverse, with Akan being the largest group at 42 percent of the population, followed by Gur (18%), Northern Mandes (17%), Krous (11%), and Southern Mandes (10%). There also small minorities of Lebanese and French. The population is divided between Muslims and Christians, with Roman Catholics being most numerous among the latter. The northern part of Côte d'Ivoire has mostly Muslims while the south has mostly Christians. The official language of the country is French, although many African languages are spoken more commonly.

Further Reading

McGovern, Mike. *Making War in Côte d'Ivoire.* Chicago: University of Chicago Press. 2011.

Mundt, Robert J. *Historical Dictionary of the Ivory Coast/Côte d'Ivoire.* Lanham, MD: Scarecrow Press, 1987.

YANGON

Yangon is the former capital and largest city of Burma, a country in Southeast Asia that is also known as Myanmar (official name: Republic of the Union of Myanmar). The city is also known as Rangoon, and is sometimes still thought of as the country's capital because the official move of capital functions to the new city of Naypyidaw is not fully complete, and because the Burmese administration that declared Naypyidaw to be the new capital of Burma in 2006 is itself in transition. Yangon is located in the southern part of the main body of Burma (that is not the part of Burma that is on the Malay Peninsula), in the large and fertile delta of the Ir-rawaddy River (also spelled Ayeyarwady), Burma's central river. Its specific river channels are the Yangon and Bago Rivers. The population of Yangon is about 4.4 million. The city is Burma's principal industrial and commercial center.

Historical Overview. Yangon was founded in the early 11th century by the Mon people who lived in lower Burma at the time. The early city was named Dagon and was centered around the Shwedagon Pagoda, a sacred Buddhist shrine that dates back to at least the sixth century. In 1755, Dagon was captured by King Alaungpaya, founder of the Konbaung Dynasty, and was renamed Yangon, meaning "run out of enemies" or "end of strife." The British captured Yangon in the First Anglo Burmese War (1824–1826) but did not hold it, and then recaptured it and all of lower Burma in 1852 in the Second Anglo-Burmese War. They designated the city to be the capital of British Burma. After the Third Anglo-Burmese War, Britain controlled upper Burma as well, and Rangoon, as the British had renamed the city, became capital of a larger colonial territory. The British laid out fine neighborhoods on a grid plan and built many new institutions, including Rangoon General Hospital and the University of Rangoon. The colonial city came to be known as "the garden city of the East." After World War I, an independence movement took root among students at the University of Rangoon, with general strikes against British rule taking place in 1920, 1936, and 1938. Rangoon was occupied by the Japanese in World War II. In 1948, Burma achieved independence. From 1962 to 1988 the country was governed by a repressive military junta led by General Ne Win. His isolationist policies impoverished the country and Yangon, as the city was named again, deteriorated. There were major antigovernment protests in 1974, 1988, and 2007, with loss of life among demonstrators. In May 2008, much of Yangon was badly damaged by Cyclone Nargis.

The movement for democracy in Burma has been led by Aung San Suu Kyi, the daughter of a leader of the independence movement against the British. Her political party won national elections in 1990 and she could have assumed the office of prime minister, but she was taken into custody instead and spent 15 of the next 21 years under house arrest at her residence in Yangon. In 2012, there were signs that the military hold on the nation is loosening and that Aung San Suu Kyi's National League for Democracy will have a larger voice in the government in the future.

Major Landmarks. Far and away, the main landmark of Yangon is the Shwedagon Pagoda. It is the most important Buddhist site in Burma and, according to legend, sits on a spot that has been sacred since the dawn of time. It is distinguished by a huge stupa made of gold plates over brick, and is one of the world's most spectacular places of worship and religious architecture. It is also one of the world's largest single depositories of both gold and teak. Other important places of worship are the Sule Pagoda, and the Botataung Pagoda, and from the time of colonialism, St. Mary's Cathedral and Holy Trinity Cathedral. Other landmarks are Aung San's house, Aung San Suu Kyi's House, the Martyrs' Mausoleum, Independence Monument, the National Museum, and the landmark Strand Hotel that was built in 1901.

Culture and Society. Burma is a rich land but its people are poor because of bad and corrupt government for most of the period since independence. There is, however, a strong prodemocracy movement that brings hope for the future. Aung San Suu Kyi has won a Nobel Prize for her brave stand against an extremely repressive military regime.

The population of Yangon is ethnically diverse. During colonial times, the majority population was from India, but now it is Bamar (Burman) people who are in the majority. The most numerous non-Burma minorities in the city are Indians, other South Asians, and Chinese, while the most numerous minority ethnic groups from within Burma are Rakhine and Karen The principal language of the city is Burmese, with English being widely understood by the educated classes. The population is devoutly Buddhist.

See also: Naypyidaw.

Further Reading

Leonard, John B. "Rangoon," *Cities: The International Journal of Urban Policy and Planning* 2, no. 1 (February 1985): 2–13.
Seekins, Donald M. *State and Society in Modern Rangoon.* New York: Routledge, 2010.
Singer, Noel F. *Old Rangoon: City of Schwedagon.* Gartmore, UK: Kiscadale Publications, 1996.

YAOUNDÉ

Yaoundé is the capital and second-largest city after the port city of Douala in the Republic of Cameroon, a nation in equatorial Africa on the Bight of Biafra of the Gulf of Guinea of the Atlantic Ocean. The city lies in the interior, in the south central part of the country, only 3°52′N of the equator. The population of Yaoundé is about 2.5 million (2012).

Historical Overview. Yaoundé was founded in 1888 during the height of European colonial exploitation in Africa by German traders as a base for dealing in ivory. Before German colonization, Cameroon's territory was divided among numerous indigenous tribal groups, the most notable of which was the Bamum Kingdom of 1394–1884 in the center of the present country. It gave way in 1884 to the German colony of Kamerun. After Germany lost World War I, Cameroon was divided between French and German administration as mandates by the League of Nations. Cameroonians struggled for independence from France in 1950s and achieved their goal in 1960 with formation of the Republic of Cameroun. British Cameroons joined in 1961, and in 1972, the country was renamed the United Republic of Cameroon. It became the Republic of Cameroon in 1984.

Major Landmarks. Yaoundé's most notable landmarks are the Cathédrale Notre Dame des Victoires, which is the seat of the city's Roman Catholic Archdiocese; the Basilique Marie-Reine-des-Apôtres built on the site of the first missionary church in Cameroon; the Cameroon Art Museum (located in a former Benedictine monastery); and the Cameroon National Museum. The Afhemi Museum was opened in 1999 by the African Arts/Handicraft and Environmental Management Institute, and displays more than 2,000 artifacts, art objects, and antiquities from around the many diverse cultures of Cameroon.

Culture and Society. Cameroon is one of Africa's most diverse countries, with as many as 230–282 folk and linguistic groups and great contrasts in natural environment and ways of living, ranging from low-lying humid tropic jungles, to highland environments, to dry lands at the edge of the Sahara Desert. As capital city, Yaoundé attracts migrants from all regions and all ethnicities, and is very diverse. The city has attracted refugees from strife-torn neighboring countries as well: the Central African Republic, Chad, and Nigeria. The official languages of the country are French and English, with French being much more prevalent, but local languages are spoken commonly on a daily basis. There is a mixture of French, English, and Pidgin English called Camfranglais that is spoken in Yaoundé and other cities. The majority of Cameroonians are Christian, particularly Roman Catholic, but in the north the main religion is Islam. Islam accounts for about 21 percent of the national total.

Indigenous Cameroonian culture is rich and vibrant. The Bamum people are known to have had a writing system of their own before the arrival of German colonialists in the 19th century. In the early 20th century, Bamum Sultan Ibrahim Njoya advanced literacy and education among his people. There is also a rich and diverse traditional music and dance culture in Cameroon, some examples of which have been popularized worldwide. The most popular sport in Cameroon is football (soccer); the national team has competed successfully in many international competitions.

Further Reading

DeLancey, Mark. *Cameroon: Dependence and Independence.* Boulder, CO: Westview Press, 1989.

Fowler, Ian and David Zeitlyn. *African Crossroads: Intersections between History and Anthropology in Cameroon.* New York: Berghahn Books, 1996.

Njoh, Ambe J. *Planning in Contemporary Africa: The State, Town Planning, and Society in Cameroon.* Burlington, VT: Ashgate, 2003.

YAREN

Yaren or Yaren District is the de facto capital of the Republic of Nauru, formerly known as Pleasant Island, a small (8.1 square miles; 21 km sq) one-island nation in the Pacific Ocean about 26 miles (42 km) south of the equator. With only 9,322 inhabitants, it is the second least populated sovereign state on earth (after Vatican City) as well as one of the smallest. The island is oval-shaped and made of phosphate rock, and is surrounded by a coral reef. Access from deep water is provided by 16 narrow channels. Yaren is in the south of the island and has a population of 1,100 (2003). It is referred to as capital only because the government of Nauru is centered there; the country actually has no official capital. The United Nations refers to Yaren as the "main district" of Nauru.

Historical Overview. Nauru has been inhabited my Micronesian and Polynesian people for more than 3,000 years. The 12 traditional clans of Nauru people are represented by the 12-pointed star on the nation's flag. The first Westerner to visit the island was the British whaler John Fearn in 1798. He named it Pleasant Island. Whaling ships came frequently to the island in the 19th century to trade for provisions. In 1888, Nauru was annexed by Germany and was incorporated into Germany's Marshall Island Protectorate. In the same year, Christian missionaries arrived from the Gilbert Islands. Phosphate was discovered in 1900, and export began in 1906–1907. After World War I, Australia, New Zealand, and Great Britain took over the phosphate reserves from Germany, and in 1923 the League of Nations give Australia a mandate over the island. From 1942 to 1945, the island was occupied by Japanese Imperial forces until surrender to Australia. Nauru gained its independence on January 31, 1968. The country prospered for a time from phosphate resources, but once the island came to be mined out in the 1980s, economic hardship set in. For a time afterward, Nauru has derived income as a tax haven. It has also been paid by the Australian government to temporarily house international refugees and asylum seekers who want to enter Australia.

Major Landmarks. Yaren district contains the Parliament House of Nauru, the administration offices, the police station, the island's largest port, and Nauru International Airport. There is also the telecommunications station and the refugee center. The interior of the island is largely waste for strip mining activity for phosphates and old industrial sites.

Culture and Society. The ethnic groups of Nauru are Nauruans (58%), other Pacific Islanders (26%), Europeans (mostly British; 8%), and Chinese (8%). The official language is Nauruan. English is also widely spoken. About one-third of the population is Protestant Christian and one-third Roman Catholic. Nauru has the largest percentage of overweight people in the world: 97 percent of men and 93 percent of women are classified as overweight or obese. The incidence rates of type 2 diabetes, heart illnesses, and diseases of the kidneys are also very high. The most popular sport on the island is Australian rules football.

Further Reading

Kendall, David. "'Doomed Island' Nauru's Short-Sightedness and Resulting Decline Are an Urgent Warning to the Rest of the Planet," *Alternatives Journal* 35, no. 1 (February 2009): 34–37.

McDaniel, Carl N. and John M. Gowdy. *Paradise for Sale: A Parable of Nature.* Berkeley: University of California Press, 2000.

YEREVAN

Yerevan is the capital and largest city in the Republic of Armenia, a small, landlocked country in the Caucuses Mountains where Eastern Europe and Western Asia come together. The city is in the west center of the country at an edge of the Ararat Plain on the Hrazdan River. On clear days, Mount Ararat in nearby Turkey can be seen from Yerevan. The population of Yerevan is about 1.1 million (2011), just over one-third of the total population of Armenia.

Historical Overview. Yerevan's early history is traced to Erebuni fortress that was built about 782 BC by the Urartian King Argishti I. Between the sixtha and the fourth centuries BC, the city was a provincial center in the First Persian Empire. In 301, Armenia was Christianized. Yerevan's first church, St. Peter and Paul, was built in the fifth century. Other churches followed. Persians and Turks alternated control over Yerevan in the Middle Ages in step with the outcomes of their various wars. Additionally, the city was besieged in 1378 by the Mongol army of Tamerlane. On June 7, 1679, Yerevan was devastated by a massive earthquake. The city was under Persian rule until 1828, after which time it was incorporated into the Russian Empire. After the 1917 Russian Revolution, Armenia became part of an independent Transcaucasian Federation, and then independent on its own as the Democratic Republic of Armenia from May 28, 1918. Independence was short lived, however, as the Bolshevik Red Army took command of Yerevan on November 29, 1920, and Armenia was incorporated into the Soviet Union a few days later on December 2. From

1920 until the collapse of the USSR in 1991, Yerevan was the capital of the Armenian Soviet Socialist Republic. The city was substantially rebuilt during the Soviet period, initially following an urban design plan put forward in 1924 by architect Alexander Tamanyan. Wide roads and monumental buildings were put in to modernize the city, but often at the cost of historic architecture, churches, and other features from Armenia's past. The Church of St. Peter and Paul was willfully destroyed by the Soviets in 1931. In 1919, during Armenia's brief period of independence, Yerevan State University was founded. Armenia became independent again on September 21, 1991.

Major Landmarks. Yerevan's oldest historic sites include the ruins of the nearly 3,000-year-old Erebuni fortress and ruins of Tsiranavor Church built in 595–602 AD in the district of Avan. Republic Square, designed by Alexander Tamanyan, is the city's main square and is considered to be an exquisite example of Soviet monumental architecture. On summer evenings, the square offers a musical fountain and light show at its Singing Fountains. Other landmarks are the Yerevan Opera House, the Soviet-style central railway station, the 18th-century Blue Mosque, and St. Atsvatsatsin of Nork Church, a new reconstruction of an historic church that had been destroyed in the Soviet period. The Armenian Genocide Memorial stands on a hill overlooking the city center and commemorates the many Armenians who were milled in 1915 by Turks. Another monument is Mother Armenia, a statue of the female personification of the nation, which was erected in 1950 in Victory Park. St. Gregory the Illuminator Cathedral was completed in 2001 to commemorate 1,700 years of Armenia Christianity. Major museums include the City Museum of Yerevan, the National Art Gallery, the Cafesjian Museum of Modern Art, and the Parajanov Museum dedicated to the work of a celebrated Soviet film director. Near the Cafesjian Museum is the Cascades, an enormous Art Deco rendition in stone of the Hanging Gardens of Babylon. The Matenadaran is an extraordinary museum that is built into solid rock in order to protect its priceless collections of ancient illuminated manuscripts.

Culture and Society. The population of Yerevan is now overwhelmingly Armenian in ethnic composition and Armenian Orthodox Christian in terms of religion. In the 19th century, however, Armenians and Christians were minorities because of the large numbers of Islamic Turks, Persians, Azeris, and others who resided in the city. The city's population fell after 1991 because of migration abroad in the context of economic crisis and suddenly opened borders, but the economy of Armenia is now much improved and since 2007, the city has started to gain population once again. The population is increasingly prosperous. Yerevan is a beautiful city with a fully developed cultural life, many fine restaurants and cafés, and an active nightlife.

Further Reading

Libaridian, Gerard. *Modern Armenia: People, Nation, State.* Piscataway, NJ: Transaction Publishers, 2007.
Miller, Donald E., Lorna Touryan Miller, and Jerry Berndt (photographer). *Armenia: Portraits of Survival and Hope.* Berkeley: University of California Press, 2003.
Payaslian, Simon. *The History of Armenia.* New York: Palgrave Macmillan, 2007.

Z

ZAGREB

Zagreb is the capital and largest city of the Republic of Croatia, an independent country at the crossroads of Central, Eastern, and Southern Europe and bordering the Adriatic Sea. The city is on the Sava River in the northwestern part of the country at the southern slopes of Medvednica Mountain. According to the census of 2011, the population of Zagreb is 686,568. The wider metropolitan area of Zagreb totals more than 1.2 million inhabitants. In addition to being national capital, Zagreb is a significant industrial center (electrical machinery, chemicals, pharmaceuticals, and textiles), and an important financial center, transportation hub, education center, and center of a vibrant cultural life. Increasingly, the city is a popular magnet for international tourism.

Historical Overview. The earliest known settlement in the vicinity of Zagreb was the first-century AD Roman town of Andaoutonia, now called Ščitarjevo, a part of the town of Velika Gorica in the Zagreb metropolitan area. Zagreb itself dates to the 11th century. It was divided at the time into two city centers, a smaller one to the east called Kaptol and the larger, western Gradec. Kaptol was a religious town and was dominated by Zagreb Cathedral, construction of which began in 1093. Gradec, on the other hand, was a secular merchants' town. In 1242, the two settlements were attacked and the cathedral was destroyed. Rebuilding followed. In the 17th century, the Ban (Croatia's governor) and the Sabor (parliament) were located in Gradec. The two communities were merged in 1850 by Ban Josip Jelačić, with Janko Kamauf becoming the first mayor.

Zagreb prospered in the 19th century as a center of Croatian national expression, and as an industrial city. The first railway was built in 1862, and later connected the city with Budapest to the north and Rijeka to the west on the Adriatic Sea, Croatia's principal port. A powerful earthquake damaged Zagreb in 1880 and was stimulus for modernization of the city with the rebuilding that followed. The city grew especially rapidly in the early 20th century, and many new districts were laid out. In 1918, after World War I, the city was part of the independent Kingdom of Slovenes, Croats, and Serbs that emerged with the demise of Austria-Hungary, a precursor of Yugoslavia. During World War II, Zagreb was capital of a fascist puppet state of Croatia. It then became part of the Second Yugoslavia after the war. Croatia declared independence in 1991 and then was engaged for four years afterward in the Croatian War of Independence against the Serbian-controlled Yugoslav People's Army. The war ended with decisive Croatian victory in August, 1995. Croatia joined the NATO military alliance in 2009 and is scheduled to become a member of the European Union in 2013.

Major Landmarks. Ban Jelačić Square is the central square of Zagreb, named after the famous governor of Slavonia, Croatia, and Dalmatia who united Kaptol and Gradec into one city. Ilica Street extends from the square and is the city's most popular shopping street. The rebuilt Zagreb Cathedral stands on a hill in Kaptol and is the city's most famous building, as well as the tallest building in Croatia. Other prominent buildings include the main rail station of Zagreb constructed in 1890–1892, the Parliament of Croatia, the 13th-century St. Mark's Roman Catholic Church, the 1895 Croatian National Theater, and the Vatroslav Lisinski Concert Hall built in 1973. Near Ban Jelačić Square is Nikola Šubić Zrinski Square in the center of which is the very beautiful building of the Croatian Academy of Sciences and Arts dating to 1866. There are many museums in Zagreb, including the Mimara Museum, the Modern Gallery, the Croatian natural History Museum, the Archaeological Museum, the Museum of Contemporary Arts, and the Museum of the City of Zagreb.

Culture and Society. More than 90 percent of the population of Zagreb is comprised of ethnic Croats. Serbs comprise about 2.4 percent of the total and are the largest minority group. The main language is Croatian, which is the official language of the country, and the principal religious group is Roman Catholic, about 88 percent of the national population. Zagreb is a center of Croatian arts and culture, and Croatian national pride. There are several universities in Zagreb, most prominently the University of Zagreb, and the city has an energetic and youthful feel. There are many events in the city, including popular concerts and festivals, and a thriving nightlife.

Further Reading

Ashbrook, John. "Politicization of Identity in a European Borderland: Istria, Croatia, and Authenticity, 1990–2003," *Nationalities Papers* 39, no. 6 (2011): 871–97.

Blau, Eve and Ivan Rupnik. *Project Zagreb: Transition as Condition, Strategy, Practice.* Barcelona, Spain: Actar D, 2007.

Hawkesworth, Celia. *Zagreb: A Cultural History.* New York: Oxford University Press, 2008.

Kent, Sarah A. "Zagreb," in Emily Gunzburger Makaš and Tanja Damljanović Conley, eds., *Capital Cities in the Aftermath of Empires: Planning in Central and Southeastern Europe,* 208–22. London: Routledge, 2010.

Appendix: Selected Historic Capital Cities around the World

Angkor

From 802 AD until 1352, Angkor was the capital of the Khmer Empire, a large territory in Southeast Asia that covered modern-day Cambodia, Laos, much of Thailand, and parts of Vietnam and Myanmar. The urban complex covered a large territory north of Tonlé Sap, the "Great Lake" in the center of Cambodia, and combined religious architecture from Hinduism, Mahayana Buddhism, and Theravada Buddhism. The word Angkor means "Holy City" in the Khmer language. The ruins of Angkor are a UNESCO World Heritage Site and have more than 1,000 temples, the most spectacular of which is Angkor Wat, an enormous complex that was built between 1113 and 1150 during the rule of the Hindu King Suryavarman II. The nearby city of Siem Reap is the base for a thriving tourism economy from around the globe.

Babylon

Babylon was one of the great capital cities of the ancient world. It was located on both banks of the Euphrates River in Mesopotamia, and flourished from about 1895 BC until it was captured by the Persians in about 539 BC. Its Hanging Gardens, attributed to King Nebuchadnezzar II and which may have been purely legendary, are regarded as one of the Seven Ancient Wonders of the World. Hammurabi's Code, written on high stone tablets during the 1792–1750 BC rule of King Hammurabi, the Lawgiver, was one of the earliest sets of written laws in the world. The ruins of Babylon are in modern-day Iraq about 55 miles (85 km) south of Baghdad. Ill-advised efforts by former Iraqi ruler Saddam Hussein to reconstruct the city caused much damage to the archaeological site.

Cahokia

Cahokia is no more than an archaeological preserve in the state of Illinois directly across the Mississippi River from St. Louis, Missouri and, in part, a St. Louis suburb. However, centuries ago it was a city of some consequence and an important ceremonial center for the native North American population of the midcontinent. Little remains except for a series of mounds and excavations of burial places and settlements underneath. In the 12th century, Cahokia had a population of as many as 20,000 inhabitants, although some generous estimates posit 50,000, and was the largest settlement on the continent. It may have been larger than any European city at the time too. How Cahokia came into being and what happened to cause its demise is one of the mysteries of the earliest American history.

Calcutta

Until 1911 when King George V declared that Delhi would be the capital of the British-ruled parts of India, British India was administered for a time from three "Presidency Towns" at three different locations of the coast: Bombay (now Mumbai) in the west on the Arabian Sea of the Indian Ocean, Madras (now Chennai) on the southeast coast; and Calcutta (now Kolkata) in the northeast at the head of the Bay of Bengal. Later, Calcutta became the main capital until the move to New Delhi. Although it was hot and muggy, the city was made into the British jewel of India and came to be referred to as the Second City of the British Empire after London. Among its masterpieces of colonial architecture were St. Paul's Anglican Cathedral completed in 1847, the 1921 Victoria Memorial Hall, a spectacular blend of architectural traditions in honor of British Queen Victoria, and the Jubilee and Howrah bridges across the Hooghly River for railroads and motor vehicles, respectively. Fort William defended the city. Beginning in 1864, the northern city of Simla (now Shimla) was made the summer capital of British India because its highlands location in the foothills of the Himalayas was much cooler.

Chang'an

Chang'an is one of the oldest cities in China and was the capital for about 10 dynasties that ruled the country from about the third-century BC until early in the 10th-century AD. In about 750 AD it reached 1 million in population, the first city in history to do so, and was the world's largest city. The city was surrounded by a wall and was built in a grid plan in line with cardinal directions. The name Chang'an means "perpetual peace" in classical Chinese. Today the city is known as Xi'an, a large modern city near the center of China that is one of the country's main tourist attractions. In addition to seeing the historic capital in the center of Xi'an, visitors travel outside the city to see the remarkable Terracotta Army, a collection of thousands of terracotta soldiers, horses, chariots, and other figures that were buried in 210–109 BC in the mausoleum of Qin Shi Huang, the first Qin Emperor of China.

Constantinople

Constantinople is the historic name for modern-day Istanbul, the largest city in Turkey. It was founded in 330 AD on the site of an historic Greek city named Byzantium by Constantine I, after whom it was named, to be the new capital of the Roman Empire. In sitting at the crossroads of Europe and Asia, and between the worlds of Christianity and Islam, the city was much contested in history and was sacked twice in the Middle Ages. After 1453 the city became the capital of the Ottoman sultans. It was endowed with beautiful mosques and other buildings, and ruled a large empire that covered much of the Middle East, Mediterranean Africa, and southeastern Europe. After the Ottoman Empire collapsed in 1922, Ankara became capital of Turkey and Constantinople was renamed Istanbul. Among the main attractions of the city are the spectacular Hagia Sophia, which was originally a church and then became a mosque, the Sultan Ahmed Mosque (also called the

Blue Mosque because of its tiles), and Topkapi Palace where sultans kept harems of young girls captured from subjugated countries.

Columbia, Texas

The Republic of Texas was an independent country from March 2, 1836, when it gained independence from Mexico until February 19, 1846, when it was incorporated into the United States. Its borders covered the present-day state of Texas, plus all or parts of neighboring Oklahoma, Kansas, Colorado, Wyoming, and New Mexico. The first Congress of the Republic of Texas convened in Columbia (now West Columbia) near the Gulf of Mexico coast, making that small town the republic's first capital. In 1837, the capital was moved to the new city of Houston, and then in 1839 to another new city, originally named Waterloo but then renamed Austin after Stephen F. Austin, the "Father of Texas." Austin is now the state capital.

Cusco

Cusco, also spelled Cuzco, is located in the Andes Mountains in southeastern Peru, and was capital of the Inca Empire from the 13th century until in 1532. It was conquered by the Spanish in 1532. The ancient city was unusual in being shaped like a puma, a sacred animal to the Incas. It is also known for intricate stonework that does not use mortar, but withstand strong earthquakes nevertheless. After the Spanish gained control, they superimposed the form of a typical Spanish colonial city on the original plan of Cusco and built churches and government buildings around a central plaza on the foundations of earlier Inca structures. The city is recognized as a UNESCO World Heritage Site, and is a popular destination for foreign tourists as well as a gateway to Machu Picchu, the remains of an iconic 15th-century Inca city situated on a barely accessible mountain top about 50 miles (80 km) to the northwest.

Fatehpur Sikri

Meaning City of Victory, Fatehpur Sikri was built in 1571–1585 by the great Mughal Emperor Akbar to be his capital. It was located in north-central India, about 25 miles (40 km) from Agra, the city in which a future Mughal ruler, Shah Jahan, would build the Taj Mahal some six decades later. It was built as one of the largest and most technologically sophisticated cities in the history of the world, and included a novel system of water wheels to bring water to the high ridge where the city was built. From there, water was distributed by force of gravity among various districts, and to fountains, baths, and pools. Fatehpur Sikri was surrounded on three sides by high walls, and on the fourth by a lake. The Buland Darwaza was the spectacular 177-ft-high (54 m) main gate, Jama Masid was the largest mosque, and Akhbar's palace was enormous. The entire complex was constructed of beautiful red sandstone in a coordinated, highly ornate architectural style that we now call Mughal. As impressive and expensive as Fatehpur Sikri was, its life was short.

The city was abandoned just after it was built because Akbar needed to fight tribes in today's Afghanistan and moved his capital to Lahore. The site of Fatehpur Sikri today is eerily empty. It is a UNESCO World Heritage Site and a popular visitor attraction.

Great Zimbabwe

Great Zimbabwe was developed over the 11th through 14th centuries as the capital of the Kingdom of Zimbabwe in what is today the Republic of Zimbabwe. It was eclipsed in about 1430 by the Kingdom of Mutapa and was then abandoned in about 1450. Europeans became aware of the site during 19th-century colonialism and began archaeological study in 1871. The racist government of British Rhodesia, as the colonial state was known, refused to acknowledge that this large and impressive city had been built by Africans, and engaged in a campaign of censorship of archaeological findings. With the birth of independent Zimbabwe, remains of the ancient capital were recognized as an important symbol of African achievement and African history reclaimed from European misrepresentations. Famous soapstone bird carvings from Great Zimbabwe are symbols of the independent African nation that is named after the ancient kingdom.

Honolulu

Honolulu is the capital and largest city (population 337,000; metropolitan area 953,000) in the state of Hawaii, but prior to the overthrow of the Hawaiian monarchy in 1893 and annexation of the Hawaiian Islands by the United States in 1898, it was the capital of the Hawaiian Kingdom. The capital had been moved to Honolulu from the Maui city of Lahaina in 1845 by King Kamehameha III, and was then developed into a truly modern city. Earlier, under the rule of Kamehameha I, the capital was in Waikīkī near Honolulu from 1804 to 1809 and in what is now downtown Honolulu from 1809 to 1812. Major landmarks of Honolulu as a royal capital include St. Andrew's Cathedral; Iolani Palace, once the king's residence and now a museum; and Ali'iōlani Hale, formerly the seat of government of the Kingdom of Hawaii and the Republic of Hawaii and now the state supreme court building. The word Honolulu means "sheltered bay" in the Hawaiian language.

Königsberg

The historic city of Königsberg, meaning the King's City and named in honor King Ottokar II of Bohemia, is now called Kaliningrad, the city named after Bolshevik revolutionary Mikhail Kalinin, and is an exclave of the Russian Federation on the Baltic Sea that borders Poland and Lithuania. The Soviet Union grabbed this German-speaking territory when World War II ended, annexed it, and then russified it. The city had been founded in 1255 by Teutonic Knights during the Northern Crusades, and then was capital of the Prussian Empire until Berlin took over in 1701. It was a beautiful and prosperous city with grand palaces, monuments, and churches. It was also known for learning and intellectual life, particularly as the city of Immanuel Kant, a leading philosopher of the Enlightenment in the late 18th century, and as the city whose pattern of seven bridges in the center posed a famous mathematics question that led to development of the sciences of graph theory and topology. Königsberg was badly damaged in World War II. For the Russians, it is now a strategic port and navy base.

Kyoto

"Kyoto" is the Japanese word for "capital city" and was the seat of Japan's emperors from 794 AD until 1869 when the imperial court was transferred to Tokyo. During various periods during this time, power did not always reside with the emperor, as there were periods of military rule during which other cities such as Kamakura and Edo (now named Tokyo) were more powerful. Kyoto, however, is the acknowledged ancient capital of Japan and is known for many historical sites such as the old Imperial Palace, beautiful old temples and pagodas, and the most perfect of Japanese gardens. When the city was founded, it was known as Heian-Kyō (the capital of peace and tranquility) and was built as a scaled replica of the ancient Chinese capital Chang'an. That grid plan is still a feature of modern-day Kyoto, now a modern city of some 1.5 million inhabitants. The Gion district preserves the culture of geishas, tea ceremony, and Japanese performance art.

Lahore

Lahore is an ancient city in northern Pakistan near the border of India and is the country's second-largest urban center, as well as an extremely important cultural center and repository of history. It was the capital of various kingdoms in the beginning of 11th century; and in the 16th century it became the capital of the powerful and culturally rich Mughal Empire. From 1802 to 1849, the city was the capital of the Sikh Empire. Later it became the capital of Punjab under the British Raj. The city is replete with beautiful historic mosques and temples, old fortifications, and historic neighborhoods and marketplaces with narrow, winding streets. It is also a modern city. The population of Lahore today is estimated at more than 10 million.

Nanjing

Meaning "southern capital" and formerly known in English as Nanking, Nanjing is one of China's oldest cities and was the national capital at various times in Chinese history. It first became a capital in 229 AD, and was then capital on and off for various dynasties into the 17th century. In 1912, it became capital of the Republic of China under Dr. Sun Yat-sen. The Japanese invaded the city in 1937 and committed atrocities against the civilian population that took as many as 300,000–350,000 lives. Afterwards, the national government of China was relocated inland to the city of Chungking (now Chongqing), while Nanjing became the capital of a Japanese puppet government. Defeat of Japan in World War II brought the capital back to Nanjing until the Communist government was formed in 1949 and proclaimed Beijing to be capital. Nanjing city is now one of China's leading industrial and educational centers, and has a population of about 8 million.

Persepolis

Persepolis was an ancient capital city of the Achaemenid Empire believed to have existed during 550–330 BC, with the oldest known remains dating to about 51 BC. The city was founded by Persian King Cyrus the Great and was developed in large part by a successor King Darius the Great and his son King Xerxes the Great. It was captured in 330 BC by the armies of Alexander the Great, the Greek ruler. The site of Persepolis is in what is today the southern part of Iran, and is now a UNESCO World Heritage Site. There are ruins of great temples, ancient gates to the city, and many bas reliefs that are both beautiful and informative about the society that they portray.

Philadelphia

Washington, DC, has been the capital of the United States since 1800, but prior to then eight other cities served as the meeting place for Congress and can, therefore, be considered to have been the capital. Philadelphia, located on the west bank of Delaware River in the Commonwealth of Pennsylvania, can be considered to be the first capital because it hosted the First Continental Congress in Carpenter's Hall in 1774 and the Second Continental Congress in Independence Hall in 1775–1776. Important landmarks of early U.S. history in Philadelphia are in Independence National Historic Park. Other early U.S. capitals are Baltimore and Annapolis in Maryland, Lancaster and York in Pennsylvania, Trenton and Princeton in New Jersey, and New York City.

St. Petersburg

St. Petersburg is a city on the Baltic Sea coast of Russia that was capital of the Russian Empire from 1703 until soon after the Russian Revolution when the seat of government was returned to Moscow. It was founded by Russian Tsar Peter the Great and was named after his patron saint. During the Soviet Period, the city was named Leningrad. The purpose for founding St. Petersburg was to bring Russia into closer contact with the capital cities of Europe, and to give the country a strategic port on the Baltic Sea. The city was built from the ground up at enormous cost to the Russian treasury and lives of conscripted workers, and was made into one of the most beautiful and most opulent cities in the world. The center of St. Petersburg is a UNESCO World Heritage Site. The Hermitage Museum houses one of the world's richest collections of art in a glorious palace that was built by the Empress Catherine the Great. Many of the leading events of the 1917 Russian Revolution took place in this city as well.

Shahjahanabad

Shahjahanabad, meaning city of Shah Jahan, was built by the Mughal Emperor Shah Jahan in 1639 and was the capital of Mughal rulers until the dynasty ended in 1857. It is at the site of Delhi, the capital of India, and is often referred to as the walled city of Old Delhi. It faces the Jamuna River and has as a focal point the enormous 17th-century Red Fort, so-called because its walls are of red sandstone. Jama Masjid, completed in 1656, is the principal mosque of Old Delhi. Chandni Chowk, meaning "moonlit market" was the principal marketplace of the old city. It was designed by Jahan Ara, the daughter of Shah Jahan, and still functions although the original canals that ran through the area in order to reflect moonlight have been filled in and are now streets. It was Shah Jahan who built the spectacular Taj Mahal, one of the so-called Wonders of the World, in Agra, south of Shahjahanabad as a mausoleum for his wife, the Empress Mumtaz Mahal.

Tenochtitlan

Tenochtitlan was founded in 1325 and was the capital of the Aztec Empire in what is today the country of Mexico in the 15th century until destruction by Spanish conquistadores in 1521. It had been the largest city in the Americas and was known for monumental temples, impressive engineering, and sustainable urban design. The city was built on an island in Lake Texcoco and was connected to other parts of the Valley of Mexico by a series of causeways. The Spanish colonial capital Mexico City was built atop the ruins of Tenochtitlan and used its building materials. The ruins of the Aztec Templo Mayor (the great temple) were discovered in the early 20th century, with archaeological work beginning in 1978. Among the many treasures of Tenochtitlan that are now on public display in Mexico City are a stone disc 10.5 ft (3.25 m) in diameter that depicts the dismembered body of the moon goddess Coyolxauhqui and the even bigger Aztec Calendar Stone.

Thebes

Thebes is located in the valley of the Nile River in Egypt approximately 300 miles (nearly 500 km) south of Cairo, the national capital, and was the capital of Egypt some 3,000–4,000 years ago. Its ruins are in the modern city of Luxor. The city has been inhabited for nearly 5,000 years. The Luxor Temple and the Karnak Temple complex are two of the most important relics of the ancient civilization, as is the Theban Necropolis across the Nile on the west bank. Thebes is a UNESCO World Heritage Site and, along with the great pyramids located much closer to Cairo, is a leading destination for foreign tourists to Egypt.

Selected Bibliography

Abdoumaliq, Simone. *City Life from Jakarta to Dakar: Movement at the Crossroads*. New York: Routledge, 2010.

Abdoumaliq, Simone and Abdelghani Abouhani, eds. *Urban Africa: Changing Contours of Survival in the City*. Dakar, Sengal: Codesria Books, 2005.

Anderson, David M. and Richard Rathbone, eds. *Africa's Urban Past*. Oxford, UK: James Curry, 2000.

Askew, Marc and William S. Logan, eds. *Cultural Identity and Urban Change in Southeast Asia: Interpretive Essays*. Geelong, Australia: Deakin University Press, 1994.

Bekker, Simon and Göran Terborn, eds. *Capital Cities in Africa: Power and Powerlessness*. Cape Town, South Africa: HSRC Press, 2012.

Benevodo, Leonardo. *The European City*. Oxford, UK: Blackwell, 1993.

Bishop, Ryan, John Phillips, and Wei Wei Yeo, eds. *Postcolonial Urbanism: Southeast Asian Cities and Global Processes*. New York: Routledge, 2003.

Biswas, Rameh Kumar. *Metropolois Now!* New York: Springer, 2000.

Brunn, Stanley D., Maureen Hayes-Mitchell, and Donald J. Ziegler. *Cities of the World: World Regional Urban Development*. Lanham, MD: Rowman & Littlefield Publishers, 2010.

Chen, Xiangming, Anthony M. Orum, and Krista E. Paulsen. *Introduction to Cities: How Place and Space Shape Human Experience*. New York: Wiley-Blackwell, 2012.

Claessens, Francois and Leen van Duin, eds. *The European City: Architectural Interventions and Transformations*. Delft, The Netherlands: Delft University Press, 2004.

Clark, Peter. *European Cities and Towns, 400–2000*. Oxford, UK: Oxford University Press, 2009.

Costa, Frank J., Ashok K. Dutt, Laurence J. C. Ma, and Allen G. Noble, eds. *Urbanization in Asia: Spatial Dimensions and Policy Issues*. Honolulu: University of Hawai'i Press, 1989.

Czaplicka, John, Nida Gelazis, and Blair Ruble, eds. *Cities after the Fall of Communism: Reshaping Cultural Landscapes and European Identity*. Washington, DC: Woodrow Wilson Center Press, 2009.

Dagens, Bruno. *Angkor: Heart of an Asian Empire*. London: Thames and Hudson, 2004.

Daum, Andreas W., ed. *Berlin-Washington, 1800–2000: Capital Cities, Cultural Representation, and National Identities*. Cambridge, UK: Cambridge University Press, 2005.

Davis, Mike. *Planet of Slums*, New York: Verso, 2007.

DeFrantz, Monika. *Capital City Cultures: Reconstructing Contemporary Europe in Vienna and Berlin*. New York: P.I.E. Peter Lang, 2011.

Douglass, M. "Local City, Capital City, or World City?" *Pacific Affairs* 78, no. 4 (2006): 543–58.

Drakakis-Smith, David. *Third World Cities*. London: Routledge, 2000.

Driver, Felix and David Gilbert. *Imperial Cities: Landscape, Display, and Identity*. Manchester, UK: Manchester University Press, 1999.

Eldredge, H. W., ed. *World Capitals: Toward Guided Urbanization*. Garden City, NY: Doubleday, 1975.

Evers, Hans-Dieter and Rüdiger Korff. *Southeast Asian Urbanism: The Meaning and Power of Social Space*. New York: St. Martin's Press, 2000.

Fisher, M. *Provinces and Provincial Capitals of the World*. New York: Scarecrow Press, 1967.

Forstall, R. L., R. P. Greene, and J. B. Pick. "Which Are the Largest? Why Lists of Major Urban Areas Vary So Greatly," *Tijdschrift voor Economische en Social Geografie* 100, no. 3 (2009): 277–97.

Friedmann, J. "The World City Hypothesis," *Development and Change* 17, no. 1 (1986): 69–83.

Freund, Bill. *The African City: A History*. Cambridge, UK: Cambridge University Press, 2007.

Garlake, P. S. *Great Zimbabwe Described and Explained*. Harare: Zimbabwe Publishing House, 1982.

Gold, John R. and Margaret M. Gold, eds. *Olympic Cities: City Agendas, Planning, and the World Games, 1896–2012*. London: Routledge, 2007.

Gottman, Jean. "The Role of Capital Cities," *Ekistics* 44, no. 264 (1977): 240–47.

Gottman, Jean. "Capital Cities," *Ekistics* 50, no. 299 (1983): 88–93.

Gottman, Jean. "The Study of Former Capitals," *Ekistics* 314–315 (1983): 541–46.

Griffiths, Ieuan Ll. *The African Inheritance*. London: Routledge, 1995.

Gugler, Josef. *Cities in the Developing World: Issues, Theory, and Policies*. Oxford: Oxford University Press, 1997.

Gunther, John. *Twelve Cities*. New York: Harper & Row, 1967.

Hall, Peter. *Cities in Civilization*. New York: Pantheon, 1998.

Hall, Peter. "The Changing Role of Capital Cities: Six Types of Capital City," in Taylor, John, Jean G. Lengellé, and Caroline Andrew, eds., *Capital Cities: International Perspectives/Les Capitales: Perspectives Internationales*, 69–84. Ottawa: Carleton University Press, 1993.

Hall, Richard W. *African Cities and Towns before the European Conquest*. New York: Norton, 1977.

Hall, Thomas. *Planning European Capital Cities: Aspects of Nineteenth Century Urban Development*. London: E & F.N. Spon, 1997.

Hardoy, Jorge E. "Ancient Capital Cities and New Capital Cities of Latin America," in Taylor, John, Jean G. Lengellé, and Caroline Andrew, eds., *Capital Cities: International Perspectives/Les Capitales: Perspectives Internationales*, 99–128. Ottawa: Carleton University Press, 1993.

Hitti, Philip Khuri. *Capital Cities of Arab Islam*. Minneapolis: University of Minnesota Press, 1973.

Huyssen, Andreas, ed. *Other Cities, Other Worlds: Urban Imaginaries in a Globalizing Age*. Durham, NC: Duke University Press, 2008.

Jaffe, R., ed. *The Caribbean City*, 162–88. Kingston: Ian Randle, 2008.

Jefferson, Mark. "The Law of the Primate City," *Geographical Review* 29 (April, 1939).

Jenks, Michael, Daniel Kozak, and Pattaranan Takkanon, eds. *World Cities and Urban Form: Fragmented, Polycentric, Sustainable?* London: Routledge, 2008.

Kasarda John D. and Allen M. Parnell, eds. *Third World Cities: Problems, Policies and Prospects*. Newbury Park, CA: Sage, 1983.

King, Anthony D. "Cultural Hegemony and Capital Cities," in Taylor, John, Jean G. Lengellé, and Caroline Andrew, eds., *Capital Cities: International Perspectives/Les Capitales: Perspectives Internationales*, 251–83. Ottawa: Carleton University Press, 1993.

Knox, Paul L. and Linda M. McCarthy. *An Introduction to Urban Geography*. Upper Saddle River, NJ: Prentice Hall, 2005.

LeGates, Richard T. and Frederic Stout, eds. *The City Reader*. New York: Routledge, 2011.

Linn, Johannes F. *Cities in the Developing World: Policies for Their Equitable and Efficient Growth*. Oxford, UK: Oxford University Press, 1983.

Lo, Fu-Chen and Yue-Man Yeung, eds. *Emerging World Cities in Pacific Asia*. Tokyo: United Nations University Press, 1996.

Locatelli, Francesca and Paul Nugent, eds. *African Cities: Competing Claims on Urban Spaces*. Leiden, The Netherlands: Brill, 2009.

Makaš, Emily Gunzburger and Tanja Damljanović Conley, eds. *Capital Cities in the Aftermath of Empires: Planning in Central and Southeastern Europe*. London: Routledge, 2010.

Massey, Doreen. *World City*. Cambridge, UK: Polity Press, 2007.

McGee, T. G. *The Southeast Asia City: A Social Geography of the Primate Cities of Southeast Asia*. New York: Praeger, 1967.

Monaco Books. *Capital Cities of Europe*. Munich, Germany: Monaco Books, 2010.

Myers, Garth A. *African Cities: Alternative Visions of Urban Theory and Practice*. London: Zed Books, 2011.

Noe, S. V. "Shahjahanabad: Geometrical Bases for the Plan of Mughal Delhi," *Urbanism Past and Present* 9, no. 2/18 (1984): 15–25.

Olds, Kris. "Globalization and the Production of New Spaces: Pacific Rim Projects in the Late 20th Century," *Environment and Planning A* 27 (1995): 1713–743.

Pacione, Michael. *Urban Geography: A Global Perspective*. Abingdon, UK: Routledge, 2005.

Palen, J. John. *The Urban World*. New York: McGraw Hill, 1997.

Potter, R. B. *The Urban Caribbean in an Era of Global Change*. Aldershot, UK: Ashgate, 2000.

Prud'homme, Remy. "New Trends in Cities of the World," in Richard V. Knight and Gary Gappert, eds., *Cities in a Global Society*, 44–57. London: Sage, 1989.

Querrian, Anne. "The Metropolis and the Capital City," *Zones* 1, no. 2 (1986): 219–22.

Rapoport, Amos. "On the Nature of Capitals and Their Physical Expression," in Taylor, John, Jean G. Lengellé, and Caroline Andrew, eds., *Capital Cities: International Perspectives/Les Capitales: Perspectives Internationales*, 31–67. Ottawa: Carleton University Press, 1993.

Redfield, R. and M. S. Singer. "The Cultural Role of Cities," *Economic Development and Cultural Change*," 3 (1954): 35–73.

Rimmer, Peter and Howard Dick. *The City in Southeast Asia: Patterns, Processes and Policy*. Singapore: National University of Singapore Press, 2009.

Rodwin, Lloyd. *Nations and Cities*. New York: Viking, 1970.

Rooney, Dawn F. *Angkor: An Introduction to the Temples*. Hong Kong: Odyssey, 1999.

Roth, Klaus and Ulf Brunnbauer, eds. *Urban Life and Culture in Southeastern Europe: Anthropological and Historical Perspectives*. Ethnologica Balkanica, 10, 2006.

Rowat, Donald C. "Ways of Governing Federal Capitals," in Taylor, John, Jean G. Lengellé, and Caroline Andrew, eds., *Capital Cities: International Perspectives/Les Capitales: Perspectives Internationales*, 149–71. Ottawa: Carleton University Press, 1993.

Rowat, Donald C. *The Government of Federal Capitals*. Toronto: University of Toronto Press, 1973.

Rüland, Jürgen, ed. *The Dynamics of Metropolitan Management in Southeast Asia*. Singapore: Institute of Southeast Asian Studies, 1996.

Sassen, Saskia. *The Global City: New York, London, Tokyo*. Princeton, NJ: Princeton University Press, 2001.

Sheraton, Mimi. *City Portraits: A Guide to 60 of the World's Great Cities*. New York: Harper & Row, 1962.

Sheridan, Greg. *Cities of the Hot Zone: A Southeast Asian Adventure.* Crow's Nest Australia: Allen & Unwin, 2003.

Simon, David. Cities, *Capital, and Development: African Cities in the World Economy.* London: Belhaven, 1992.

Sonne, W. *Representing the State: Capital City Planning in the Early Twentieth Century.* Munich, Germany: Prestel, 2003.

Sutcliffe, Anthony. "Capital Cities: Does Form Follow Values" in Taylor, John, Jean G. Lengellé, and Caroline Andrew, eds., *Capital Cities: International Perspectives/Les Capitales: Perspectives Internationales,* 195–212. Ottawa: Carleton University Press, 1993.

Taylor, John, Jean G. Lengellé, and Caroline Andrew, eds. *Capital Cities: International Perspectives/Les Capitales: Perspectives Internationales.* Ottawa: Carleton University Press, 1993.

Taylor, Peter et al., eds. *Global Urban Analysis: A Survey of Cities in Globalization.* London: Earthscan: 2011.

Therborn, Göran. "Monumental Europe: On the Iconography of European Capital Cities," *Housing, Theory, and Society* 19 (2002): 26–47.

Therborn, Göran. "Eastern Drama: Capital Cities of Eastern Europe, 1830–2006," *International Review of Sociology* 16, no. 2 (2006): 209–42.

Therborn, Göran and K. C. Ho. "Capital Cities and their Contested Roles in the Life of Nations," *City: Analysis of Urban Trends, Culture, Theory, Policy, Action*" 13, no. 1 (March, 2009): 53–62.

Vale, Lawrence. *Architecture, Power, and National Identity.* New Haven, CT: Yale University Press, 1992.

Waley, Paul. *Japanese Capitals in Historical Perspective: Place, Power and Memory in Kyoto, Edo and Tokyo.* London: Routledge, 2003.

Ward, Philip. *Japanese Capitals: A Cultural, Historical, and Artistic Guide to Nara, Kyoto, and Japan.* New York: Hippocrene Books, 1987.

Whitfield, Peter. *Cities of the World: A History in Maps.* Berkeley: University of California Press, 2005.

Wright, Herbert. *Instant Cities.* London: Black Dog Publishers, 2009.

Zukin, Sharon. *Landscapes of Power: From Detroit to Disney World.* Berkeley: University of California Press, 1991.

Index

Note: Page numbers in **boldface** reflect main entries in the book.

About the Author

Roman Adrian Cybriwsky, PhD, is professor of geography and urban studies at Temple University, Philadelphia, PA. He has taught as well at Temple University's campus in Tokyo, Japan, and as a Fulbright Scholar at the National University of Kyiv Mohyla Academy in Kyiv, Ukraine. His recent books include *Historical Dictionary of Tokyo*; *Roppongi Crossing: The Demise of a Tokyo Nightclub District and the Reshaping of a Global City*; and *Tokyo: The Shogun's City at the 21st Century*. His newest book is about Kyiv and will appear soon. Professor Cybriwsky holds a doctorate in geography from The Pennsylvania State University.